Applied Animal Endocrinology

E. James Squires

*Department of Animal and Poultry Science, University of Guelph,
Guelph, Ontario, Canada*

CABI Publishing

CABI Publishing is a division of CAB International

CABI Publishing
CAB International
Wallingford
Oxon OX10 8DE
UK

CABI Publishing
875 Massachusetts Avenue
7th Floor
Cambridge, MA 02139
USA

Tel: +44 (0)1491 832111
Fax: +44 (0)1491 833508
E-mail: cabi@cabi.org
Website: www.cabi-publishing.org

Tel: +1 617 395 4056
Fax: +1 617 354 6875
E-mail: cabi-nao@cabi.org

©CAB International 2003. All rights reserved. No part of this publication may be reproduced in any form or by any means, electronically, mechanically, by photocopying, recording or otherwise, without the prior permission of the copyright owners.

A catalogue record for this book is available from the British Library, London, UK.

Library of Congress Cataloging-in-Publication Data

Squires, E. James
  Applied animal endocrinology/E. James Squires.
    p. cm
  Includes bibliographical references (p.   ).
  ISBN 0-85199-594-2 (alk. paper)
  1. Veterinary endocrinology. 2. Livestock–Reproduction. I. Title.
SF768.3.S68 2003
636.089´64–dc21

2003003596

ISBN 0 85199 594 2

Typeset by MRM Graphics Ltd, Winslow, Bucks
Printed and bound in the UK by Cromwell Press, Trowbridge

Contents

Preface		x
Dedication		xii
Acknowledgements		xii
Abbreviations		xiii
1 Hormone and Receptor Structure and Function		**1**
1.1	Introduction	1
	What is a hormone?	1
	Why are hormones necessary?	1
	How do hormones function?	2
	What effects are due to hormones?	3
	How is hormone action selective?	4
	Types of hormones	5
	Location of endocrine glands	5
1.2	Synthesis, Release and Metabolism of Hormones	5
	Synthesis of protein hormones	5
	Synthesis of steroid hormones	7
	Synthesis of eicosanoids	9
	Synthesis of thyroid hormones	11
	Hormone release	11
	Metabolism and excretion of hormones	12
1.3	Receptors and Hormone Action	13
	Extracellular receptors	13
	Second messenger systems	14
	Adenylate cyclase–cAMP–protein kinase A pathway	14
	Guanyl cyclase–cGMP-dependent protein kinase pathway	15
	Genomic actions of cAMP	16
	Calcium-dependent phospholipase C–protein kinase C system	16
	Interaction of cAMP and Ca^{2+} pathways	18
	Tyrosine kinase receptors: catalytic receptors	18
	Cytokine receptors	20
	Receptor serine kinase	20
	Termination of hormone action	20
	Intracellular receptors	20
	Structural and functional domains of nuclear receptors	22
	Binding sites of the hormone–receptor complex on DNA	22
	Organization of nuclear chromatin and the nuclear matrix	23

		Chromatin	23
		Nuclear matrix	24
		Identification of DNA regulatory sequences	25
		Identification of DNA-binding proteins	26
		Integration of peptide and steroid hormone actions	26
	1.4	Pituitary–Hypothalamic Integration of Hormone Action	27
		Structure–function relationship of pituitary and hypothalamus	27
		Posterior pituitary hormones	28
		Anterior pituitary hormones	28
		Hypothalamic control of pituitary hormone secretion	29
		Hypothalamic releasing and release-inhibiting hormones	30
		Control of hormone release	31
	Questions for Study and Discussion		32
	Further Reading		33

2 Endocrine Methodologies — 35

	2.1	Methods for Studying Endocrine Function	35
		Model systems	35
		Whole animal model	35
		Isolated organs or tissues	37
		In vitro models	38
		Use of inhibitors and agonists	39
		Use of antibodies	40
		Immune response	40
		Detection and purification of antibodies	41
		Monoclonal antibodies	41
		Use of antibodies to identify the site of hormone synthesis or target tissue	42
	2.2	Measurement of Hormones and Receptors	42
		Assay of hormones	42
		Types of hormone assays	43
		Bioassays	43
		Chemical assays	44
		Liquid chromatography	45
		Gas chromatography	46
		Electrophoresis	47
		Competitive binding assays	47
		Assay requirements	47
		Measurements of hormone–receptor binding	49
		Competition binding	50
	2.3	Methods for the Production of Hormones	51
		Steroids and non-protein hormones	51
		Protein and peptide hormones	52
		Determination of amino-acid sequence	53
		Peptide and protein synthesis	54
		Non-peptide mimics of peptides	55
		Production of recombinant proteins	56
	2.4	Manipulation of Endocrine Function	57
		Hormone delivery methods	57
		Types of sustained-release devices	58
		Pulsatile release of hormone	59
		Hormone residues	60

		Immunomodulation of hormone action	61
		Types of immunoglobulins	61
	Transgenic animals		62
		Uses for transgenic animals	62
		Production of transgenic animals	63
	Questions for Study and Discussion		64
	Further Reading		65
3 Manipulation of Growth and Carcass Composition			**66**
3.1	Overview		66
	Effects on growth, feed efficiency and lean yield		67
3.2	Anabolic Steroids and Analogues		68
	Mechanism of action		70
		Direct effects	71
		Indirect effects	72
	Delivery systems		72
	Safety issues		74
3.3	Use of Intact (Uncastrated) Male Pigs		75
	Advantages and problems of intact male pigs		75
	Effects of sex steroid hormones		76
	Description of boar taint		77
	Measurement of boar taint		78
	Use of tainted meat in processed products		79
	Metabolism of androstenone and skatole		80
	Factors affecting boar taint		81
		Diet and management	82
		Genetic factors	82
	Immunological methods to control boar taint		83
3.4	Somatotrophin		83
	Control of ST release		85
	Mechanism of action of ST		88
		Direct effects	88
		ST receptors	88
		Metabolic effects	88
		Indirect effects	89
	Delivery/dose effects		89
	Safety/quality aspects		90
3.5	β-Adrenergic Agonists		90
	Mechanism of action		91
	β-AA structures		91
	β-AA receptors		91
	Physiological responses to β-AA		93
	Delivery/dose		94
	Safety aspects		94
	Alternative approaches for using growth promoters		95
3.6	Thyroid Hormones		96
	Synthesis and metabolism		96
	Metabolic effects		97
	Effects on growth and development		98
3.7	Dietary Polyunsaturated Fatty Acids		99
	Mechanism of action		99
	Linoleic acid, linolenic acid and γ-linolenic acid		100

		Applications	102
		Conjugated linoleic acid	102
		Metabolic effects of CLA isomers	103
		Mechanism of action of CLA	104
		CLA preparations	104
	3.8	Leptin	105
		Leptin receptors	105
		Involvement in energy metabolism and reproduction	106
		Direct effects	106
		Applications	107
	3.9	Cholecystokinin and Appetite	107
		Mechanism of action	107
		Applications	108
	3.10	Antibiotics, Antimicrobials and Other Factors	108
	3.11	Dietary Chromium and Insulin	109
		Insulin	109
		Glucagon	110
		Mechanisms of action	110
		Physiological effects	111
		Dose	111
		Safety issues	112
	3.12	Effects of Stress on Meat Quality	112
		Pale, soft, exudative and dark, firm, dry meat	112
		Porcine stress syndrome	114
		Testing for PSS	115
		Endocrine factors that affect PSS pigs and PSE pork	117
		Manipulations to reduce the incidence of PSE	118
	Summary and Conclusions		118
	Questions for Study and Discussion		119
	Further Reading		120
4	**Endocrine Effects on Animal Products**		**124**
	4.1	Mammary Gland Development and Milk Production	124
		Introduction	124
		Mammary gland development	124
		Involution and the dry period	126
		Model systems for studying mammary gland development and function	126
		In vitro cell culture systems	126
		Whole animal studies	126
		Hormones and mammary gland development (mammogenesis)	127
		Hormones and initiation of lactogenesis	128
		Maintenance of lactation (galactopoiesis)	129
		Hormonal effects	129
		Milk removal	130
		Effect of bST	130
		Mechanism of action	131
		Delivery	131
		Safety concerns of bST use	132
		Factors affecting milk composition	132
		Milk protein	132
		Milk fat	133
		Metabolic diseases related to lactation	134

		Ketosis	134
		Milk fever	134
		Hormones involved	134
		Predisposing factors	135
		Treatment and prevention	135
	4.2	Egg Production	136
		Sexual development	136
		Hormonal effects	137
		Genetic effects	137
		Regulation of follicular development and egg production	138
		Application	140
		Eggshell formation	141
		Shell matrix	142
		Calcium metabolism	143
		Applications	145
	4.3	Wool Production and Endocrine Defleecing	145
		Introduction	145
		Defleecing methods	145
		Model systems used to study function of follicles	147
		Growth factor effects on hair and wool follicles	147
		Insulin-like growth factors	147
		Fibroblast growth factors	148
		Transforming growth factor-β	149
		EGF family of growth factors	149
		EGF receptor	150
		Effects of EGF on follicles	150
		Other effects of EGF	151
		Summary of growth factors affecting fibre growth	151
	Questions for Study and Discussion		151
	Further Reading		152
5	**Endocrine Manipulation of Reproduction**		**154**
	5.1	Manipulation of Reproduction in Mammals	154
		Sexual differentiation and maturation	154
		Differentiation of the gonads and ducts	155
		Differentiation of the brain	155
		Sex differentiation in cattle, sheep and pigs	156
		Sex-determining genes	157
		Regulation of meiosis in germ cells	157
		Regulation of the oestrous cycle	158
		Overview of the oestrous cycle	159
		Follicular development	159
		Oestrus and ovulation	161
		Luteal phase	161
		Pregnancy	162
		Parturition	162
		Puberty and seasonality	162
		Regulation of LH production	163
		Regulation of steroidogenesis	164
		Manipulation of the oestrous cycle	164
		Hormone preparations for manipulating reproduction	165
		Use of hormone agonists to control fertility	170

	Methods for the detection of oestrus	171
	Oestrus behaviour	171
	Milk progesterone	172
	Induction and synchronization of oestrus	172
	Strategies for synchronizing oestrus	173
	Prostaglandin $F_{2\alpha}$-based systems	173
	GnRH and the Ovsynch® protocol	173
	Progestin-based systems	174
	Superovulation and embryo transfer	175
	In vitro production of embryos	176
	Maintenance of pregnancy	177
	Induction of abortion/parturition	177
	Postpartum interval	178
	Cystic ovarian disease	178
	Effects of nutrition	179
	Effects of stress	179
	Inducing puberty	180
	Advancing cyclicity in seasonal breeders	180
	Immunological control of reproduction	181
5.2	**Endocrine Manipulations in Aquaculture Fish**	182
	Control of reproduction	182
	Sex reversal	182
	Hormonal treatments for sex reversal	183
	Indirect methods	183
	Induction of spawning	185
	Effects on growth and nutrient utilization	186
	Applications	187
	Stress and effects on the immune system	188
	Applications	189
	Questions for Study and Discussion	189
	Further Reading	189
6	**Effects on Animal Behaviour, Health and Welfare**	**192**
6.1	Control of Broodiness in Poultry	192
	Applications	193
6.2	Applications of Pheromones	194
	Types of pheromones	194
	Chemistry of pheromones	194
	Pheromone production and release	195
	Detection of pheromones	196
	Vertebrate pheromones	196
	Rodents	197
	Pigs	198
	Cattle	198
	Sheep and goats	199
	Fish	199
	Other	199
	Insect pheromones	200
	Applications	202
	Pest control	202
	Population monitoring	202
	Mating disruption	202

		Mass trapping	203
		Insect management	204
		Reproduction control in mammals	204
6.3	Effects of Stress		204
	What is stress?		204
	Hormonal responses to stress		205
		SA system	205
		HPA axis	206
		CRH and CRH receptors	207
		Role of various hormones in stress responses	208
	Assessment of stress		210
		Behavioural and physiological measures	210
		Hormonal measures	210
		Issues related to sampling	211
	Effects of stress on the immune system and disease resistance		212
		Overview	212
		Stress effects on the immune system	214
	Effects on reproduction		215
	Effects on growth performance		217
	Summary		218
6.4	Endocrine Applications in Toxicology		218
	Endocrine disruptors or modulators		218
	Assessment of endocrine disruptor activity		219
		In vitro assays	220
		In vivo assays	220
	Sources of endocrine disruptors		221
		Plant-derived endocrine modulators	221
		Xenobiotic endocrine modulators	223
	Indirect mechanisms of action		224
		Effects on hormone metabolism	224
		Effects on thyroid function	225
		Effects on adrenal function	225
		Effects on CNS function and behaviour	225
	Summary		226
	Questions for Study and Discussion		227
	Further Reading		227

Index 230

Preface

The purpose of this book is to provide information on a number of different endocrine systems that affect animal production, and to describe how these systems can be manipulated or monitored to advantage. A number of excellent endocrinology texts are already available that describe the function of various endocrine systems, but these texts, for the most part, deal with human or comparative endocrinology. This book focuses on commercial animals, and endocrine systems that can affect growth and carcass composition, the production of animal products (milk, eggs and wool), reproduction efficiency and animal health, behaviour and welfare are described. Detailed information on the mechanism of action of the endocrine systems is covered, and an attempt is made to integrate knowledge from similar topics by focusing on common mechanisms and themes (for example, see the discussion of dietary polyunsaturated fatty acids (PUFA) in Chapter 3, Section 3.7). This information is used to understand potential methods for altering these systems and, hopefully, to stimulate ideas for the development of new methods.

The first two chapters cover the essential background information in endocrinology, my version of 'Endocrinology 101', but also include information on the production of hormones and the methods for manipulating endocrine systems. In the remaining chapters, endocrine systems that affect some aspect of animal production are described. Each chapter includes an overview of the problem or application, followed by a description of the endocrine systems affecting the problem and a discussion of how these systems can be manipulated or monitored to advantage.

In Chapter 1, the structure and function of hormones and receptors are covered. The main concepts of endocrinology are reviewed in sufficient depth to provide the necessary background material for the rest of the book. An attempt was made to avoid excessive detail and to introduce some potential applications to heighten interest. An initial overview of hormones and endocrinology is followed by a discussion of the synthesis, release and metabolism of hormones, the intracellular and extracellular mechanisms of hormone action and the integration of hormone action at the level of the hypothalamus and pituitary.

Chapter 2 covers various endocrine methodologies. The methods that are used to study how endocrine systems function are described, followed by methods for measuring hormones and receptors. Methods used for the production of hormones are then described and, finally, a number of methods for manipulating hormone function are covered.

In Chapter 3, endocrine systems that affect growth rate, feed efficiency and carcass composition are described. This includes anabolic steroids and analogues, use of uncastrated (intact) male pigs and the problem of boar taint, somatotrophin, β-adrenergic agonists, thyroid hormones, dietary PUFA (linoleic, linolenic, γ-linolenic acid (GLA) and conjugated linoleic acid (CLA)), leptin, control of appetite by cholecystokinin (CCK), antibiotics, antimicrobials and other factors, dietary chromium and insulin, and the effects of stress on meat quality.

In Chapter 4, the endocrine effects on animal products other than meat are covered. These include endocrine effects on mammary gland development and milk production (including the regulation of mammogenesis, lactogenesis and galactopoiesis), the effects of bovine somatotrophin (bST), the factors affecting milk composition, and metabolic diseases related to lactation. This is followed by a discussion of endocrine effects on egg production, including those on ovary sexual development in chickens, and the regulation of follicular development and eggshell formation. Finally, wool production and endocrine defleecing are covered.

Chapter 5 is concerned with the endocrine manipulation of reproduction. In the first section, sexual differentiation and maturation of mammals are covered, followed by the regulation of the oestrous cycle, pregnancy and parturition. Methods for manipulating reproduction are then discussed, including manipulation of the oestrous cycle, pregnancy, the postpartum interval, inducing puberty and advancing cycling in seasonal breeders. The next section covers endocrine manipulations in aquaculture, including control of reproduction, effects on growth and nutrient utilization and the effects of stress.

In Chapter 6, the applications of endocrinology in monitoring and manipulating animal behaviour, health, performance and welfare are described. The control of broodiness in turkeys and applications of pheromones in vertebrates and insects are discussed first. This is followed by a section on the effects of stress, including the hormonal responses to stress and the effects of stress on immune function, reproduction and growth performance. The endocrine applications in toxicology are covered in the final section, which illustrates that changes in endocrine function can be used as indicators of endocrine disruptors in the environment and food chain.

This text is aimed at senior undergraduate and graduate students in animal science and veterinary medicine, although others in academia and industry who are interested in applications of endocrinology in animal production systems should also find it useful. It is based on a course, 'Applied Endocrinology', that has been taught at the Department of Animal and Poultry Science, University of Guelph, for the past 15 years. It is my hope that it will help to integrate knowledge of endocrine function in commercial animals and stimulate new ideas and applications for improving animal production, health and welfare. Constructive criticism and comments will be most appreciated.

E. James Squires
University of Guelph
jsquires@uoguelph.ca

Dedication

To Gail, Allison, Victoria and Kimberly, and in loving memory of Dorothy Cavell Squires.

Acknowledgements

Special thanks are due to Davis Brooks for library assistance, Victoria Squires and Kimberly Squires for their artistic skills, Laurie Parr for her tremendous help in all aspects of preparing the manuscript and Brian McBride, Tina Widowski, Jim Atkinson and Heidi Engelhardt for valuable comments and reviewing the manuscript.

Abbreviations

AA	arachidonic acid
β-AA	β-adrenergic agonists
Ab	antibody
ACTH	adrenocorticotrophic hormone
ADG	average daily gain
ADH	antidiuretic hormone
ADI	acceptable daily human intake
Ag	antigen
AI	artificial insemination
AIS	androgen insensitivity syndrome
ALA	α-linolenic acid
AMH	anti-Müllerian hormone
AP-1	activating protein-1
AR	androgen receptor
ARA	androgen receptor co-activator
ATP	adenosine triphosphate
BCIP/NBT	5-bromo-4-chloro-3-indoyl phosphate/nitro blue tetrazolium
BMP	bone morphogenic protein
BMR	basal metabolic rate
bST	bovine somatotrophin
BW	body weight
cAMP	cyclic adenosine monophosphate
CAD	cation–anion difference
CAP	6-chloro-8-dehydro-17-acetoxy-progesterone
CBG	corticosteroid binding globulin
CBP	corticoid-binding protein
CCK	cholecystokinin
cDNA	complementary DNA
CFTR	cystic fibrosis transmembrane conductance regulator
CIDRs	controlled internal drug-releasing devices
CL	corpus luteum
CLA	conjugated linoleic acid
CNS	central nervous system
ConA	concanavalin A
COX	cyclooxygenase
CRC	calcium release channel
CREB	cAMP-responsive-element binding protein

CRF	corticotrophin releasing factor
CRH	corticotrophin releasing hormone
CRHR	corticotrophin releasing hormone receptor
CrNic	chromium nicotinate
CRP	C-reactive protein
CrPic	chromium picolinate
cST	chicken somatotrophin
CV	coefficient of variation
DAG	diacylglycerol
DDT	dichlorodiphenyltrichloroethane
DES	diethylstilbestrol
DFD	dark, firm, dry (meat)
DGLA	dihomo-γ-linolenic acid
DHA	docosahexaenoic acid
DHEA	dehydroepiandrosterone
DHEAS	dehydroepiandrosterone sulphate
DHT	5α-dihydrotestosterone
E_2	oestradiol
EBI	ergosterol biosynthesis inhibiting (fungicides)
eCG	equine chorionic gonadotrophin
ECM	extracellular matrix
ECP	oestradiol cypionate
EDC	endocrine disruptor chemical
EGF	epidermal growth factor
EGFR	epidermal growth factor receptor
ELISA (or EIA)	enzyme-linked immunosorbent assay
EPA	eicosapentaenoic acid OR Environmental Protection Agency
ER	oestrogen receptor
EROD	7-ethoxyresorufin-O-deethylase
FA	fatty acid
FAS	fatty acid synthase
FEBP	fetoneonatal oestrogen binding protein
FFA	free fatty acid
FGA	fluorogestone acetate
FGF	fibroblast growth factor
FGFR	fibroblast growth factor receptor
FIA	fluorescence immunoassay
FID	flame ionization detector
FSH	follicle stimulating hormone
GC	gas chromatography
GDF-9	growth differentiation factor 9
GDP	guanosine diphosphate
GFP	green fluorescent protein
GH	growth hormone, somatotrophin
GHBP	growth hormone binding protein
GHRH	growth hormone releasing hormone
GH-RIH	growth hormone release-inhibiting hormone
GHRP	growth hormone releasing peptide
GHS	growth hormone secretagogue
GHS-R	growth hormone secretagogue receptor
GI	gastrointestinal
GIP	gastric inhibitory peptide

GLA	γ-linolenic acid
GLUT4	glucose transporter 4
GnRH	gonadotrophin releasing hormone
GnRH-A	GnRH analogue
GR	glucocorticoid receptor
GRIPI	glucocorticoid receptor interacting protein I
GTF	glucose tolerance factor
GTP	guanosine triphosphate
HAT	hypoxanthine–aminopterin–thymidine (medium)
HB-EGF	heparin-binding epidermal growth factor
hCG	human chorionic gonadotrophin
9-HDA	9-hydroxydec-2-enoic acid
HDL	high-density lipoprotein
H-FABP	fatty acid binding protein isolated from heart
HGF	hepatocyte growth factor
HGPRT	hypoxanthine guanine phosphoribosyl transferase
HIT	histidine triad
HMG	high-mobility group (proteins)
HMI	3-hydroxy-3-methylindolenine
HMOI	3-hydroxy-3-methyloxindole
HNF	hepatic nuclear factor
hnRNA	heterogeneous nuclear RNA
HOB	methyl *p*-hydroxybenzoate
HPA	hypothalamic–pituitary–adrenal
HPI	hypothalamic–pituitary–interrenal (axis)
HPLC	high performance liquid chromatography
HREs	hormone-responsive elements
17β-HSD	17β-hydroxysteroid dehydrogenase
hST	human somatotrophin
HVA	4-hydroxy-3-methoxyphenylethanol
IBMX	isobutylmethylxanthine
IFN	interferon
IGF-I	insulin-like growth factor-I
IGFBP	insulin-like growth factor binding protein
IL	interleukin
IP_3	inositol-1,4,5-phosphate
IP_4	inositol-1,3,4,5-tetrakisphosphate
IPM	integrated pest management
IRM	integrated resource management
IVF	*in vitro* fertilization
IVM	*in vitro* maturation
JAK	Janus kinase
KGF	keratinocyte growth factor
KLH	keyhole limpet haemocyanin
LA	linoleic acid
LDL	low-density lipoprotein
LH	luteinizing hormone
LHRH	luteinizing hormone releasing hormone
LMWCr	low molecular weight chromium-binding substance
β-LPH	β-lipotrophin
LPS	lipopolysaccharide
LT	leukotriene

LUC-NE	nerve fibres of the locus ceruleus that secrete noradrenaline
MALDI-TOF	matrix-assisted laser desorption ionization–time of flight mass spectrometry
MAP	6-methyl-17-acetoxy-progesterone
MAS	meiosis activating substance
MCH	melanin concentrating hormone
MCSF	macrophage colony stimulating factor
MDGI	mammary-derived growth inhibitor
MGA	melengestrol acetate
mGnRH	mammalian GnRH
MH	malignant hyperthermia
MHC	major histocompatibility complex
MIF	melanotrophin release-inhibiting factor
MIH	MSH inhibiting hormone
MIS	Müllerian inhibiting substance
MOET	multiple ovulation and embryo transfer
3MOI	3-methyloxindole
MPS	meiosis preventing substance
MR	mineralocorticoid receptor
MRF	MSH releasing factor
MRI	magnetic resonance imaging
MRL	maximum residue level
MS	mass spectrometry
MSH	melanocyte stimulating hormone
MUP	major urinary protein
NDF	Neu differentiation factors
NEFA	non-esterified fatty acids
NF-κB	nuclear factor-κB
NGF	nerve growth factor
NOEL	no observed effects limit
NOS	nitric oxide synthase
NPY	neuropeptide Y
NSAIDs	non-steroidal anti-inflammatory drugs
OC17	ovocleidin-17
9-ODA	9-oxodec-2-enoic acid
ODP	odorant-binding protein
OECD	Organization for Economic Cooperation and Development
$1,25(OH)_2D_3$	1,25-dihydroxyvitamin D_3
OPN	osteopontin
OR	odour receptor
oST	ovine somatotrophin
PCBs	polychlorinated biphenyls
PCPA	*p*-chlorophenylalanine
PCR	polymerase chain reaction
PDGF	platelet-derived growth factor
PDI	potential daily intake
PGC	primordial germ cell
PGD_2	prostaglandin D_2
PGE_2	prostaglandin E_2
$PGF_{2\alpha}$	prostaglandin $F_{2\alpha}$
PIF	prolactin inhibiting factor
PIH	prolactin inhibiting hormone

PKCI	protein kinase C inhibitor
PLC	phospholipase C
PMA	phorbol 12-myristate 13-acetate
PMSG	pregnant mare serum gonadotrophin
POA	preoptic area
POMC	pro-opiomelanocortin
PPARγ	peroxisome proliferator-activated receptor γ
PPRE	peroxisome proliferator responsive element
PR	progesterone receptor
PRF	prolactin releasing factor
PRH	prolactin releasing hormone
PRID	progesterone-releasing intrauterine device
PRL	prolactin
PSE	pale, soft, exudative (meat)
PSS	porcine stress syndrome
pST	porcine somatotrophin
PST	phenol sulphotransferase
PTH	parathyroid hormone
PTHrP	parathyroid hormone related protein
PTK	phosphotyrosine kinase
PTU	6-propyl-2-thiouracil
PUFA	polyunsaturated fatty acid
PYY	peptide YY
QMP	queen mandibular pheromone
QTL	quantitative trait locus
RAR	retinoic acid receptor
REM	rapid eye movement
RER	rough endoplasmic reticulum
RFLP	restriction fragment length polymorphism
RFN	reddish-pink, firm, non-exudative (meat)
RIA	radioimmunoassay
RSE	reddish-pink, soft, exudative (meat)
rT_3	reverse T_3
RXR	9-*cis* retenoic acid receptor
SA system	sympathetic nervous system activation of the adrenal medulla
SARMs	selective androgen receptor modulators
SCD	stearoyl-CoA desaturase
SDS	sodium dodecyl sulphate
SF-1	steroidogenic factor 1
sGnRH-A	salmon GnRH analogue
SH2	src homology region 2
SHBG	sex hormone binding globulin
SLA	swine lymphocyte antigen
SPF	specific-pathogen-free
SR	sarcoplasmic reticulum
SREBP	sterol regulatory elements binding protein
SS	somatostatin
ST	somatotrophin, growth hormone
StAR	steroid acute regulatory protein
STATs	signal transducers and activators of transcription
$3,3'\text{-}T_2$	diiodotyrosine

T_3	triiodothyronine
T_4	thyroxine
TAT	transactivator protein
TBA	trenbolone acetate
TBT	tributyltin
TCD	thermal conductivity detector
TCDD	tetrachlorodibenzo-*p*-dioxin
TDF	testis-determining factor
TF	transcription factor
Tfm	testicular feminization
TGFβ	transforming growth factor-β
Th	helper T cells
TMB	3,3′,5,5′-tetramethylbenzidine
TNF-α	tumour necrosis factor-α
TP-1	trophoblast protein-1
TPA	tetradecanoylphorbol acetate
TR	thyroid hormone receptor
TRH	thyrotrophin releasing hormone
TSH	thyroid stimulating hormone (thyrotrophin)
TXA_2	thromboxane A_2
UCP	uncoupling protein
VDR	1,25-hydroxy vitamin D_3 receptor
VEGF	vascular endothelial growth factor
VIP	vasoactive intestinal peptide
VNO	vomeronasal organ
ZP	zona pellucida

1

Hormone and Receptor Structure and Function

1.1 Introduction

What is a hormone?

A **hormone** is a chemical messenger that coordinates the activities of different cells in a multicellular organism. Bayliss and Sterling first used the term in 1904 to describe the actions of secretin, which is a hormone produced by the duodenum to stimulate the flow of pancreatic juice. The classical definition of a hormone is: a chemical substance that is synthesized by particular **endocrine glands** and then enters the bloodstream to be carried to a **target tissue,** which has specific **receptors** that bind it. Other mechanisms of hormone delivery also exist (Fig. 1.1). **Neuroendocrine** hormones are synthesized by nervous tissue and carried in the blood to the target tissue; for example, the various releasing factors that are produced in the hypothalamus, which travel to the anterior pituitary via the hypothalamus–pituitary blood portal system. **Neurocrine** hormones are released into the synaptic cleft by neurones that are in contact with the target cells. **Paracrine** hormones diffuse to neighbouring cells, while **autocrine** hormones feed back on the cell of origin in a form of self-regulation. At the other extreme, pheromones are produced by one animal and released into the environment to be received by other animals (see Section 6.2).

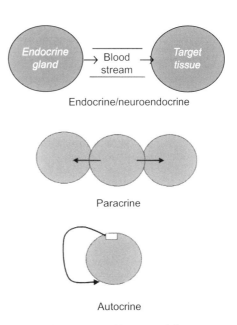

Fig. 1.1. Mechanisms of hormone delivery.

Why are hormones necessary?

Hormones are involved in maintaining homeostasis – consistency of the internal environment that is maintained for the benefit of the whole organism. Homeostasis was first recognized by Claude Bernard in the 19th century, who noted that the internal environment (i.e. the fluid bathing cells) had to be regulated independently of external environment. Being able to regulate and maintain its internal environment gives the animal freedom from changes in the external

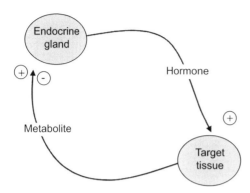

Fig. 1.2. Feedback system to regulate hormone production.

environment, allowing it to live in changing or harsh environments. However, there are metabolic costs associated with maintaining homeostasis. For example, maintenance of a constant body temperature allows animals to function in cold environments, while cold-blooded animals (poikilotherms) only function during warm temperatures. The added energy costs of maintaining deep body temperature above that of the environment means that warm-blooded animals have a higher energy requirement for maintenance than do poikilotherms.

Homeostasis is maintained by **negative feedback**. For example, an endocrine tissue produces a hormone that affects the production of a metabolite by the target tissue. The metabolite then interacts with the endocrine gland to reduce the production of the hormone. This forms a cyclic system in which the levels of the metabolite are maintained at a particular level. The set point of the system can be altered to affect the levels of the metabolite by altering the sensitivity of the target tissue to the hormone or the sensitivity of the endocrine gland to negative feedback from the metabolite (Fig. 1.2).

In addition to maintaining homeostasis, hormones can also be used to drive change in an organism. In this case, levels of hormone increase to some peak, and this occurs by **positive feedback**. Positive feedback amplifies the response, so the tissue must be desensitized or turned over to stop the response. An example of this response is the surge of luteinizing hormone (LH) that leads to ovulation (see Section 5.1). LH produced by the pituitary gland stimulates the developing ovarian follicle to produce oestrogen, which stimulates the hypothalamus to produce gonadotrophin releasing hormone and increase LH production by the pituitary. This produces a surge of LH, which decreases only after the follicle ovulates (Fig. 1.3).

How do hormones function?

Hormones cause a trigger effect to modulate the activity of the target tissue. The effects of hormones are seen long after levels of the hormone return to basal values. In contrast, nervous signals are short lasting and more immediate. However, nervous signals can regulate hormone production as well. Hormones are present in trace amounts in plasma, usually ranging from 10^{-9} to 10^{-6} g ml^{-1}. They are present at all times in order to maintain receptors in the target tissue and keep the tissue primed for a

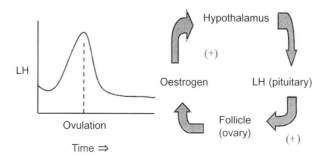

Fig. 1.3. Positive feedback system leading to the LH surge and ovulation.

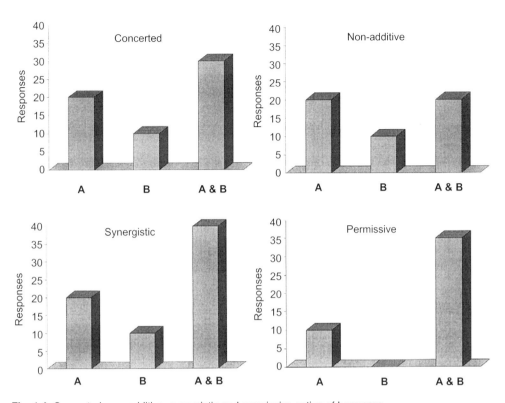

Fig. 1.4. Concerted, non-additive, synergistic and permissive action of hormones.

response. Hormones are secreted in variable amounts according to need, and there is a constant turnover by inactivation and excretion of the hormone.

The combined effects of more than one hormone on a biological response can occur in a number of different ways (Fig. 1.4). The actions of different hormones are **concerted** or additive if they cause the same response and the combined effect of the hormones is simply the sum of the separate actions of the individual hormones. This suggests that the two hormones act by different mechanisms. In some cases, two hormones can cause the same effect but the effects due to the different hormones are **non-additive**. This implies that the two hormones may act by the same common mechanism. The effects of two different hormones are **synergistic** when the combined effect of the two hormones together is more than the sum of the separate effects of the individual hormones. Some hormones, for example steroid hormones and thyroid hormones, can have a **permissive** action and have no effect on their own but must be present for another hormone to have an effect. A permissive hormone could act by increasing the number of receptors or affecting the activity of the second messenger system for the second hormone. For example, oestradiol has a permissive action for progesterone by inducing the expression of progesterone receptors in the oviduct.

What effects are due to hormones?

Hormones cause changes in cellular metabolism, but they do not make a cell do something it was not previously capable of. Hormones do not directly cause changes in gene structure but can activate genes to influence **gene expression** and ultimately protein synthesis. Hormones can alter **catalytic rates of enzymes**, by mechanisms such as the phosphorylation and dephosphorylation of proteins. Hormones can also alter **membrane permeability** to affect transport processes and ion movements, alter muscle contraction, exocrine secretion and water permeability.

These general mechanisms can cause a variety of effects in the animal. Hormones can:

- cause morphological changes, such as the differences in body shape between males and females;
- act as mitogens to accelerate cell division or alter gene expression to trigger differentiation of cells;
- stimulate the overall rate of protein synthesis or the synthesis of specific proteins;
- be involved in stimulating smooth muscle contraction; for example, oxytocin stimulates contraction of the myoepithelium in the mammary gland for milk ejection;
- affect exocrine secretions; for example, secretin, a peptide hormone from intestinal mucosa, stimulates pancreatic secretions;
- control endocrine secretions, and a number of trophic hormones from the anterior pituitary can stimulate or inhibit hormone secretion from target organs;
- regulate ion movements across membranes and control permeability to water; for example, antidiuretic hormone (ADH, vasopressin) increases water reabsorption by the kidney; and
- have a dramatic effect on behaviour, such as sex-related behavioural characteristics, maternal behaviour, nesting activity and broodiness (see Chapter 6).

How is hormone action selective?

The method of hormone delivery to the target cells and the presence of specific receptors in the target cells determine the selectivity of hormone action. For example, the hypophyseal–portal system linking the hypothalamus to the pituitary delivers releasing hormones from the hypothalamus directly to the target cells in the anterior pituitary. Smaller quantities of hormone are needed since there is less dilution of the hormone in selective delivery systems compared to hormones that reach their target via the peripheral circulation. Many hormones are linked to **carrier proteins** in the blood, which stabilize the hormone and increase its half-life in the circulation. For example, sex hormone binding globulin is synthesized in the liver and binds testosterone and oestradiol in the circulation with a high affinity. However, the main factor that determines the sensitivity of a particular tissue to a hormone is whether or not the tissue contains the specific receptor for the hormone – the tissue will not respond to the hormone unless it has enough of the specific receptor.

Receptors are specific proteins present in target cells that bind a particular hormone and initiate a response. Receptors are normally present in small numbers (10,000 molecules per cell). There are two general types, cell-surface receptors (Fig. 1.5) and intracellular receptors (Fig. 1.6). Peptide and protein hormones generally do not enter the cell, but interact with cell-surface receptors. For some cell-surface receptors, a second messenger system is needed to transmit the hormone response signal from the outside to the inside of the cell. This involves the activation of a protein kinase, which phosphorylates specific proteins within the cell to alter their function. Steroid hormones and thyroid hor-

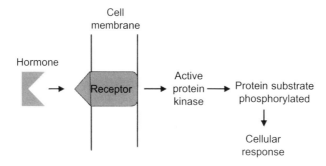

Fig. 1.5. Action of hormones via cell-surface receptors.

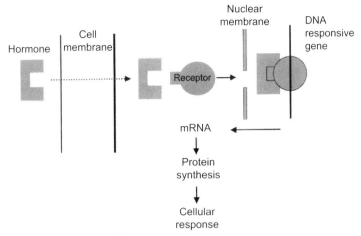

Fig. 1.6. Action of hormones via intracellular receptors.

mones enter the cell to interact with intracellular receptors and regulate gene expression.

Types of hormones

The major structural groups of hormones are:

- steroids;
- proteins, polypeptides and glycoproteins;
- amino-acid derivatives (especially derivatives of tyrosine); and
- fatty acids and derivatives, such as prostaglandins.

The structures of some non-protein hormones are given in Fig. 1.7.

Location of endocrine glands

The location of the key endocrine glands is given in Fig. 1.8. Table 1.1 lists the hormones produced by these glands and their functions. Applications involving many of these hormones are covered in this text and the relevant sections are listed in Table 1.1.

1.2 Synthesis, Release and Metabolism of Hormones

Synthesis of protein hormones

Peptide and protein hormones consist of a linear chain of amino acids. As with any protein, the specific sequence of the different amino acids determines the primary structure and nature of the protein. The amino-acid sequence information for a protein is contained in the sequence of bases (A,C,G,T) in the coding region of the gene that codes for the protein. A three-base sequence codes for one amino acid; this is known as the genetic code. This code is copied from DNA into mRNA by transcription and the mRNA is used to direct protein synthesis by the process of translation (Fig. 1.9).

Signal peptides are short sequences of 15–30 hydrophobic amino acids located at the amino-terminal (beginning) of proteins. The presence of a signal sequence (S) directs the newly synthesized protein into the endoplasmic reticulum and then to export from the cell. Other proteins enter the cytosol and from there are directed to the mitochondria (M) or nucleus (N), or to other sites within the cell. Proteins move between the various compartments by vesicular transport. The uptake of proteins by particular vesicles is controlled by the sorting signal sequences in the proteins (Fig. 1.10). A program (SignalP) for identifying signal peptides and their cleavage sites has been described by Nielsen *et al.* (1997) and can be accessed online (http://www.cbs.dtu.dk/).

Newly synthesized protein hormones containing signal sequences are known as **prehormones**. Some peptide hormones are synthesized as part of a larger precursor, called a **prohormone**. Examples of prohor-

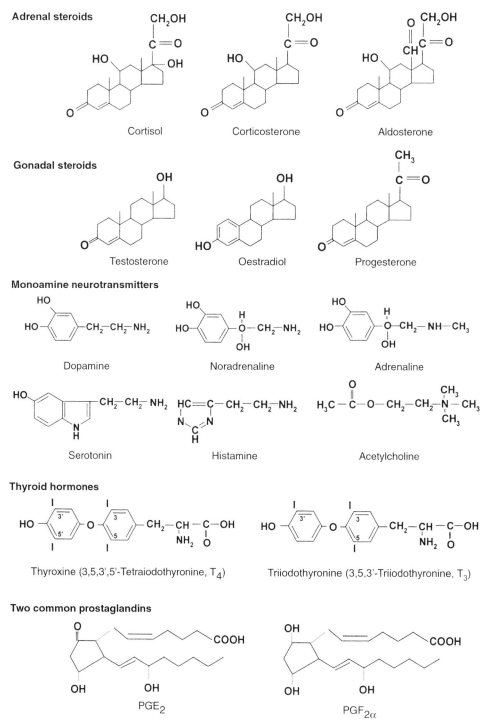

Fig. 1.7. Structures of representative non-protein hormones.

Fig. 1.9. Transcription of DNA to RNA and translation of RNA into protein.

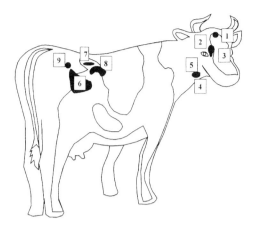

Fig. 1.8. The location of key endocrine glands in cattle: 1, pineal; 2, hypothalamus; 3, pituitary; 4, thyroid; 5, parathyroid; 6, pancreas; 7, adrenal; 8, kidney; 9, ovary (testis in males).

mones include proparathyroid hormone, the precursor of parathyroid hormone and proinsulin, which is the precursor of insulin. Pro-opiocortin is the precursor of several trophic peptide hormones produced in the anterior pituitary. The newly synthesized prohormone with a signal peptide is known as a **preprohormone** (Fig. 1.11).

A number of hormones, including thyroid stimulating hormone (TSH), follicle stimulating hormone (FSH) and LH, have sugar units attached to the amino-acid side-chains, and are known as **glycoproteins**. After protein synthesis, the preprohormone moves from the endoplasmic reticulum to the Golgi apparatus, where sugar residues are attached to asparagine, serine and other amino-acid side-chains in a process called glycosylation. These sugar units can form complex branched chains. From the Golgi apparatus, the proteins are packaged into secretory vesicles and the active hormone is generated by cleavage of the prohormone sequences. The secretory granules fuse with the plasma membrane to release their contents by exocytosis when the cell is stimulated. For more information on the mechanisms of protein secretion, see the review by Blázquez and Shennan (2000).

Synthesis of steroid hormones

Steroid hormones are produced in the gonads and the adrenal gland. The gonadal

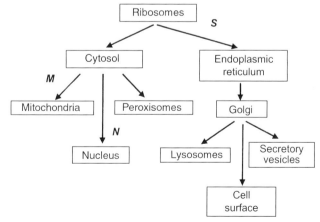

Fig. 1.10. Role of signal peptides in directing the movement of proteins within cells. A typical signal sequence (**S**) is: M-M-S-F-V-S-L-L-L-V-G-I-L-F-W-A-T-E-A-E-Q-L-T-K-C-E-V-F-Q- (a patch of hydrophobic amino acids is underlined). The signal (**M**) for importing into the mitochondria is: M-L-S-L-R-Q-S-I-R-F-F-K-R-A-T-R-T-L-C-S-S-R-Y-L-L-. The signal (**N**) for importing into the nucleus is: P-P-K-K-K-R-K-V-.

Table 1.1. Summary of hormones produced by various endocrine glands and their functions.

Endocrine glands	Hormones produced	Physiological response	Relevant book sections
Hypothalamus	TRH	TSH and PRL by anterior pituitary	See Section 1.4 for interaction between hypothalamus and pituitary
	GnRH	LH and FSH	
	CRH	ACTH, β-endorphin, stress	
	GHRH and GH-RIH	GH	
	PRF and PIF	PRL	
	MRF and MIF	MSH	
Anterior pituitary (adenohypophysis)	GH	Somatomedin by liver	Section 3.4
	PRL	Mammary gland/lactogenesis	Section 4.1
	TSH	Thyroid hormone	Section 3.6
	FSH	E_2, follicular growth/spermatogenesis	Sections 4.2, 5.1
	LH	E_2 and P_4, ovulation/androgen	Section 5.1
	ACTH	Adrenal steroids	Sections 3.12, 6.3
	MSH	Melanogenesis	
	β-endorphin	Analgesic	
Posterior pituitary (neurohypophysis)	Oxytocin	Milk ejection	Section 1.4
	Vasopressin	Antidiuretic hormone	
Pineal	Melatonin	Seasonality, gonad function	Section 5.1
Parathyroid	PTH	Calcium and phosphorus metabolism	Sections 4.1, 4.2
Thyroid	T_4 and T_3	Metabolic rate	Section 3.6
Adrenal cortex (outer part)	Cortisol, corticosterone, aldosterone	Carbohydrate metabolism, sodium retention	Sections 3.12, 6.3
Adrenal medulla	(Nor)Epinephrine	Alarm reactions	Sections 3.5, 6.3
Gonads	Androgens, oestrogens	Sexual development/behaviour	Sections 3.2, 3.3, 4.1, 5.1, 5.2
	Progestins	Pregnancy	
	Inhibin	FSH	
	Relaxin/oxytocin	Parturition	
Pancreas	Insulin, glucagon	Blood glucose	Section 3.11
Gastrointestinal tract	Gastrin, GIP, secretin	HCl and bicarbonate	Section 3.9
	CCK	Pancreatic enzymes	
	Motilin, neurotensin	Gastric activity	
	VIP	Blood flow	
Kidney	Erythropoietin	Blood cell formation	
Various tissues	Eicosanoids	Smooth muscle	Section 3.7
	Growth factors	Growth and differentiation	Sections 4.1, 4.3

steroids include the **progestins**, **oestrogens** and **androgens**. Progesterone is a major progestin, which prepares the lining of the uterus for implanting of the ovum and is involved in the maintenance of pregnancy. Oestrogens, such as oestradiol, are involved in the development of female secondary sex characteristics and in the ovarian cycle. The androgens are involved in the development of male secondary sex characteristics. Testosterone is a major androgen. The adrenal cortex produces **glucocorticoids** and **mineralocorticoids**. Cortisol, a major glucocorticoid, promotes gluconeogenesis and fat and protein degradation. Aldosterone, a major mineralocorticoid, increases absorption of sodium, chloride and bicarbonate by the kidney to increase blood volume and blood pressure.

The synthesis of steroid hormones occurs on the smooth endoplasmic reticulum and in the adrenal mitochondria. Cholesterol is the precursor of all steroid hormones and is present as low-density lipoprotein (LDL) in plasma. Many of the steps in the biosynthesis of steroids involve an electron transport chain in which cytochrome P450 is the termi-

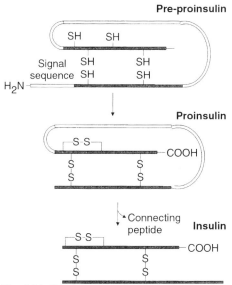

Fig. 1.11. Structure of insulin illustrating the signal peptide and pro-sequence.

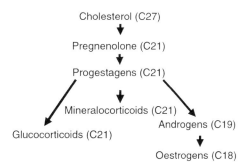

Fig. 1.12. Overall pathways of steroid hormone synthesis (see Fig. 1.7 for structures of these steroids).

nal electron acceptor and carries out hydroxylation reactions. The overall scheme is shown in Fig. 1.12.

The conversion of cholesterol to pregnenolone (Fig. 1.13) involves removal of the C_6 side chain from cholesterol by hydroxylation at C_{20} and C_{22} and cleavage of this bond by **desmolase** (P450 side-chain cleavage). This step occurs in adrenal mitochondria and is stimulated by adrenocorticotrophic hormone (ACTH).

Pregnenolone is then converted to progesterone by oxidation of the 3-hydroxy to a 3-keto group and isomerization of the Δ5 double bond to a Δ4 double bond. Progesterone is converted to cortisol by hydroxylation at C_{17}, C_{21} and C_{11}. Progesterone is converted to aldosterone by hydroxylation at C_{21} and C_{11}, and oxidation of the C_{18} methyl to an aldehyde (Fig. 1.14).

Progesterone is converted into androgens by hydroxylation at C_{17} and cleavage of the side-chain to form androstenedione (an androgen). The 17-keto group is reduced to a hydroxyl to form testosterone. Androgens are converted into oestrogens by loss of the C_{19} methyl group and aromatization of the A ring. The formation of oestrogens from androgens is catalysed by the aromatase enzyme CYParom (CYP19) (Fig. 1.15).

Synthesis of eicosanoids

The **eicosanoid hormones** include **prostaglandins, prostacyclins, thromboxanes** and **leukotrienes**. They are locally produced within cell membranes and have autocrine and paracrine effects. They stimulate inflammation, regulate blood flow and blood pressure, affect ion transport and modulate synaptic transmission. They are synthesized from 20 carbon fatty acids, such as

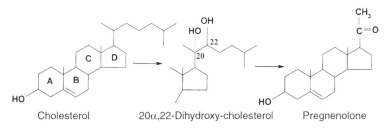

Fig. 1.13. Conversion of cholesterol to pregnenolone.

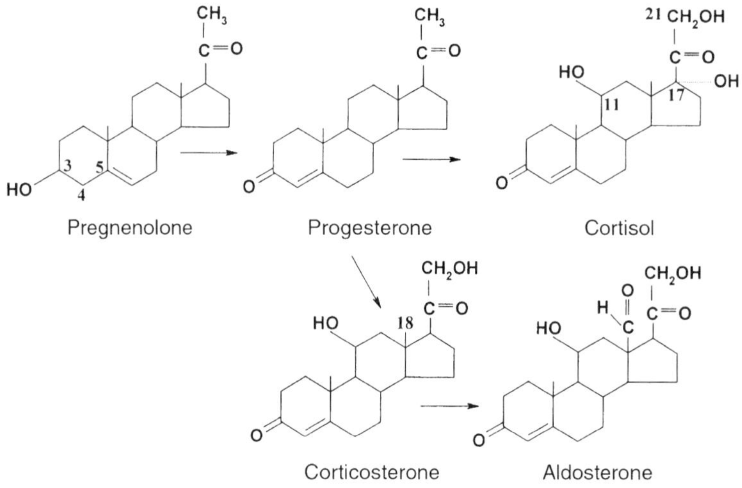

Fig. 1.14. Metabolism of pregnenolone to aldosterone.

arachidonic acid (20:4) derived from membrane lipids (Fig. 1.16).

The enzyme cyclooxygenase (COX) catalyses the first step in the conversion of arachidonate to prostaglandins and thromboxanes. Non-steroidal anti-inflammatory drugs (NSAIDs), such as aspirin, ibuprofen and acetaminophen, inhibit COX and reduce

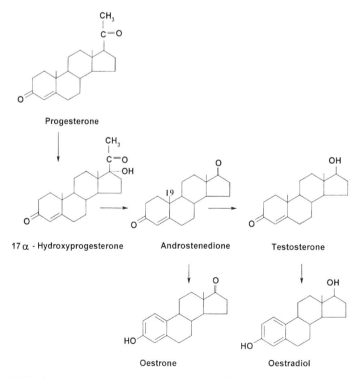

Fig. 1.15. Metabolism of progesterone to androgens and oestrogens.

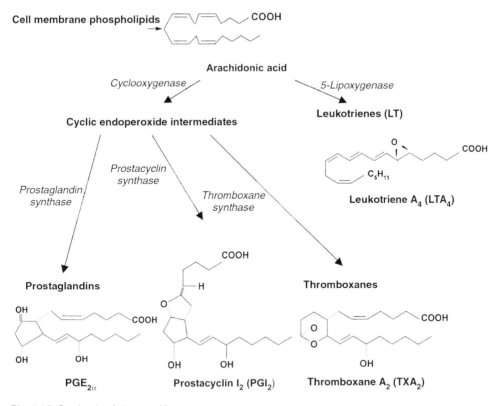

Fig. 1.16. Synthesis of eicosanoids.

the production of prostaglandins and thromboxanes. Prostaglandin E_2 (PGE$_2$) and $F_{2\alpha}$ (PGF$_{2\alpha}$) control vascular smooth muscle activity. Prostaglandin I_2 (PGI$_2$) is produced by the blood vessel wall and is the most potent natural inhibitor of blood platelet aggregation. Thromboxanes such as TXA$_2$ are produced by thrombocytes (platelets) and are involved in the formation of blood clots and the regulation of blood flow to the clot. Leukotrienes are made by leukocytes and are extremely potent in causing vasocontraction and inducing vascular permeability.

Synthesis of thyroid hormones

Synthesis of **thyroid hormones** occurs in the thyroid gland. The synthesis is stimulated by thyrotrophin (TSH) released from the anterior pituitary. Thyrotrophin is released in response to thyrotrophin releasing hormone produced from the hypothalamus. Thyroid hormones are synthesized by iodination of tyrosine residues in the thyroglobulin protein. Proteases in lysosomes degrade thyroglobulin to release triiodothyronine (T_3) and thyroxine (T_4) (Fig. 1.17).

Hormone release

Steroids are not stored but released immediately to diffuse out of the cell. Protein and peptide hormones are stored in granules within the gland and are released in response to various stimuli. **Trophic hormones** can stimulate hormone release: for example TSH, which stimulates the release of thyroxine. The trophic hormones FSH and LH stimulate the synthesis and release of gonadal steroids, while ACTH stimulates the synthesis and release of adrenal steroids (see Section 1.4). Hormones can be released in response to nervous stimuli from environmental cues such as light, smell, sound and temperature.

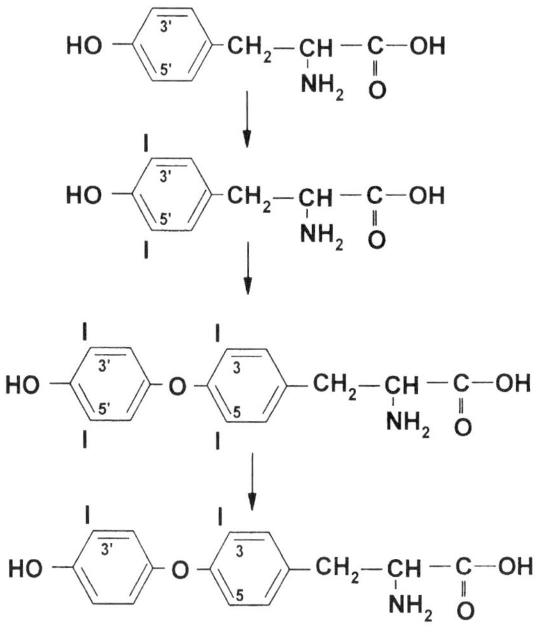

[Figure of thyroid hormone synthesis structures: Tyrosine → Diiodotyrosine → Thyroxine → Triiodothyronine]

Fig. 1.17. Synthesis of thyroid hormones.

This **neuroendocrine transduction** illustrates the integration of the nervous and endocrine systems. Hormones are also released in response to levels of various metabolites. For example, intracellular glucose levels control glucagon and insulin secretion, amino acids stimulate somatotrophin release and increase uptake of amino acids, while extracellular Ca^{2+} regulates parathyroid hormone and calcitonin secreting cells. These effects are all examples of **stimulus–response coupling**.

Metabolism and excretion of hormones

Hormones must be metabolized rapidly and removed so that feedback mechanisms can operate and hormones can regulate cellular functions. Removal or inactivation follows exponential decay kinetics (Fig. 1.18). The half-life of the hormones in the circulation is a measure of the longevity of hormone action. Many synthetic hormones and hormone analogues are designed to have a longer half-life and thus be effective for longer periods of time than natural hormones.

Peptide hormones are degraded by peptidases, such as the cathepsins in lysosomes, which split the peptide bonds in the molecule. **Exopeptidases** degrade protein from the carboxy-terminal end or the amino-terminal end. **Endopeptidases,** such as trypsin and chymotrypsin, degrade the protein at internal sites with some specificity. Trypsin hydrolyses peptide bonds where the carboxyl group is from lysine or arginine, while for chymotrypsin the carboxyl group in the peptide bond comes from phenylalanine, trypto-

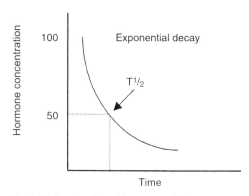

Fig. 1.18. Inactivation of hormones follows exponential decay.

phan or tyrosine. Deamination or reduction of disulphide bonds (e.g. insulin) can also inactivate proteins. This occurs in kidney, liver and in target cell lysosomes.

Steroid hormones are bound to protein carriers in blood, such as serum albumin or steroid-binding globulin, which are necessary since steroids are lipophilic. Binding to protein carriers also increases the half-life of steroids. Physiologically, only 5–10% of the hormone is present in the free or unbound form.

Steroids are degraded by a two-phase process in the liver and in the kidney. This process inactivates the steroids and makes them more water soluble for excretion. In phase one, enzymes such as cytochrome P450 (CYP) add functional groups such as hydroxyl groups. These metabolites are then conjugated to glucuronic acid or sulphates by transferase enzymes (Fig. 1.19). These more water-soluble metabolites are excreted by the kidney in the urine or by the liver in the bile salts.

1.3 Receptors and Hormone Action

Hormones interact with receptors located either on the cell surface or inside the cell to initiate their effects on the target tissue. Binding of hormones to cell-surface receptors activates intracellular enzyme systems to alter cell function. Hormones that cross the cell membrane act by binding to intracellular receptors. The hormone–receptor complex then interacts with DNA to affect expression of specific genes.

Extracellular receptors

Extracellular receptors are large macromolecules located on the outer surface of the plasma membrane in target tissues. For example, the insulin receptor has a molecular mass of 200–400 kDa, consisting of two α-subunits of 130 kDa and two β-subunits of 90 kDa linked by disulphide bonds. Usually, there are separate receptors for each hormone and the function of the cell (i.e. the cell type) dictates whether a particular receptor will be present on a cell and the number of receptors present. **Experimental evidence** that a hormone receptor is located on the cell surface includes:

1. Demonstrating that antibodies against the receptor can block hormone action;
2. Limited proteolysis of intact cells would be expected to destroy the receptor and remove the hormone response;
3. Coupling the hormone to a large molecule that cannot enter the cell and demonstrating that the effect of the hormone is still present; and
4. Demonstrating that the receptor is present in a plasma membrane preparation produced by subcellular fractionation (100,000 g pellet).

Hydrophobic regions on the receptor protein interact with lipid in the membrane. The receptor can be solubilized with detergents and purified by affinity chromatography using the hormone bound to a column matrix. Receptors can be glycoproteins and contain carbohydrate residues. **Experimental tools** to demonstrate this are:

1. Treat the receptor preparation with neuraminidase or β-galactosidase to remove the sugar residues. This inhibits binding of the hormone.
2. Concanavalin A (ConA; a protein from jack bean that binds to a D-glucosyl residues) can

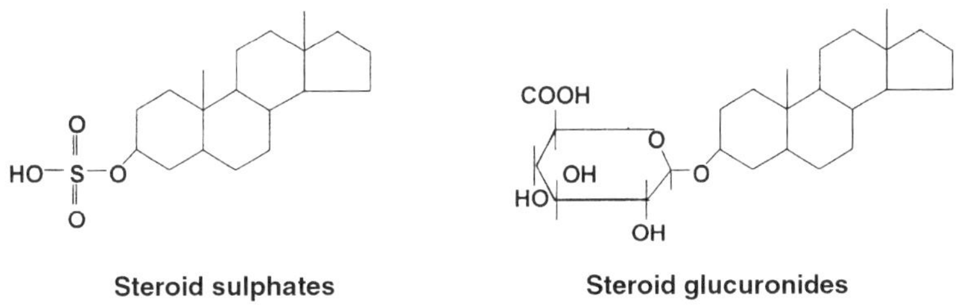

Steroid sulphates   **Steroid glucuronides**

Fig. 1.19. Structure of steroid sulphates and glucuronides.

be used to inhibit hormone binding. In addition, ConA can be used for affinity chromatography of glycoproteins (see Section 2.2, Chemical assays).

Second messenger systems

Some hormones interact with a **cell-surface receptor** and stimulate the synthesis of intracellular **second messenger** compounds. The hormone does not enter the cell to elicit a response but stimulates one of two main pathways:

1. The adenylate cyclase–cAMP–protein kinase A pathway or the related guanylate cyclase–cGMP-dependent protein kinase pathway; and
2. The calcium-dependent phospholipase C–protein kinase C pathway.

In the first system, hormone binding to the receptor activates the enzyme adenylate cyclase or guanylate cyclase, which synthesize either cAMP or cGMP. These second messengers activate protein kinase A. In the second system, binding of the hormone to the receptor activates phospholipase C, which splits phosphatidylinositol in the cell membrane to inositol phosphate and diacylglycerol. The inositol phosphate increases levels of intracellular calcium, which, together with the diacylglycerol, activates protein kinase C.

Both protein kinase A and protein kinase C can phosphorylate and activate various intracellular proteins to alter cellular metabolism. These proteins are inactivated by removing the phosphate using the enzyme phosphoprotein phosphatase.

Adenylate cyclase–cAMP–protein kinase A pathway

The enzyme adenylate cyclase catalyses the formation of cAMP from ATP. cAMP activates protein kinases which phosphorylate intracellular proteins to alter their activity (Fig. 1.20). The formation of cAMP is an amplification step that increases the effective hormone concentration, since one adenylate cyclase enzyme catalyses the formation of many cAMP molecules. The enzyme phosphodiesterase degrades cAMP to AMP by the following reaction:

ATP $\rightarrow$ cAMP $\rightarrow$ AMP
   Adenylate   Phosphodiesterase
   cyclase

Several properties of cAMP make it suitable as a second messenger. It is derived from ATP

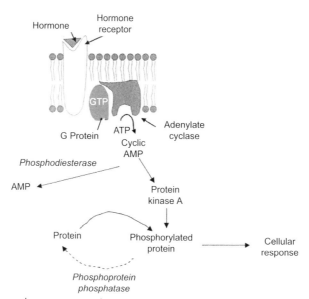

Fig. 1.20. Cyclic AMP second messenger system.

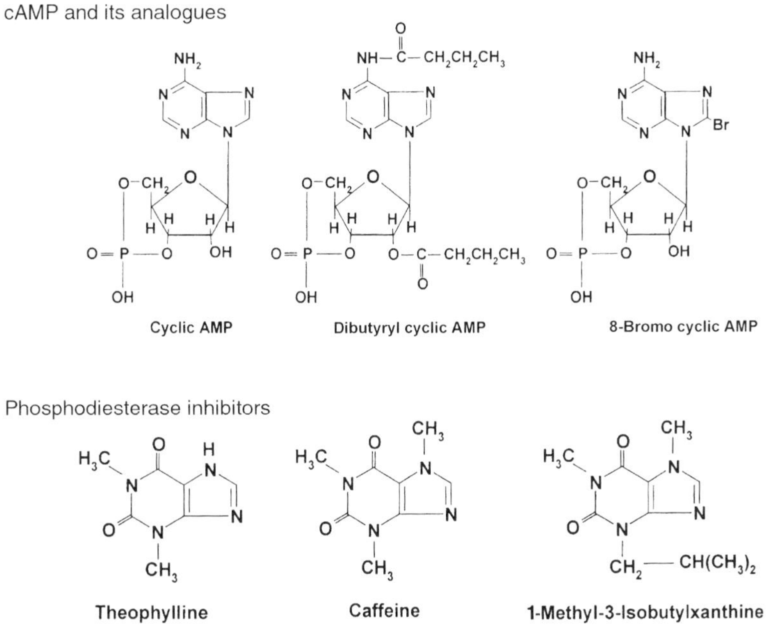

Fig. 1.21. cAMP and its analogues and phosphodiesterase inhibitors.

but is chemically stable. ATP is ubiquitous and cAMP is formed from it in a single reaction. Since cAMP is not a metabolic precursor, but an allosteric regulator, it can be controlled independently of metabolism. cAMP is a small and easily diffusable molecule and it has a number of functional groups that allow specific binding to regulatory subunits of protein kinases.

Experimentally, the involvement of cAMP as a second messenger can be determined if **physiological levels** of hormone increase cAMP in cells and cAMP production precedes the physiological effect. The hormone should also stimulate adenylate cyclase activity in broken cells. Treatment with exogenous cAMP or its analogues, such as dibutyryl cAMP and 8-bromo cAMP (Fig. 1.21) should produce the hormone response. Phosphodiesterase inhibitors, such as theophylline, caffeine or isobutylmethylxanthine (IBMX), will decrease cAMP clearance and potentiate the response. Adenylate cyclase can be activated by treatment with the diterpene, forskolin (Fig. 1.22), which binds directly with the catalytic subunit to permanently activate it.

A number of different hormones act via the cAMP second messenger system (Table 1.2). The substrates for cyclic AMP-dependent protein kinases include triglyceride lipase, which is involved in the regulation of lipolysis, phosphorylase *b* kinase in the regulation of glycogenolysis, cholesterol ester hydrolase in the regulation of steroidogenesis and fructose 1,6-diphosphatase in the regulation of gluconeogenesis. These latter enzymes are all activated by phosphorylation. Enzymes that are inactivated by phosphorylation include pyruvate kinase in the regulation of glycolysis and gluconeogenesis, glycogen synthase in the regulation of glycogen synthesis, and 3-hydroxy-3-methylglutaryl-CoA reductase in the regulation of cholesterol biosynthesis.

Guanyl cyclase–cGMP-dependent protein kinase pathway

The cGMP system is similar to the cAMP system, but may act in opposition to cAMP. This is known as the **yin–yang hypothesis**. For example, activation of the cAMP-dependent kinases results in smooth muscle relaxation,

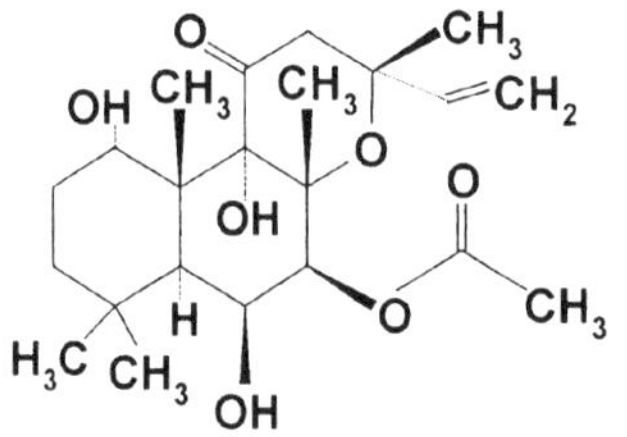

Forskolin

Fig. 1.22. Forskolin, an activator of adenylate cyclase.

while activation of the cGMP-dependent kinases results in smooth muscle contraction. Levels of cGMP are normally 10–50 times lower than those of cAMP.

GENOMIC ACTIONS OF CAMP. Activating protein kinase A and subsequent phosphorylation of intracellular proteins can cause immediate cellular responses, such as modification of metabolic pathways and regulation of ion flows and muscle contraction. However, cAMP can also have effects on gene transcription by protein kinase A activation of the cAMP-responsive-element binding protein (CREB), or modification of the structural proteins in chromatin. Activated CREB binds to specific cAMP-responsive elements in the regulatory regions of certain genes to activate gene expression.

Calcium-dependent phospholipase C–protein kinase C system

The primary intracellular effector in this pathway is calcium, which activates a calcium-dependent protein kinase C. Hormone binding activates phospholipase C, which catalyses the hydrolysis of phosphatidylinositol-4,5-bisphosphate to produces inositol-1,4,5-phosphate (IP_3) and diacylglycerol (DAG) (Fig. 1.23). Inositol-1,4,5-phosphate and its metabolite, inositol-1,3,4,5-tetrakisphosphate (IP_4), increase intracellular Ca^{2+} by activating calcium channels at the endoplasmic reticulum and cell surface. Diacylglycerol activates protein kinase C by increasing its affinity for Ca^{2+}, and protein kinase C then phosphorylates cellular proteins to regulate their activity (Fig. 1.24).

In both of these systems, the receptor is discrete from the enzyme system that it activates; both 'float' in the cell membrane lipid. The receptor interacts with the adenylate cyclase or phospholipase C via an intermediate G-protein. The G-protein is activated by binding GTP and inactivated when the GTPase activity converts the GTP to GDP. G-protein function can be studied using a non-hydrolysable form of GTP or cholera toxin, both of which inhibit the GTP-ase activity. The *Ras* oncogene codes for a permanently active G-protein, which might explain its role in the development of cancer. G-proteins act to couple cell-surface receptors for hormones, neurotransmitters, odorants and photons of light to effector molecules such as ion channels or enzymes that generate second messenger molecules.

An example of the stimulus response coupling due to this system is the effect of gonadotrophin releasing hormone (GnRH). GnRH is produced by the hypothalamus and causes LH and FSH release by the anterior pituitary. GnRH increases cellular Ca^{2+} levels and affects inositol metabolism. LH release is Ca^{2+} dependent and an increase in intracellular Ca^{2+} causes a release of LH.

Calcium also binds to **calmodulin** to form an active complex. Calmodulin is a heat-stable globular protein of molecular

Table 1.2. Hormones that act via the adenylate cyclase–cAMP–protein kinase A pathway.

Glucagon	Chorionic gonadotrophin
Vasopressin	Parathyroid hormone
Thyrotrophin (TSH)	Calcitonin
Adrenocorticotrophic hormone (ACTH)	Luteinizing hormone releasing hormone (LHRH)
Luteinizing hormone (LH)	Thyrotrophin releasing hormone (TRH)
Follicle stimulating hormone (FSH)	Secretin

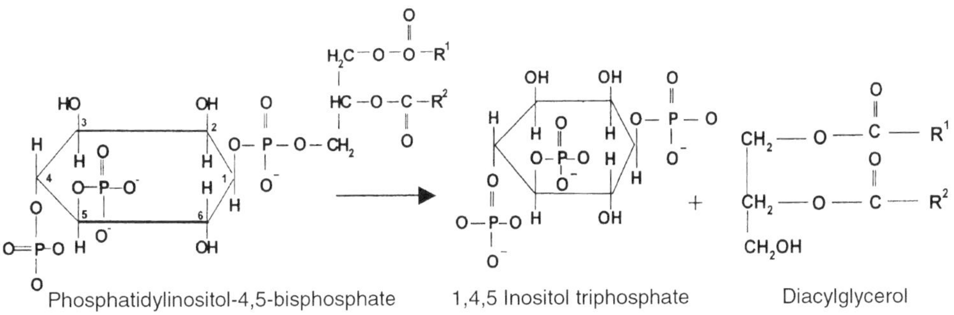

Fig. 1.23. Action of phospholipase C.

mass 16 kDa and is a calcium-dependent regulatory protein found in all eukaryotic cells. It controls intracellular Ca^{2+} and binds four Ca^{2+} ions per molecule to form an active complex. This complex acts as an allosteric regulator of protein kinase C and other enzymes. It also controls the activity of cellular filamentous organelles (via actin and myosin) responsible for cell motility, exoplasmic flow (hormone secretion) and chromosome movement.

Experimental tools used to determine the involvement of the calcium-dependent phopholipase C–protein kinase C system are:

1. Increase intracellular Ca^{2+} levels using Ca^{2+}-selective ionophores (A23187) or liposomes loaded with Ca^{2+}.
2. Decrease intracellular Ca^{2+} by chelating

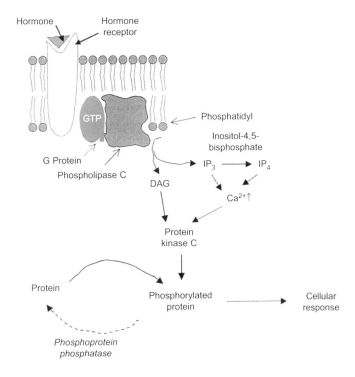

Fig. 1.24. Calcium-dependent protein kinase C second messenger system.

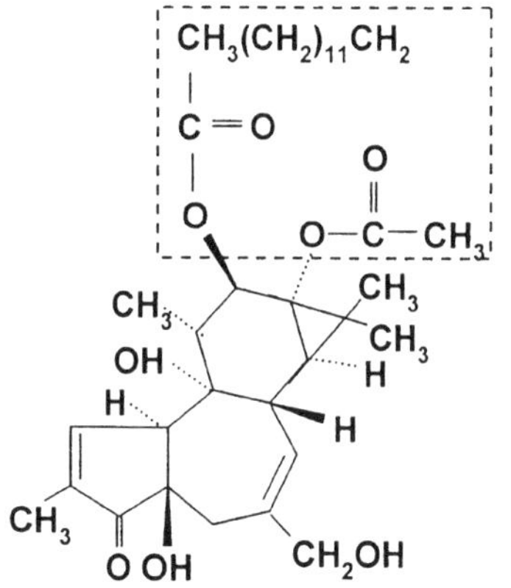

Fig. 1.25. Tetradecanoylphorbol acetate (TPA) or phorbol 12-myristate 13-acetate (PMA). The outlined area has a structure similar to that of diacylglycerol.

with EGTA, using Ca^{2+}-channel blockers, or inorganic Ca^{2+} antagonist (La^{3+}).
3. Use phorbol esters (tetradecanoylphorbol acetate (TPA)), which resemble diacylglycerol (Fig. 1.25, see structure inside dashed lines), to activate protein kinase C.
4. Inhibit phospholipase C with U73122; the inactive analogue, U73343, is used as a positive control.

INTERACTION OF cAMP AND Ca^{2+} PATHWAYS. There is a considerable amount of 'cross-talk' between the different secondary messenger systems. Ca^{2+} binds to calmodulin and this complex can bind to phosphodiesterase to activate it and decrease cAMP levels. Protein kinase A, which is activated by cAMP, can phosphorylate some Ca^{2+} channels and pumps and alter their activity to affect intracellular calcium levels. Protein kinase C can be phosphorylated by protein kinase A to change its activity. Protein kinase C and protein kinase A can phosphorylate different sites on the same protein, so that its activity is regulated by both cAMP and Ca^{2+}.

Tyrosine kinase receptors: catalytic receptors

The tyrosine kinase receptors do not use a second messenger system to activate a separate protein kinase, but have a kinase domain as part of the receptor structure. The activated receptor phosphorylates tyrosine residues in its kinase domain and can then phosphorylate other proteins, and is thus called a **tyrosine kinase** receptor. The receptor consists of a transmembrane domain, an extracellular domain for hormone recognition and a cytoplasmic domain that transmits the regulatory signals and contains ATP binding sites. The cytoplasmic domain has a C-terminal tail with autophosphorylation sites. The phosphorylated receptor acts as a kinase enzyme and phosphorylates substrates. These phosphorylated substrates transmit several regulatory signals into the cell.

There are three main classes of tyrosine kinase receptors (Fig. 1.26). The class I receptor (e.g. epidermal growth factor (EGF) receptor) is a monomeric transmembrane protein with intracellular and extracellular domains on the same molecule. The extracellular domain contains two cysteine-rich repeat regions. The class II receptor (e.g. insulin receptor) is a heterotetrameric receptor in which the two α-subunits and the two β-subunits are linked by disulphide bonds. The class III receptor (e.g. receptors for platelet-derived growth factor (PDGF) or nerve growth factor (NGF)) is a monomeric protein with cysteine residues over the extracellular domain. The intracellular domain has unique amino acid inserts in the middle of the kinase domain.

Hormone binding causes dimerization of the monomeric receptors (Fig. 1.27). The kinase domains in the monomers are then phosphorylated and activated by their partner. The kinases can then phosphorylate and activate other proteins. The receptors also interact with proteins containing SH_2 domains that bind to phosphotyrosines. The amino-acid sequence next to the phosphotyrosines specifies which protein, containing an SH_2 domain, can bind to receptor. The SH_2 domain protein can be attached to a different enzymatic domain or be a linker molecule that binds to other enzyme molecules that could not normally bind to the receptor. In this way, linker molecules can bind a number of different specific molecules to bring them together to produce the desired biological effect. For example, the appropriate kinase and phosphatase can be held in position so

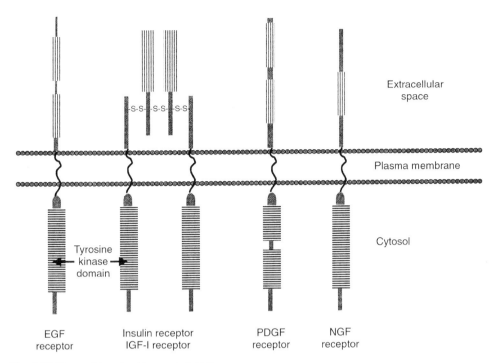

Fig. 1.26. Types of tyrosine kinase receptors.

that a protein (receptor, ion channel, etc.) is activated by the kinase in the presence of the appropriate signalling molecule and deactivated by the phosphatase when the signalling molecule is absent. The role of SH_2 domain proteins is described in a popular press article by Scott and Pawson (2000).

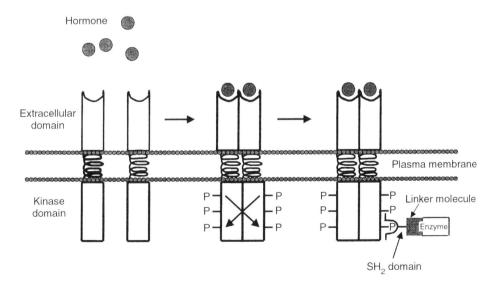

Fig. 1.27. Mechanism of action of tyrosine kinase receptors.

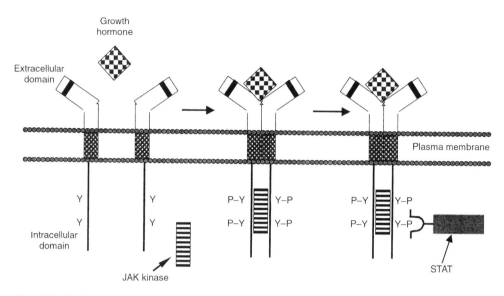

Fig. 1.28. Mechanism of action of cytokine receptors.

CYTOKINE RECEPTORS. Receptors for cytokines (growth hormone (GH), prolactin, erythropoietin, interferons and interleukins) do not have intrinsic kinase activity (Fig. 1.28). The GH receptor exists as a monomer when it is not bound to hormone. The binding of hormone causes dimerization of the receptor and binding of Janus kinase (JAK) tyrosine kinase, which phosphorylates the receptor to activate it. Phosphotyrosines act as docking sites for intracellular signalling molecules, such as STATs (signal transducers and activators of transcription), which activate various genes.

RECEPTOR SERINE KINASE. Receptor serine kinases include receptors for transforming growth factor-β (TGFβ), activin and inhibin. These peptide growth factors are involved in the control of cell proliferation and differentiation. The existence of these receptors was demonstrated by chemical cross-linking of radiolabelled hormone to cell-surface proteins on responsive cells.

The binding of hormone results in formation of a heterodimer of receptors I and II. Serine residues on RI are phosphorylated by RII to activate the complex (Fig. 1.29).

Termination of hormone action

After hormones interact with receptors, they cluster together, and this triggers vesicularization of membrane and endocytosis. The receptors may then be degraded by lysomal enzymes or the receptor can be recycled. The hormone at the cell surface can be degraded by serum enzymes. The cyclic nucleotides are degraded by phosphodiesterases and the phosphorylated proteins are dephosphorylated by phosphoprotein phosphorylase.

Intracellular receptors

Steroid and thyroid hormones operate via intracellular receptors. In 1966, Toff and Gorski identified oestrogen receptors, using radiolabelled hormone, and showed that the receptor was only present in target cells. Receptors for glucocorticoids (GR), mineralocorticoids (MR) and androgens (AR) are found predominantly in the cytoplasm, while receptors for thyroid hormone (TR), oestrogen (ER), progesterone (PR), retinoic acid (RAR) and 1,25-hydroxy vitamin D_3 (VDR) are predominantly in the nucleus. Receptors for steroid hormones act as transcription factors to regulate the transcription of target genes.

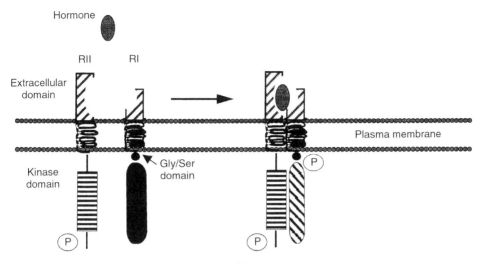

Fig. 1.29. Mechanism of action of receptor serine kinases.

Steroid hormone receptors move between the nucleus and cytoplasm and, in the absence of hormone, are bound to the 90 kDa heat-shock protein complex (Hsp90). (TR, RAR and VDR do not bind Hsp90.) Binding of the hormone to the receptor results in release of the Hsp90 complex and translocation of the hormone–receptor complex to the nucleus. A dimer of the hormone–receptor complex then interacts with hormone-responsive elements on specific genes to affect DNA transcription. This exposes template sites on DNA, either directly or by influencing pre-existing repressor molecules, to increase the initiation sites for RNA polymerase and increase transcription (Fig. 1.30). These actions of steroid hormones occur over a much longer time frame (hours) compared to peptide hormones.

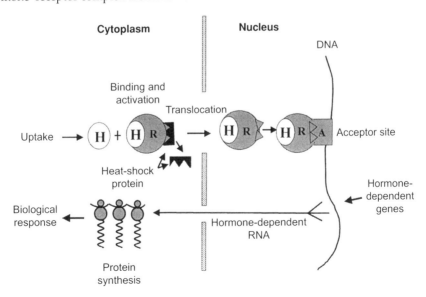

Fig. 1.30. Mechanism of action of steroid hormones.

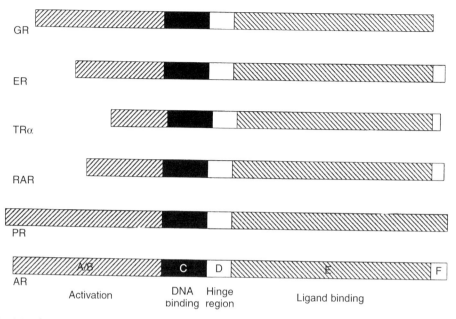

Fig. 1.31. Structural domains of intracellular receptors.

Steroid hormones can also affect the extent of mRNA degradation. The effects of steroid hormones can be determined **experimentally** by:

1. Measuring steady-state levels of mRNA by Northern blotting. Levels would increase with increased synthesis or decreased degradation of mRNA.
2. In order to separate synthesis from degradation, a nuclear run-on assay can be used to measure the rate of mRNA synthesis.
3. Actinomycin D (an inhibitor of RNA transcription) and puromycin (an inhibitor of protein synthesis) can be used to inhibit the effects of steroid hormones.

Steroid hormones can affect the response to other hormones. This can be through synthesis of receptors or protein kinases to increase hormone response or phosphoprotein phosphatases, which are antagonistic to cyclic nucleotide actions.

Structural and functional domains of nuclear receptors

Nuclear receptors have common structural domains involved in DNA binding and hormone binding (Fig. 1.31). Other regions are involved in dimerization of the receptor and translocation of the receptor to the nucleus.

The **ligand-binding domain** is the region where the hormone binds to the receptor. Sequence diversity in this region gives the specificity of the receptor for the hormone. The **DNA-binding domain** of the receptor is comprised of 60–70 amino acids arranged in 'zinc fingers'. This region is where the hormone–receptor complex binds to hormone-responsive genes to stimulate transcription (Fig. 1.32).

Binding sites of the hormone–receptor complex on DNA

Hormone target genes contain specific **hormone-responsive elements** (HREs), which are involved in binding to the hormone–receptor complex. Identification of these sequences in a gene suggests that hormones regulate the gene. The sequences are usually in the 5′ regions, but may also be in the introns or the 3′ region. All HREs have a similar structure, consisting of pairs of hexamers arranged as inverted palindromes, with one side of the palindrome having the complementary

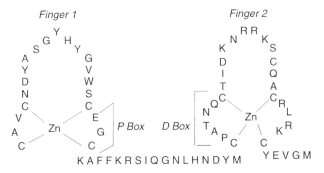

Fig. 1.32. DNA-binding domains of intracellular receptors. The P-box sequences define the specificity of receptor binding to DNA. The sequences are: ER, EGCKA; TR, EGCKG; RAR, EGCKG; GR, GSCKV; PR, GSCKV.

sequence to the other side. The complementary sequences are separated by a defined number of intervening bases. Steroid receptors bind to the HREs as homodimers, with the axis of symmetry over the centre of the palindrome. Receptors for thyroid hormone (TR) and vitamin D (VDR) bind to their HREs as heterodimers along with the RXR receptor (named for binding to 9-*cis* retinoic acid). Recent evidence has shown that the polyunsaturated fatty acid docosahexaenoic acid (DHA) binds to the RXR receptor *in vivo*, which may explain the vital role of DHA in development.

Only two types of consensus hexamers are used. The steroid receptors (GR, MR, AR and PR) bind to 5'-AGAACA-3', while the ER, RXR, TR and VDR bind to 5'-AGGTCA-3'. The residues in the P box of the DNA-binding domain of the receptor determines which type of HRE consensus sequence is recognized by the receptor. The spacing between the pairs of hexamers in the palindrome varies from one to five nucleotides for different receptors.

Organization of nuclear chromatin and the nuclear matrix

CHROMATIN. The DNA in eukaryotic cells is tightly associated with protein in the form of chromatin (Fig. 1.33). Chromatin exists as a string of nucleosomes, which are composed of 146 bp of DNA wrapped around a core of eight histones. Histones are small, basic proteins that are rich in arginine, lysine and his-

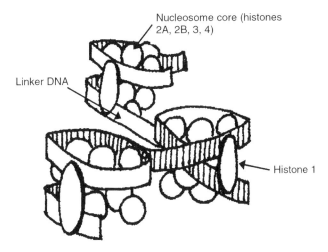

Fig. 1.33. Nucleosome structure of chromatin.

tidine. The histone core of the nucleosome consists of two each of H2A, H2B, H3 and H4, and this complex is coated with positive charges that interact with the negatively charged DNA. Histones H3 and H4 are very similar between species, with a difference of only two amino acids in H4 and four amino acids in H3 found between pea and calf thymus. An additional histone (H1) is involved in the packaging of the individual nucleosomes to form condensed chromatin. There are no changes in the histones that are related to cell function and they are not involved in binding of the hormone–receptor complex to chromatin.

DNA that is highly condensed is often inactive, with transcription taking place in the more loosely packed regions of the genome, which can be detected by their hypersensitivity to nucleases or chemical probes. The histones in active chromatin are highly acetylated, which removes a positive charge on lysine and reduces the binding to DNA and the packing density of the chromatin. The binding of a steroid hormone–receptor complex produces hypersensitive sites in chromatin allowing the binding of other transcription factors to initiate gene transcription.

Chromatin also contains non-histone proteins or acidic proteins. The acidic proteins are rich in aspartic acid and glutamic acid. Many types of acidic proteins are present, and their composition varies between tissues. One group of these, the non-histone high-mobility group (HMG) proteins, has been implicated in the regulation of transcription by enhancing the binding of steroid hormone receptors to their target sites on DNA. In particular, HMG-1 and HMG-2 bind to specific DNA structures and they can also modify and stabilize the structure of DNA, to aid in the assembly of protein complexes at the promoter. They thus act to increase the sequence-specific binding of steroid hormone receptors to DNA and increase transcription of steroid hormone responsive genes. However, HMG-1/2 do not stimulate binding of non-steroid nuclear receptors, such as VDR, RAR and RXR (for further details, see Boonyaratanakornkit et al., 1998).

The increase in initiation sites caused by binding of steroid hormone–receptor complexes to DNA can be demonstrated using inhibitors of free RNA polymerase, such as rifampicin or α-amanitin. In this experiment (Fig. 1.34), RNA polymerase is added to chromatin, which then binds to available initiation sites. Rifampicin is then added, which binds to and inhibits the excess RNA polymerase not bound to chromatin. Nucleotides are then added to start transcription and, after one copy is made, RNA polymerase is inhibited. The number of copies of RNA transcribed is a measure of the number of initiation sites on the chromatin.

NUCLEAR MATRIX. The nuclear matrix is a three-dimensional skeletal network that exists in the nucleus. It is isolated by extracting nuclei with Triton X-100 detergent, followed by treatment with DNase I and 2 M NaCl. The matrix allows the nucleus to be

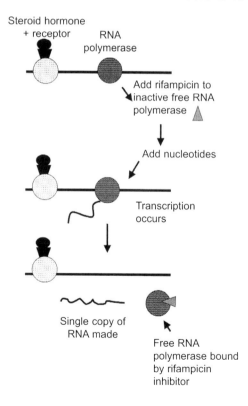

Fig. 1.34. Experiment to illustrate that steroid hormone receptor binding increases RNA polymerase initiation sites in target genes.

organized into different domains for DNA replication, transcription and RNA splicing. The nuclear matrix is analogous to the cytoskeleton, which is involved in the intracellular transport of molecules in the cytosol, for example the movement of hormone–receptor complexes to the nucleus.

Specific, high-affinity steroid receptors have been found to be associated with actively transcribed genes that are also associated with the nuclear matrix. The interaction of receptors with the nuclear matrix is hormone dependent. The nuclear matrix may thus provide a mechanism whereby the hormone–receptor complex binds to specific DNA sequences. Acceptor proteins for receptors have been identified associated with the matrix, and specific domains of the receptors are involved in matrix binding. Release of receptors from the matrix may be ATP dependent. Nuclear matrix proteins of 10–17 kDa have been identified that cause specific high-affinity binding of the progesterone and oestrogen receptor to genomic DNA. These acceptor proteins are expressed only in tissues that are responsive to progesterone.

The nuclear matrix is also involved in the organization of DNA, DNA replication, heterogeneous nuclear RNA (hnRNA) synthesis and processing. It may play a role as a communication network from one part of the cell to another, in the transport of molecules (e.g. chromosomes during replication), and as a framework for reactions. For more information on nuclear acceptor sites for the progesterone receptor, see Spelsberg *et al.* (1996).

IDENTIFICATION OF DNA REGULATORY SEQUENCES. DNA sequences that are important in regulating gene expression can be identified using an *in vitro* system (Fig. 1.35). The **experimental method** involves:

1. Linking the regulatory sequences of interest to a test gene and transfecting the gene construct into a cell line. For measuring hormone-responsive elements, the cell line must also contain the receptor for the hormone.
2. The activity of the test gene is measured in the presence of hormone and receptor.
3. The sequences of interest are then gradually deleted and the effect of the deletion on expression of the test gene is measured.

Deletion of important regulatory sequences will dramatically affect the level of gene expression. For example, deletion of sequences important in activation of the gene by a hormone–receptor complex will eliminate the stimulation of gene transcription by the hormone. Deletion of repressor regions will result in increased activity of the test gene.

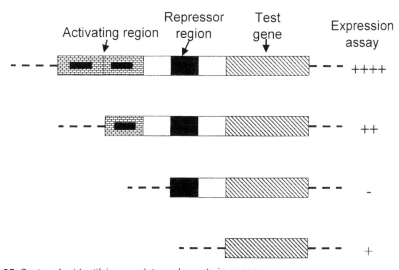

Fig. 1.35. System for identifying regulatory elements in genes.

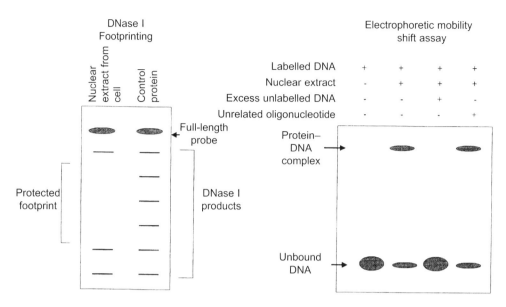

Fig. 1.36. DNase footprinting and electrophoretic mobility shift assays.

IDENTIFICATION OF DNA-BINDING PROTEINS. The binding of protein factors to DNA can be determined by DNase footprinting and electrophoretic mobility shift assays (Fig. 1.36). These techniques can be used to determine sequences that are involved in hormone/receptor binding.

DNase footprinting assays begin by labelling a double-stranded DNA sequence that contains the region of DNA (promoter) suspected of binding protein factors. The labelled DNA is then incubated with a source of binding protein, such as an extract of nuclei from cells in which the promoter functions. A control incubation is carried out with a source of protein that does not bind the DNA. This is followed by limited digestion with DNase I to introduce approximately one nick in each DNA molecule. The DNA fragments are separated by denaturing polyacrylamide gel electrophoresis to generate a 'ladder' of DNA fragments. The region of the DNA probe bound by the protein is protected from DNase digestion and this results in a hole or 'footprint' in the ladder.

The electrophoretic mobility shift assay is used to confirm that a small region (25 bp) of DNA binds to protein. A labelled DNA probe is incubated with a source of binding protein, such as a nuclear extract, and the mixture is separated by non-denaturing gel electrophoresis to preserve DNA–protein complexes. The unbound DNA probe has a low molecular weight and moves rapidly through the gel, while the DNA–protein complex is of much larger size and is 'shifted' to a higher region of the gel. The specificity of binding of protein to the labelled probe is confirmed by the absence of the labelled higher molecular weight band when an excess of the same DNA that is unlabelled is included. An unrelated oligonucleotide should have no effect.

Integration of peptide and steroid hormone actions

The classic descriptions of peptide and steroid hormone actions described above are not the whole story. Steroid hormones can also cause rapid effects by acting at the cell surface and not only longer-term effects on gene transcription by binding to intracellular receptors. For example, plasma sex hormone binding globulin (SHBG) is involved in the transport of androgens and oestrogens in the blood, and it regulates the amount of free sex steroids that are released to the target cells. In addition, a SHBG receptor is present on the surface of target cells. SHBG can bind to its receptor at the cell surface and then bind free steroids, to activate a cAMP second messenger system. In this way, the SHBG modulates the effects of sex steroids acting on receptors within the cells. Steroids that bind to the SHBG but do not activate the second messenger system act as antagonists. The actions of SHBG have been reviewed (Fortunati, 1999). In addition to gonadal steroids, rapid, non-genomic effects for glucocorticoids, mineralocorticoids, vitamin D_3 and thyroid hormone are also known. For further information on the rapid non-genomic effects of steroid hormones, see the review by Falkenstein et al. (2000).

Steroid hormones can also affect the activity of protein hormones by stimulating the synthesis of protein hormone receptors on the cell surface, or by affecting the synthesis of protein kinases or other intracellular proteins that are involved in the action of peptide hormones. As mentioned previously, production of cAMP by the action of protein hormones can also have effects on gene transcription. This occurs by phosphorylation and activation of the cAMP-responsive-element binding protein (CREB) by protein kinase A or by modification of the structural proteins in chromatin. Many hormone-responsive genes have specific cAMP-responsive elements in their regulatory regions. Activated CREB binds to these regions to activate gene expression. For more information, see the texts by Stryer (1995) and Nelson and Cox (2000).

1.4 Pituitary–Hypothalamic Integration of Hormone Action

Structure–function relationship of pituitary and hypothalamus

The hypothalamus is a part of the brain located below the third ventricle above the median eminance. The pituitary gland or hypophysis is located below the hypothalamus in a hollow pocket of the sphenoid bone known as the 'sella turcica' and is linked to the hypothalamus. The pituitary gland consists of two distinct lobes, the posterior pituitary or neurohypophysis and the anterior pituitary or adenohypophysis. The posterior pituitary is nervous tissue that develops as an outgrowth of the diencephalon. It receives hormones that are made in the magnocellular neurones in the hypothalamus and are transported along the axons to the posterior pituitary. The anterior pituitary is glandular tissue, and is subdivided into pars distalis and pars intermedia. It is the 'master gland', which produces a number of trophic releasing hormones that stimulate hormone release by target tissues (Fig. 1.37).

The hypothalamus is innervated with many neurones from other parts of the body, and receives signals from cells such as baroreceptors and osmoreceptors and other environmental cues. It produces releasing hormones and release inhibiting hormones, which are delivered to the anterior pituitary by the hypothalamic–pituitary portal system. Release from the posterior pituitary is by direct nervous stimulation, from other neurones in the hypothalamus to the neurones that produce the posterior pituitary hormones. The release of hormones from the hypothalamus and pituitary is pulsatile because of the pulsatile firing of nerves.

Releasing factors or release inhibiting factors produced by specific neurones in the hypothalamus regulate release of hormones from the anterior pituitary. These factors are produced in very small amounts and are

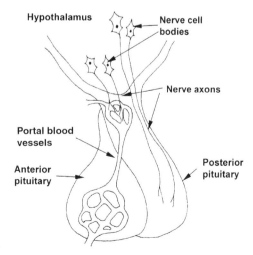

Fig. 1.37. Organization of the pituitary and hypothalamus.

delivered from the hypothalamus to the anterior pituitary by the **hypothalamic–hypophyseal portal system** (Fig. 1.37).

Release of hormones by the posterior pituitary is under direct nervous control. The posterior pituitary gland has nerve endings with bulbous knobs that lie on the surfaces of capillaries. Vasopressin or ADH (antidiuretic hormone) and oxytocin are secreted into the capillaries by exocytosis from the neurones that produce them (Fig. 1.38).

Posterior pituitary hormones

Oxytocin and **vasopressin** are the two hormones that are released from the posterior pituitary. Oxytocin causes contraction of smooth muscles. These include the myoepithelial cells for milk let down in the mammary gland, and in the myometrium for the contraction of the uterus for parturition. Vasopressin, also known as antidiuretic hormone, stimulates reabsorption of water from the distal tubular kidney to maintain blood osmolarity when blood volume or blood pressure is decreased. Both vasopressin and oxytocin are polypeptides containing nine amino acids with a disulphide bridge between two cysteines in the molecule (Fig. 1.39).

These two hormones are almost identical, except that in vasopressin, phenylalanine and arginine replace isoleucine and leucine of the oxytocin molecule. They are synthesized as **preprohormones** in the cell bodies of specific neurones in the hypothalamus. The prohormones are cleaved to active hormones

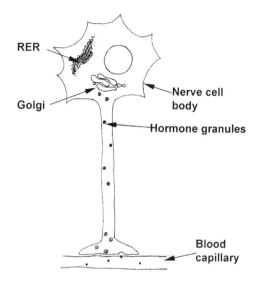

Fig. 1.38. Hormone production and axonal transport by nerve cells. RER, rough endoplasmic reticulum.

during fast axonal transport from the cell body down the axon in the posterior pituitary and before they are released into the circulation (Fig. 1.38). They are released in response to changes in osmotic or barometric pressure, pain, fright or stress, adrenal insufficiency, hypoxia or cardiac failure.

Anterior pituitary hormones

The anterior pituitary produces a number of trophic hormones that cause hormone release from target tissues (Table 1.3). Hormones of

Table 1.3. Hormones of the anterior pituitary.

Hormone	Number of amino acids	Structure
GH	191	Single-chain proteins with 2–3 disulphide bonds
PRL	199	
TSH	211	Glycoproteins, 2 subunits with common α-subunit
FSH	210	
LH	204	
ACTH	39	Short, linear-chain peptides
MSH	13/22	
β-LPH	91	

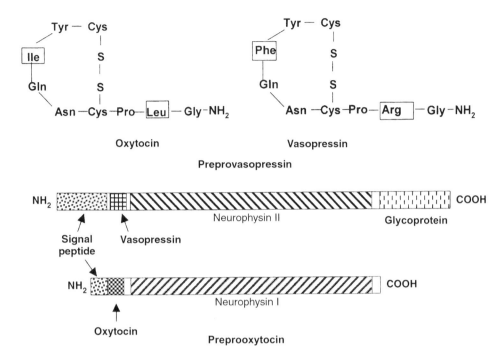

Fig. 1.39. Structures of oxytocin, vasopressin and their preprohormones.

the anterior pituitary are proteins or glycoproteins. They have longer half-lives than releasing hormones produced by the hypothalamus. They are made in specific cells (thyrotrope, gonadotrope, corticotrope, melanotrope, somatotrope, mammotrope) in the anterior pituitary gland.

Hypothalamic control of pituitary hormone secretion

The control of pituitary hormone release by the hypothalamus (Fig. 1.40) has been demonstrated by a number of **experiments**. These include:

1. Placing a mechanical barrier between the hypothalamus and the pituitary, or ectopic transplantation of the pituitary gland (to another site that has a good blood supply, see Section 2.1). This will either decrease or enhance secretion of certain pituitary hormones.

2. Pharmaceutical agents that affect neurotransmitter production, or electrical stimula-

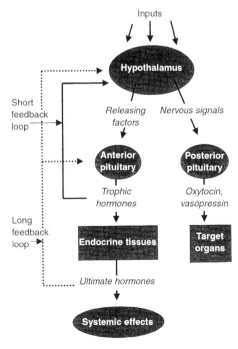

Fig. 1.40. Regulation of pituitary hormone release.

Table 1.4. Hypothalamic hormones.

Releasing hormone	Trophic hormone	Target tissue and hormone	Ultimate effect
TRH (thyrotrophin releasing hormone)	TSH (thyrotrophic hormone)	↑ T4 and T3 from thyroid	Regulates metabolic rate
	PRL (prolactin)	Mammary gland	Synthesis of milk
GnRH (gonadotrophin releasing hormone)	LH (luteinizing hormone)	Ovary	Affects ovulation
	FSH (follicle stimulating hormone)	Ovary	Affects gonadal development
CRH or CRF (corticotrophin releasing hormone)	ACTH (adrenocorticotrophic hormone)	↑ Adrenocortical steroids from adrenal cortex	Response to stress
	β-LPH (β-lipotrophin degrades to β-endorphin [Met]enkephalin		Analgesia in stress neurotransmitter
GHRH and GH-RIH (growth hormone releasing hormone, and growth hormone release-inhibiting hormone)	GH (somatotrophin))	↑ Somatomedin production by liver	Increased growth
MRF, MIH (melanocyte stimulating hormone releasing and inhibiting hormones)	MSH (melanocyte stimulating hormone)	Melanocytes	↑ Skin pigmentation
PRH and PIH (prolactin releasing and inhibiting hormones)	PRL (prolactin)	Mammary gland	Synthesis of milk

tion of discrete hypothalamic sites, can affect the release of pituitary hormones.
3. Hypothalamic extracts injected into an intact animal affect pituitary secretions.
4. Releasing factors have been localized to specific hypothalamic neurones, and levels of these factors are higher in the hypothalamic–pituitary portal system than in the systemic circulation.
5. Antisera to hypothalamic releasing hormones lead to enhanced or inhibited pituitary hormone secretion (for example, see immunization against GnRH, Section 3.3).

Hypothalamic releasing and release-inhibiting hormones

The function of the releasing and inhibitory hormones from the hypothalamus is to control the secretion of the anterior pituitary hormones. For each type of anterior pituitary hormone there is usually a corresponding hypothalamic releasing hormone; for some of the anterior pituitary hormones there is also a corresponding hypothalamic inhibitory factor (Table 1.4).

The hypothalamic releasing hormones are generally made as a larger prohormone, which is cleaved later to form active peptide hormones. Using labelled antibodies, it can be demonstrated that specific neurones make different releasing hormones. The releasing hormone precursors are made in cell bodies and transported down the axons to the nerve endings for storage. They are released by exocytosis of granule contents (excretion of substances through the cell membrane) into the hypothalamic–pituitary portal blood system in response to electrical signals from other neurones. Chemical signals for release include metabolite levels and other hormones such as steroid

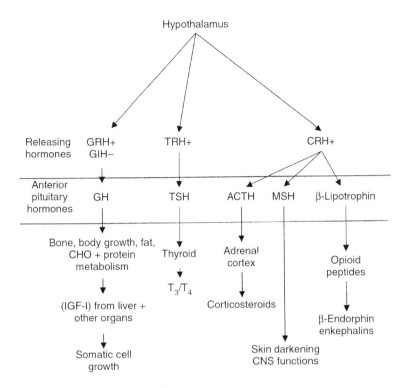

Fig. 1.41. Overview of regulation of metabolic hormones.

hormones or small molecules that can cross the **blood–brain barrier**.

The releasing and inhibiting hormones produced by the hypothalamus are all simple peptides or proteins. They range in size from three amino acids for TRH (pyro-E-H-P-amide), ten amino acids for GnRH (pyro-E-H-W-S-Y-G-L-R-P-G-amide), 14 amino acids for GH-RIH, 41 amino acids for CRH and 44 amino acids for GHRH.

Some hormones (e.g. GH, MSH, PRL) are under tonic inhibition by release inhibiting hormones. These hormones are needed early in life, but as the hypothalamus matures and becomes more active, the hormones that are 'driven' by releasing hormones increase, while GH and MSH decrease.

Control of hormone release

Electrical (nervous) stimulus in response to environmental or internal signals causes the release of releasing hormone by the hypothalamus. The releasing hormone is produced in nanograms and is delivered to the pituitary by the portal system. This causes the release of trophic hormone by the anterior pituitary in microgram amounts. The trophic hormone then causes release of the milligram amounts of the ultimate hormone by the target gland. Note that each step in this process produces an **amplification** of the response. The overall scheme of the regulation of metabolic hormones is given in Fig. 1.41 and the regulation of reproductive hormones is given in Fig. 1.42.

Hormone secretion from the pituitary gland occurs in an episodic or rhythmic manner. This is regulated by the biological clock in the suprachiasmatic nucleus of the hypothalamus. This may prevent the down regulation of receptors that would occur in response to a continuous level of hormone secretion. For example, the levels of cortisol are highest in the morning and decrease in the afternoon and evening. Secretion of somatotrophin is more pulsatile in females than in males. The pulse frequency, pulse

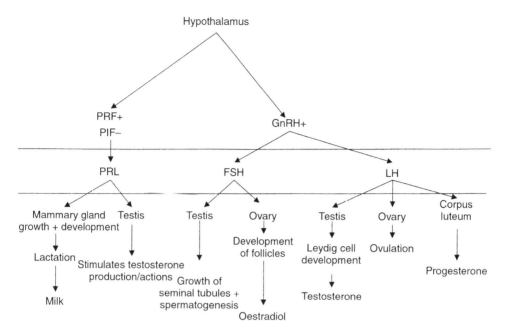

Fig. 1.42. Overview of regulation of reproductive hormones.

amplitude and average hormone levels can be calculated using a pulse detection algorithm, 'Pulsar', developed by Merriam and Wachter in 1982. The nature of this pulsatile release is very important. For example, differences in the frequency of GnRH release by the hypothalamus differentially affect the subsequent release of LH and FSH by the pituitary.

The release of hormones by the anterior pituitary is regulated by feedback control. There is a short feedback loop from the anterior pituitary to the hypothalamus and a long feedback loop of the ultimate hormone on the CNS, hypothalamus or anterior pituitary (Fig. 1.40).

Questions for Study and Discussion

Section 1.1 Introduction

1. What is the definition of a hormone?
2. Why are hormones necessary?
3. How do hormonal systems function and what are the effects of hormones?
4. How are hormones selective?
5. Why are there both intracellular and extracellular receptors?
6. What are the different chemical structures of hormones?
7. Where are the different hormones produced and what are their physiological roles?

Section 1.2 Synthesis, Release and Metabolism of Hormones

1. Describe the mechanism for the synthesis of protein hormones and the process of movement and processing of newly synthesized protein in the cell.

2. Describe the synthesis of steroid hormones.
3. Describe the synthesis of eicosanoid hormones and thyroid hormones.
4. Discuss factors that regulate hormone release and hormone metabolism and excretion.

Section 1.3 Receptors and Hormone Action

Protein hormones
1. Outline the chemical nature of receptors.
2. Describe the cAMP and cGMP second messenger systems.
3. Outline the calcium/diacylglycerol–protein kinase C-dependent signalling pathway.
4. Describe methods to determine whether the cAMP- or calcium-dependent pathways are involved in any system.
5. Outline the mechanisms of action of tyrosine kinase receptors, cytokine receptors and serine kinase receptors.

Steroid hormones
1. Discuss the actions of steroid hormones in the cell. What is the time frame for steroid hormone and peptide hormone action?
2. Describe the structure of steroid hormone receptors.
3. Discuss the potential role of the nuclear matrix and nuclear chromatin in hormone action.
4. Describe experiments to identify DNA sequences involved in hormonal regulation of gene expression.

Section 1.4 Pituitary–Hypothalamic Integration of Hormone Action

1. Illustrate the position of the brain, hypothalmus and pituitary gland. (a) Outline the integration of hormonal systems with the hypothalamus. (b) Describe the integration of hypothalamic and pituitary function (nervous/pituitary–hypothalamic portal system).
2. What is the nature of the hypothalamic releasing and release inhibiting hormones (function and structure)?
3. What hormones are produced by the posterior pituitary gland (structure, synthesis and function)? What stimulates their release and how is this regulated?
4. What are the properties of the hormones produced by the anterior pituitary gland? How is their release regulated?

Further Reading

General

Endocrine Society (annual publication) *Introduction to Molecular and Cellular Research* [syllabus of annual course]. Endocrine Society, Bethesda, Maryland.

Griffin, J.E. and Ojeda, S.R. (2000) *Textbook of Endocrine Physiology*, 4th edn. Oxford University Press, New York.

Hadley, M.E. (2000) *Endocrinology*, 5th edn. Prentice Hall, Englewood Cliffs, New Jersey.

Nelson, D.L. and Cox, M.M (2000) *Lehninger Principles of Biochemistry*, 3rd edn. Worth Publishers, New York.

Norman, A.W. and Litwack, G. (1997) *Hormones*, 2nd edn. Academic Press, Toronto.

Stryer, L. (1995) *Biochemistry*, 4th edn. W.H. Freeman, New York.

Hormone synthesis and metabolism

Blázquez, M. and Shennan, K.I.J. (2000) Basic mechanisms of secretion; sorting into the regulated secretory pathway. *Biochemistry and Cell Biology* 78, 181–191.

Nielsen, H., Engelbrecht, J., Brunak, S. and von Heijne, G. (1997) Identification of prokaryotic and eukaryotic signal peptides and prediction of their cleavage sites. *Protein Engineering* 10, 1–6.

Hormone receptors

Boonyaratanakornkit, V., Melvin, V., Prendergast, P., Altmann, M., Ronfani, L., Bianchi, M.E., Taraseviciene, L., Nordeen, S.K., Allegretto, E.A. and Edwards, D.P. (1998) High-mobility group chromatin proteins 1 and 2 functionally interact with steroid hormone receptors to enhance their DNA binding *in vitro* and transcriptional activity in mammalian cells. *Molecular and Cellular Biology* 18, 4471–4487.

Falkenstein, E., Tillmann, H.-C., Christ, M., Feuring, M. and Wehling, M. (2000) Multiple actions of steroid hormones – a focus on rapid nongenomic effects. *Pharmacological Reviews* 52, 513–555.

Fortunati, N. (1999) Sex hormone binding globulin: not only a transport protein. What news is around the corner? *Journal of Endocrinological Investigation* 22, 223–234.

Freedman, L.P. (ed.) (1998) *Molecular Biology of Steroid and Nuclear Hormone Receptors*. Birkhauser, Boston, Massachusetts.

Scott, J.D. and Pawson, T. (2000) Cell communication: the inside story. *Scientific American* June, 72–79.

Spelsberg, T.C., Lauber, A.H., Sandhu, N.P. and Subramaniam, M. (1996) A nuclear matrix acceptor site for the progesterone receptor in the avian c-*myc* gene promoter. *Recent Progress in Hormone Research* 51, 63–97.

Hypothalamus and pituitary

Merriam, G.R. and Wachter, K.W. (1982) Algorithms for the study of episodic hormone secretion. *American Journal of Physiology* 243, E310–E318.

2
Endocrine Methodologies

2.1 Methods for Studying Endocrine Function

It is necessary to first characterize an endocrine system and understand how it functions. This knowledge of the endocrine system can then be used to improve or monitor animal performance, health and welfare. In broad terms, this involves identifying the endocrine tissue and target organs and determining the physiological effects of hormone and their significance to the animal. The detailed **information needed** includes:

1. The chemical structure of the hormone (for example protein, steroid, amino acid or fatty acid derivative);
2. The details of the biosynthesis, storage and secretion, and the stimulus for secretion of the hormone;
3. The transport of the hormone and the mechanism of hormone action in the target cell; and
4. The pathways of metabolism, inactivation and excretion of the hormone.

Model systems

A number of *in vivo* and *in vitro* **model systems** can be used to study endocrine systems. Models range from the whole animal, to isolated perfused organs, tissue slices, isolated cells in culture and subcellular fractions (Fig. 2.1). The model system of choice for a particular study depends on what information is needed and what level of organization is needed to achieve the objectives of the study.

For example, the overall effects of an endocrine system on animal performance would need to be studied in a whole animal, while the mechanism of action of the hormone might best be studied in an isolated cell system.

Whole animal model

Experiments using the **whole animal** have the advantage of measuring *in vivo* production and uptake of hormones by various organs. The true '*in vivo*' function and effect on the animal can be established in the whole animal model. The net effects on the whole animal are determined and the interplay

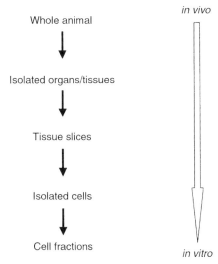

Fig. 2.1. Model systems for studying endocrine function.

among individual organs can be examined. However, whole animal experiments have the disadvantages of high cost for animals and reagents and high biological variability among different animals. Biological variability can be reduced somewhat by using highly inbred lines of animals with a similar genetic make-up.

The classical methods used to study endocrine tissues are whole animal studies in which organs are removed and replaced. The approach is to remove the organ, note the effects on the animal and then replace the organ. Sham-operated animals, in which the surgery is performed up to the point that the organ would be removed but the animal is left intact, act as controls (Fig. 2.2).

Organs can be removed **surgically**; for example, **adrenalectomy** or removal of the adrenals, which would result in lowered plasma adrenal steroids and lower plasma sodium, along with other effects. In the case of paired organs, such as the adrenals and gonads, removal of one organ causes the other to increase in size to compensate for the lost capacity. This can be from hypertrophy (increase in cell size) or hyperplasia (increase in cell number). Note that hyperplasia can be distinguished from hypertrophy by an increase in the DNA content of the tissue. Another example of surgical removal of organs is **hypophysectomy** or removal of the pituitary (hypophysis). Hypophysectomy demonstrates that the endocrine system is under pituitary control. For example, in the winter flounder *Pseudopleuronectes ameri-canus*, a hypophysectomized fish turns grey due to a lack of melanocyte stimulating hormone (MSH) from the pituitary.

Organs can also be inactivated by treatment with chemicals or by removing the nervous stimulation of the tissue. Examples of **chemical inactivation** include treatment with alloxan, which destroys the islet cells in the pancreas that produce insulin. Cobalt chloride treatment destroys the glucagon-secreting cells. This is particularly useful when surgical removal is impossible, as in the islets of Langerhans, which are scattered throughout the pancreas. An example of removal of nervous stimulation to an organ, called **deafferentation**, is vagotomy or disruption of the vagus nerve to the gastrin-producing cells of the stomach. Normally the vagus nerve stimulates gastrin secretion in anticipation of food or distension of the stomach. Gastrin stimulates HCl and pepsinogen secretion into the stomach.

To confirm that the endocrine response is due to a particular tissue, the tissue can be placed back in the animal and a reversal of the effects from removal of the tissue should be seen. In these studies, it is necessary to control for rejection of the implanted tissue and the use of highly inbred strains may minimize these problems. Tissues can be transplanted to an **ectopic site**, which is well supplied with blood, for example beneath the kidney capsule, or within the orbit of the eye. A variant of this is **parabiosis**, in which the blood supplies of two animals are connected together. One animal is treated and the effects in the other animal are noted to demonstrate that blood-borne factors are causing the effect. An **extract of the gland** can also be injected to check for the presence of the hormone by measuring the endocrine response. The extract is then purified to obtain separate compounds with hormonal activity.

In whole animal studies where the effects of various hormones are studied *in vivo*, the following issues should be addressed:

1. What animal would be used? Is a smaller but still relevant animal model available (e.g. sheep instead of cattle)? Inbred strains or specific genetic crosses?

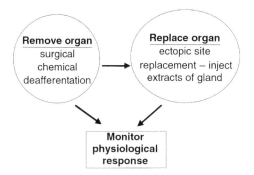

Fig. 2.2. Classical methods for studying hormone action in whole animals.

2. What dose of hormone would be used? Is this a physiological or pharmacological level?
3. What is the appropriate physiological state (e.g. age, maturity, reproductive status)?
4. What is the appropriate route of administration (in the feed, by injection – implant or multiple injection)?
5. What are the potential interactions or side-effects of the treatment?
6. What variable will be measured?
7. What are the advantages and disadvantages of this model?

Isolated organs or tissues

An isolated perfused organ system is useful for determining the overall metabolic effects on an organ, for example the effects of insulin/glucagon on glucose metabolism. It could also be used to study hormone production by an organ in response to blood-borne signals such as tropic hormones or levels of metabolites. The blood vessels supplying and draining the organ or tissue are cannulated and the vascular system is perfused with a modified blood cocktail that is continuously oxygenated and recirculated (Fig. 2.3). The system has the advantages of lack of effects from endogenous hormones and neural influences, as it isolates the effects of other tissues in the animal. The system is also easily manipulated and sampled. The disadvantages of the system are that surgical skill is required, and since it is not *in vivo*, it can sometimes be difficult to maintain organ viability.

Another version of this is the use of isolated organs or tissues *in situ*. In this model, catheters are placed in the blood vessels supplying and draining the organ or tissue, and the organ is left in the circulation and not removed from the animal. Blood samples can be taken from the arterial and venous catheters to determine what effects the organ has on levels of hormones or metabolites. Test compounds can be supplied to the organ via the arterial catheter and the effects on the organ noted. An example of this would be the *in vivo* catherization of the vessels supplying the mammary gland in dairy cattle. Separate catheters are placed in the external pudental arteries supplying the two sides of the mammary gland. This split-udder design allows one side of the mammary gland to be used for a treatment, while the other side is used as a control (see Section 4.1).

With the organ perfusion/infusion model, the following factors should be considered:

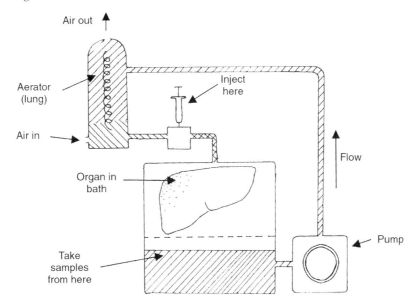

Fig. 2.3. Organ perfusion system.

1. What is the appropriate animal (species, strain, etc.) to use to study the effect?
2. How will the hormone be given in the perfusion/infusion?
3. What dose will be used, and how long will the treatment be carried out?
4. What is the appropriate physiological state (age, maturity, reproductive status, etc.)?
5. What dependent variables will be measured?
6. What are the advantages and disadvantages of this experimental approach?

In vitro models

In vitro models are useful for studying the details of the endocrine effects on specific tissues or cell types, but they suffer from several limitations. First, the tissue being studied tends to be in a catabolic state when studied *in vitro*. Thus, although particular biochemical pathways can be studied by *in vitro* methods, the results are qualitative rather than quantitative. Secondly, the results can also be dramatically affected by the composition of the incubation medium. The presence of particular hormones or metabolites can alter the activity of metabolic pathways. Thirdly, the concentration of hormones used may be pharmacological rather than physiological and thus the results obtained *in vitro* can differ dramatically from the results *in vivo*.

The next lower level of organization is the use of thin **tissue slices** that are incubated in an appropriate medium containing necessary metabolites and oxygen for the cells to remain viable. This has the advantage of being technically easier than perfusion, not requiring blood cells, and allowing for the study of different areas of the same organ. However, the slices must be very thin so that oxygen can penetrate through to the cells on the interior of the slice, and there are many damaged cells on the surface of the slice.

Isolated cells have the advantage of being a more defined system that can be used to study hormone actions on particular genes at the cellular level, including the effects of cellular growth factors or studying the control of gene expression (hormone-responsive elements, etc., see Section 1.3). Various inhibitors can be used to determine the mechanism of action of hormones within the cell. Cell culture uses fewer animals, and usually many treatments can be performed from one preparation of cells. Various types of cells can be isolated, so that the contribution of different cell types to the overall metabolism of a tissue can be studied. However, cell culture is not an *in vivo* system and the culture conditions are critical (for example, the presence of serum), so that artefacts can be produced inadvertently if the conditions are not controlled carefully.

Isolated cells can be prepared by treating a tissue with collagenase to break down the connective tissue and release the cells. The viability of the cells is determined by their exclusion of the dye Trypan Blue and the cells can be used in a **primary cell culture system**. The disadvantages of primary tissue culture are that cell viability and receptor/enzyme functions change over the time that the cells are in culture. An alternative to primary cell culture is to use established **immortalized cell lines** that have been derived from carcinomas from a particular tissue and species of animal. Many of these can be obtained commercially (for example from the American Type Culture Collection). Other cell lines have been immortalized experimentally by transforming primary cells with a tumour-inducing virus or chemicals. An example of this is the MAC-T3 cell line derived from mammary alveolar cells transfected with large T antigen. Immortalized cell lines produce a stable long-term system. However, they may not have the same properties as freshly isolated primary cells, so results obtained using these cells should be interpreted with caution.

With the cell culture model system, the following points should be considered:

1. What is the appropriate cell type? Is a primary cell culture or immortalized cell line appropriate?
2. What are the appropriate conditions for cell culture? Can a defined medium be used or is serum required? What culture surface is required (e.g. collagen-coated plates or plastic)? Is a feeder cell line required in co-culture?
3. What is the appropriate duration of treat-

ment? Is cell viability and response to hormones maintained (e.g. receptors maintained)?
4. What are the appropriate end-point measurements (e.g. specific proteins produced, cell growth, etc.)?
5. What is the appropriate dose of hormone for culture conditions (physiological versus pharmacological)?
6. What are the advantages and disadvantages of this experimental model?

Various **subcellular fractions** can be prepared by homogenizing cells or tissue in buffer and then centrifuging the homogenate at different speeds (differential centrifugation). These fractions can be used to study specific processes; for example, membrane transport using membrane vesicles, or enzyme activities, or structure/function relationships of molecules involved in endocrine responses. The advantages of subcellular fractions are their usefulness for detailed biochemical studies and that they represent a well-defined system. However, this system can produce artefacts, particularly if high levels of compounds are used, so the results may not reflect what is found in an *in vivo* system.

Use of inhibitors and agonists

A number of **inhibitors** can be used to determine the mechanism of action of hormones, particularly in studies using cell culture or subcellular fractions. These include inhibitors of the synthesis of DNA, RNA and protein, metabolic inhibitors, receptor agonists and antagonists and compounds that affect membrane transport.

Inhibitors of **RNA synthesis** include actinomycin D, which intercalates with the double helix of DNA to prevent transcription, and rifampicin, which inhibits the initiation of transcription. Inhibitors of **protein synthesis** include cycloheximide and puromycin. Cycloheximide inhibits the peptidyl transferase activity of the 60S ribosomal subunit. Puromycin resembles the aminoacyl terminus of aminoacyl-tRNA and joins to the carboxyl group of the growing peptide chain, causing it to dissociate from the ribosome, resulting in premature termination of the growing amino-acid chain.

Inhibitors of **microtubules** and **microfilaments** include colchicine and cytochalasin B. Colchicine is a plant alkaloid that inhibits microtubule assembly by binding with tubulin. It inhibits insulin secretion, suggesting that microtubules may function in the secretion of this hormone. Cytochalasin B is a fungal metabolite that specifically interferes with microfilament function without affecting microtubule integrity. It inhibits secretion of a number of hormones, suggesting that filamentous organelles may be involved in the secretory process of these hormones.

A number of hormone **antagonists and agonists** are available. Cyproterone acetate antagonizes testosterone binding at androgen receptors and can be used to reduce aggression and to treat prostatic hypertrophy. Dexamethasone is a synthetic glucocorticoid agonist that is widely used. Chlorpromazine and ergot alkaloids (e.g. ergocryptine) are dopamine receptor antagonists. Spironolactone is an aldosterone receptor antagonist and saralisin is an angiotension II receptor antagonist. Thiouracil inhibits the uptake of iodide and synthesis of thyroid hormone in the thyroid gland.

The methylxanthines, theophylline and caffeine, are **phosphodiesterase inhibitors**, which elevate intracellular cAMP levels by inhibiting degradation of cAMP. A number of cAMP analogues are also available, including dibutyryl cAMP and 8-bromo cAMP. Forskolin is an activator of adenylate cyclase and phorbol ester activates protein kinase C (see Section 1.3).

Metabolic inhibitors include iodoacetic acid, which blocks glycolysis, and 2-deoxyglucose, which inhibits glucose uptake and utilization within cells. Inhibitors that affect ATP formation include dinitrophenol, which uncouples oxidative phosphorylation, and oligomycin, which prevents mitochondrial phosphorylation of ADP to ATP.

The cytotoxic agent alloxan destroys the beta cells of the pancreas that secrete insulin, while cobalt chloride destroys the alpha cells of the pancreas that secrete glucagon.

Ion transport can be affected by treatment with **ionophores**, which span biological membranes to carry ions into cells. The ionophore A23187 is used to increase intracellular Ca^{2+}, and valinomycin has a high specificity for K^+ transport. Verapamil is a calcium-channel antagonist.

Liposomes can be used as transport vehicles that fuse with the cell membrane and transfer the contents into the cytoplasm. Liposomes that incorporate specific proteins can be potentially targeted to different cells. This provides a novel method of studying the mechanisms of stimulus–secretion coupling, particularly as it relates to the release of a chemical messenger.

Use of antibodies

Antibodies (Ab) are produced by animals as a defence against foreign compounds, called **antigens** (Ag). Antibodies are widely used in endocrine studies and can be prepared by injecting hormones or receptors obtained from one species into another species.

Immune response

There are two types of immune responses, the **cell-mediated response** and the **humoral** response (see Section 6.3). The cell-mediated response is involved in the destruction of 'self' cells gone wrong, such as infected or cancerous cells, and is carried out by cytotoxic T lymphocytes. The humoral response is the generation of soluble antibodies against foreign antigens.

The humoral response is due to B lymphocytes, which are precommitted to respond to a limited number of antigens. The initial exposure to an antigen produces the **primary immune response**, in which specific B cells that respond to particular antigen divide and differentiate (Figs 2.4 and 2.5). The majority of the responding B cells become memory cells and others become antibody-producing plasma cells. Thus, there is only a small production of Ab from the primary response. The **secondary immune response** occurs upon subsequent exposure to the same antigen. This results in a rapid proliferation and differentiation of memory cells into antibody-producing plasma cells and high antibody titres (a titre is the highest dilution of serum that gives a measurable Ab response). A large number of different antibody-producing cells are generated, and each of these produces a unique antibody directed against a particular **epitope** on the surface of the antigen. Thus, the immunization process generates a mixture of antibodies, called a **polyclonal antibody** preparation.

To elicit an antibody response, the substance (Ag) must be recognized as foreign (non-self), and be large (for example, proteins) and complex in structure (not a homopolymer such as polylysine). A **hapten** is a low molecular weight compound that can react with an antibody but is unable to induce antibody formation by itself. It must be attached to a larger molecule, such as keyhole limpet haemocyanin (KLH) that will readily be recognized as foreign, to be antigenic. Protein hormones are usually antigenic since

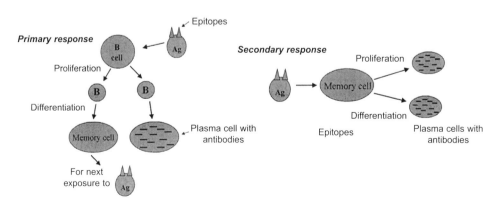

Fig. 2.4. The humoral response to foreign antigens.

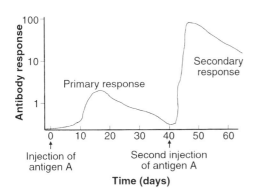

Fig. 2.5. Primary and secondary responses to antigen exposure.

their structures differ between species. Steroid hormones have a low molecular weight and their structure is the same in different species, so they need to be linked to a large protein to make them antigenic.

Normally an **adjuvant** is given with the antigen to enhance the immune response. A classic adjuvant is Freund's complete adjuvant, which is an emulsion of water in oil containing killed bacteria. It causes a slow continuous release of antigen and local irritation, which stimulates macrophage activity but can cause discomfort for the animals.

Detection and purification of antibodies

The presence of antibody in plasma can be detected using immunodiffusion (Ouchterlony) plates (Fig. 2.6). Serum containing antibodies (antiserum) is placed in one well of an agar plate and the antigen is placed in an adjacent well. The presence of an immunoprecipitate between the two wells indicates that the antibodies are present. The antibodies can be purified by affinity chromatography, using protein A to purify the IgG fraction or by linking the antigen to a support such as Sepharose to fractionate different antibodies in a polyclonal mixture.

Monoclonal antibodies

Antibodies produced *in vivo* are polyclonal, since they are produced from more than one B-cell clone. It is possible to produce antibodies from one clone (**monoclonal antibody**) to increase specificity, since a monoclonal antibody recognizes only one particular hapten (Fig. 2.7). Once a mouse has been immunized with the antigen, the spleen is removed, and the cells are isolated. The antibody-producing cells are then fused with a myeloma cell line to produce immortalized hybridoma cells. The individual hybridoma cell clones are then purified to produce cell lines that produce one particular monoclonal antibody. The cell line that produces the monoclonal

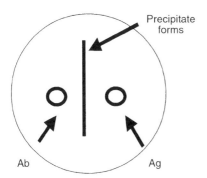

Fig. 2.6. Ouchterlony immunodiffusion experiment.

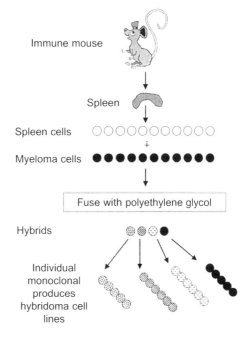

Fig. 2.7. Production of monoclonal antibodies.

antibody with the desired specificity is then used to produce large amounts of antibody.

A selection process is used to isolate the hybridoma cells from the unfused myeloma cells and spleen cells. The myeloma cell line is missing the enzyme hypoxanthine guanine phosphoribosyl transferase (HGPRT) from the purine salvage pathway. The cells are grown in HAT medium, containing hypoxanthine, thymidine and the synthetic folic acid analogue aminopterin, which blocks the thymidylate synthetase enzyme and forces the cells to use the purine salvage pathway. The unfused myeloma cells cannot survive under these conditions and unfused spleen cells cannot last more than a few days in culture. Thus, only the hybridoma cells that express the HGPRT from the spleen cells survive in the HAT medium.

Use of antibodies to identify the site of hormone synthesis or target tissue

Antibodies against hormones can be tagged with a fluorescent, enzyme, or electron-dense probe and used to identify hormone-producing or target cells. Labelled antibodies will accumulate in cells containing the hormone, which is the antigen to that antibody. For example, in **immunocytochemistry**, labelled antibody is applied to histological sections of tissue. In **immunoenzyme histochemistry**, an enzyme that gives a colour reaction is coupled to the antibody to increase the sensitivity of detection. Antibody binding is detected using fluorescent light microscopy, or electron-dense probe via electron microscopy.

Labelled hormone can also be used to localize areas of hormone binding. Radiolabelled hormone is injected and the tissue is sampled, fixed and the label is detected by **autoradiography**. This involves exposing the tissue to X-ray film and allowing the decay of radioisotope to produce spots on the exposed film. A related technique, *in situ* **hybridization**, is used to detect the presence of mRNA for a hormone or receptor in a particular cell type. In this case, the tissue section is exposed to a radiolabelled DNA probe for the hormone or receptor, and the binding of the probe to the tissue section is determined by autoradiography.

Antibodies can also be used to neutralize the effect of a hormone or to modulate hormone action, as described in Section 2.3.

2.2 Measurement of Hormones and Receptors

Assay of hormones

Assays for hormones are used primarily to measure the levels of hormones in plasma or tissue extracts. An assay must meet several criteria and be validated to be of use. The assay should be calibrated with a standard of known activity (Fig. 2.8). The standard must be tested at a number of concentrations to demonstrate that the response in the assay is proportional to the amount of hormone. A plot of the response in the assay versus the concentration of standard is known as a standard curve. The slope of the line gives the response factor that is used to convert the

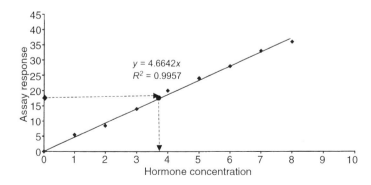

Fig. 2.8. Typical standard curve.

response in the assay that is obtained with a sample to the concentration of hormone in the sample.

Sufficient replicates should be done to demonstrate reproducibility; this is determined by a low coefficient of variation (CV = standard deviation/mean). The assay must be sensitive, usually in the picogram to nanogram range for plasma hormones. The standard and unknown should be measured at the same time and under the same conditions as the samples. There should be near 100% recovery of standard added to extract, and this recovery should be consistent between samples. A comparison of standard curves obtained with pure standard in buffer to a sample spiked with standards should show that the curves are parallel. This parallelism of standard curves needs to be validated for each tissue type (for example, plasma will be different from muscle) to demonstrate that the response in the assay is not dependent on the sample matrix (Fig. 2.9). The assay should also be simple, of low cost and suitable for routine use. The measure used in the assay must also relate to hormone activity. This is important when the assay is based on structural components of hormones that may not be related to activity.

What does all this mean? In order to use an assay, you first determine the response in the assay for different amounts of a pure reference hormone. Construct a standard curve by plotting the response in the assay versus the amount of added hormone and determine the equation of the line. When you measure the response in the assay with an unknown sample, use the equation from the standard curve to determine the amount of hormone in the unknown. To check for parallelism, prepare additional standard curves by adding the pure reference hormone to the sample matrix you are using.

Types of hormone assays

BIOASSAYS. The **bioassay** is the primary method for measuring the amount of a hormone, since it is based on measuring the physiological responses caused by the hormone. Whole animals, or tissue preparations utilizing an organ or tissue that is naturally responsive to the hormone, can be used. Some examples of bioassays are listed in Table 2.1.

The parameter or response measured in the bioassay should be specific for the hormone. An example is the pregnancy test, in which urine from the test subject is injected into rats or rabbits and the effect on the ovaries is noted. Human chorionic gonadotrophin (hCG) in the positive test causes ovaries to turn from being normally pale yellow to pink, due to increased blood flow. The response should also be directly controlled by the hormone and not dependent on the presence of other factors for a response. Otherwise, these other factors may also contribute to variability in the response parameter. The response should also be easily measured, such as an increase in weight. The parameter can be electrophysiological. For example, in measuring olfaction in fish, the trans-membrane potential in neurosecre-

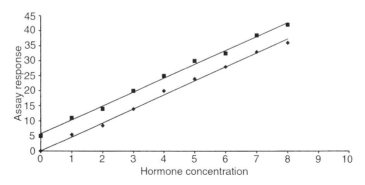

Fig. 2.9. Example of parallelism in standard curves.

Table 2.1. Some examples of bioassays (adapted from Hadley, 2000).

Hormone	Assay system	Responses monitored
In vivo systems		
Insulin	Fasted rodent	↓ Blood glucose
FSH	Immature hypophysectomized rodent	↑ Weight of follicular size of ovaries
TSH	Any vertebrate	↑ Thyroid uptake of ^{131}I and release of $[^{131}I]T_4$ and $[^{131}I]T_3$ from thyroid
Thyroxine	Larval amphibian	Metamorphic change
Somatotrophin	Tibia test (rat)	↑ Width of epiphysial (cartilage) plate
Oxytocin	Rat uterus	↑ Contraction
Parathormone	Parathyroidectomized rat	↑ Ca^{2+} in plasma
Oestrogens	Castrated or immature rodent	Vaginal cornification
Androgens	Castrated or immature rodent	↑ Weight of prostate and seminal vesicles
Prolactin	White squab pigeon	↑ Height of crop sac epithelium
Chorionic gonadotrophin	Female frog	Induced ovulation
	Galli–Mainini male frog test	Induced spermatogenesis
	Ascheim–Zondek test, mature rodent or rabbit	Formation of haemorrhagic follicles and corpora lutea
In vitro systems		
Melanotrophins	Frog skin	↑ Darkening of skin, melanosome dispersion
	Melanoma cells	↑ Tyrosinase activity, melanin production
Corticotrophin	Perfused adrenal	↑ Synthesis and release of cortisol

tory neurones in the hypothalamus is recorded with microelectrodes. The magnitude of the electrical potential is a measure of the response of the fish to various compounds.

The **disadvantages of bioassays** include:

1. A lack of sensitivity, particularly with whole animal studies, which require a lot of hormone to produce a response.
2. Poor reproducibility due to the wide variability in response that is obtained with different animals.
3. High cost and difficulty of use for animals and animal preparations.

However, in spite of these factors, any type of assay for hormones must be correlated with **hormone activity** as determined by some sort of bioassay.

In recent years, hormone activity has been assessed using *in vitro* systems consisting of hormone receptors coupled to a 'reporter' activity by genetic engineering. Binding of the hormone to the receptor turns on the reporter and this is easily measured. Examples of this approach are the oestrogen receptor coupled to luciferase, which is an enzyme from the firefly that produces a fluorescent product. This system has been used to measure binding of oestrogenic 'endocrine disruptor' compounds in the environment to the oestrogen receptor, to assess their potential biological activity (see Section 6.4). Similarly, the growth hormone secretagogue receptor (GHS-R, see Section 3.4) coupled to a fluorescent reporter gene, aequorin, has been used to screen for compounds that bind to the GHS-R. This approach is particularly useful in measuring binding between two different receptors for a particular hormone. For example, corticotrophin releasing hormone (CRH) acts via two classes of receptors, CRHR1 and CRHR2 (see Section 6.3). These receptors have been cloned and expressed in cells in culture, to be used as a bioassay system for compounds that bind specifically to one form of the receptor.

CHEMICAL ASSAYS. **Chemical assays** for hormones are based on some aspect of the structure of the particular hormone. Sometimes it is possible to take advantage of unique aspects of the hormone structure to design a simple and specific measure. For example, one assay method for thyroxine is to measure total protein-bound iodine. Uptake of ^{125}I can also be used to measure biosynthesis of thyroxine in the thyroid gland. However, chemical assays that test for functional groups on the hormone molecule that may not be involved with activity can give false results compared to biological activity.

The **solubility** and **stability** of the hormone are related to its chemical nature. If the hormone is soluble in organic solvents, it is likely a steroid or other hydrophobic molecule; while if it is soluble in water, it is more polar and probably a protein or amino-acid derivative. Protein hormones can also be inactivated by heating, changing the pH or digestion with proteases.

One form of chemical assay involves purifying the hormones by chromatography and then measuring the amount of hormone. Separation of compounds by chromatography depends on the differential partitioning of the compounds between the stationary phase (the material in the column) and the mobile phase that is flowing through the column.

Liquid chromatography. Purification by chromatography can be achieved using open-column chromatography, but **high performance liquid chromatography** (HPLC) is more commonly used as an analytical technique. HPLC uses high-resolution columns containing very uniform particles. These columns can separate very similar molecules with high efficiency, and normally operate under high pressure. The column is connected to a detector, which measures some chemical characteristic of the column effluent, such as the absorbance of ultraviolet or visible light, fluorescence, refractive index, conductivity, radioactivity or even a molecular mass (using a mass spectrometer). The typical arrangement of a chromatography system is shown in Fig. 2.10.

The types of chromatography columns include:

1. **Ion exchange** (either anion or cation), which separates molecules based on charge;
2. **Adsorption** (**normal phase**) chromatography, which is similar to thin-layer chromatography in that it works by surface adsorption of molecules to silica and is used to purify small organic molecules; and
3. **Hydrophobic interaction** (**reverse phase**), which separates by relative polarity or hydrophobicity, with the polar compounds eluting first from the column. Common reverse-phase columns are C18 and C4, which are used for separating small hydrophobic molecules such as steroids, drugs and amino acids, and phenyl, which is used for proteins.

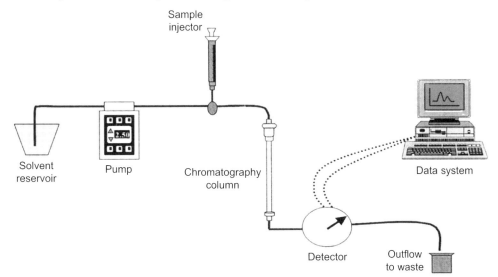

Fig. 2.10. Typical chromatography system.

In these types of chromatography, separation of the various compounds is achieved by changing the composition of the mobile phase over time. The compounds that are tightly bound to the column are removed by changing the pH, ionic strength or percentage of organic solvent in the mobile phase, usually as a gradient. A group of compounds with a similar chemistry can also be separated using an isocratic system with no changes in the mobile phase during the chromatographic run.

Size exclusion or **gel filtration** chromatography separates molecules based on their size and shape. Small molecules enter the pores in the gel matrix and are retained on the column, while the larger molecules are excluded from the gel and elute first in the initial void volume. Gel filtration is not usually of high enough resolution for analytical work, but it is useful in preparative work for large molecules such as proteins. Another type of liquid chromatography is **affinity chromatography**. This uses a ligand attached to the column matrix that binds specifically the compound of interest, giving a one-step purification procedure. An example of this would be to attach an antibody or binding protein to the column matrix and then apply a crude hormone preparation to the column. The hormone binds to the protein on the column and the other compounds pass through the column. The pure hormone is then eluted with a high salt or pH gradient. The types of liquid chromatography systems are summarized in Table 2.2.

Gas chromatography. Gas chromatography (GC) separates compounds by adsorption or partitioning between the gaseous mobile gas phase and the liquid or solid stationary phase. The partitioning is dependent on temperature, so the temperature is increased during the run to elute the low boiling compounds first and the high boiling compounds last. Compounds must be volatile to be separated by GC, so charged groups are normally derivatized to make them less polar. Common detectors for GC are thermal conductivity (TCD) and flame ionization detectors (FIDs). The FID is sensitive over a large range of concentrations, but destroys the sample. With the FID, the effluent from the column passes through a hydrogen/air flame that produces ions from organic molecules present in the effluent. The ions are collected on an electrode to produce an electrical signal. With the TCD, organic molecules in the column effluent change the thermal conductivity of the carrier gas, which changes the resistance of an electrically heated wire or thermistor. This change in resistance is converted to an electrical signal. The TCD is not as sensitive as a FID, but does not destroy the sample.

Mass spectrometry can be used to identify the molecular weight, structure and position of functional groups of small molecules. The mass spectrometer produces charged ions from the sample, consisting of parent ion and fragments of original molecule, and then sorts these ions by mass/charge ratio in a magnetic field. The relative numbers and the mass/charge ratio of each ion is characteristic of a particular compound and can be used to identify the structure of the compound. A mass spectrometer can be linked to a GC (GC-MS) or HPLC (HPLC-MS) to identify the

Table 2.2. Liquid chromatography systems.

Type	Separation principle	Usage
Ion exchange	Net charge	Proteins, charged molecules
Reverse phase	Relative hydrophobicity	C18 – small hydrophobic molecules
		Phenyl – proteins
Normal phase	Surface adsorption	Small organics
Size exclusion	Molecular size and shape	Large molecules (proteins, etc.)
Affinity	Specific binding affinity	Unlimited

*labelled hormone (fixed amount)
              +   + specific binding protein (fixed amount)
   unlabelled hormone (variable amount)
                                        ↓
                   *labelled hormone–   →   unlabelled hormone–
                       protein complex   ←   protein complex

compounds that are separated by the chromatography.

Electrophoresis. Proteins can be separated by electrophoresis, usually on polyacrylamide gels. Native or non-reducing gels separate intact protein molecules based on charge, while gels containing sodium dodecyl sulphate (SDS) separate molecules based on their molecular weight. SDS coats proteins and gives them an overall negative charge, so the migration of the protein in the gel is affected by the molecular weight and not the charge of the protein. After electrophoresis, the gels are stained to visualize all the proteins. The proteins can also be transferred or blotted to a solid membrane support and stained with a specific antibody that recognizes the protein of interest. This is known as Western blotting. A protein band can also be cut from a gel, the protein eluted from the gel, and a partial amino-acid sequence determined. This information can be used to identify the protein and to produce DNA oligonucleotide probes as a first step in cloning the protein. Capillary electrophoresis is a high-resolution analytical method useful for separating a variety of compounds. It uses just a few nanolitres of sample for capillary tubes with internal diameters of 20–100 µm.

COMPETITIVE BINDING ASSAYS. The competitive binding assay is based on specific binding of a hormone by a protein (usually a specific antibody). A fixed amount of labelled hormone and a fixed amount of a binding protein are used, along with a variable amount of unlabelled hormone. An equilibrium is established between the amount of unlabelled hormone and labelled hormone that is complexed with the binding protein (see equation at top of page).

The proportion of labelled versus unlabelled hormone bound to the antibody at equilibrium depends on the amount of unlabelled hormone present in the assay. The amount of bound label decreases with increasing amount of unlabelled hormone. Since we can only measure the amount of *labelled* hormone bound to the protein, we use this to determine how much *unlabelled* hormone was present in the assay.

The procedure involves first preparing standard curves to determine the useful concentrations of binding protein. A series of standard curves are produced using a fixed amount of binding protein for each curve, with the same amount of radioactive hormone throughout, and adding various amounts of unlabelled hormone. The amount of labelled bound hormone is plotted on the y axis (linear scale), with the concentration of unlabelled hormone on the x axis (log scale). The amount of binding protein that gives a moderate slope from 10–90% saturation is used for subsequent assays (Fig. 2.11).

The hormone concentration in a test sample is determined by adding the sample to the binding protein and labelled hormone mixture. The amount of labelled hormone bound is measured and the hormone concentration can be determined from the standard curve.

ASSAY REQUIREMENTS. Competitive binding assays require specific, high-affinity binding of hormone and the absence of non-specific interfering substances in the assay. Sometimes there are substances present in physiological fluids that affect binding of hormone to the binding protein and it may be necessary to partially purify the hormone and remove the interfering substances before assay.

Several types of binding proteins can potentially be used in competitive binding assays. Hormones bind to receptors with high affinity and specificity, and it is possible to use target-tissue plasma membrane or

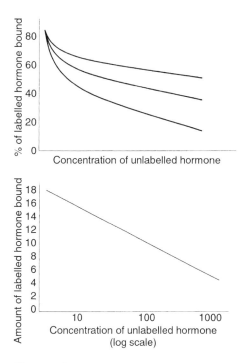

Fig. 2.11. Standard curves for competitive binding assays.

intact cells as a source of binding protein. The hormone binds to its own receptor and the specificity of binding can be checked by competition with other hormones to replace labelled hormone. However, a purified source of the binding protein is needed and this is usually accomplished with a high titre of antibodies. It is also possible to quantitate the number of receptor binding sites if a known concentration of hormone is used. This is known as Scatchard analysis (see below).

Proteins present in blood serum, such as specific globulins that bind hormones, can also potentially be used in competitive binding assays, but the binding affinity and specificity is usually not adequate. Most often, specific antibodies are used as binding proteins and the assays are then called immunoassays. Immunoassays are a popular method for assay of protein hormones. Antibodies to steroids are generated using the steroids conjugated to an antigenic protein at a site on the steroid that leaves the unique functional groups on the steroid available as haptens for antibody production.

Competitive binding assays require efficient methods to separate bound hormone from free hormone. These methods must not change the binding equilibrium, and should be simple and efficient so that a number of samples can be processed at the same time. A second antibody can be used to precipitate the primary antibody–hormone complex, leaving unbound hormone in solution. Alternatively, the free hormone can be adsorbed on a solid phase material, such as charcoal with a coating of albumin or dextran. Bound and free hormone can also be separated using column chromatography with gel filtration media such as Sephadex; convenient spin columns are available for separating different-sized molecules. A most convenient way of separating bound from free hormone is to link the binding hormone to a solid support, such as glass beads, filter paper, or the assay tube. This latter format is referred to as a solid-phase assay (Fig. 2.12).

Labels used for immunoassays can be radioactive, fluorescent or an enzyme that produces a colour change. A radioimmunoassay (RIA) can use steroids labelled with ^{14}C or ^{3}H; ^{125}I can be used to label steroids or tyrosine residues of protein by mild oxidation. A fluorescence immunoassay (FIA) uses a fluorescent label such as fluorescein isothiocyanate, which has an excitation maximum at 485 nm and emission maximum at 525 nm. In the enzyme-linked immunosorbent assay (ELISA or EIA), an enzyme such as alkaline phosphatase or horseradish peroxidase is coupled to the hormone. The enzyme catalyses the hydrolysis of substrates that produce a coloured product, such as BCIP/NBT (5-bromo-4-chloro-3-indolyl phosphate/nitro blue tetrazolium) for alkaline phosphatase and TMB (3,3',5,5'-tetramethylbenzidine) for peroxidase. The extent of colour change is used to quantitate the amount of labelled hormone–antibody complex that is present.

As an alternative to the competitive immunoassays, an **antibody sandwich ELISA** can be used to measure hormone concentrations. In this format, a capture antibody is bound to a solid support (tube or microtitre plate) and incubated with unlabelled hormone (standard or in a sample). This is then

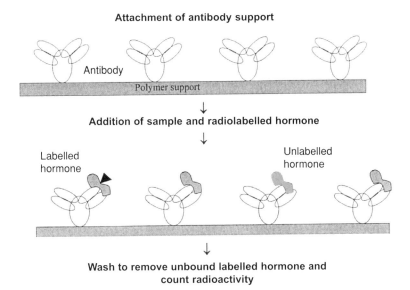

Fig. 2.12. Example of solid-phase competitive immunoassay. The steps are coupling of specific antibody to a solid support, addition of the sample and radiolabelled hormone, and washing to remove soluble compounds followed by counting the bound radioactivity.

followed by incubation with a second indicator antibody against the hormone that is conjugated to an enzyme. The activity of the enzyme is then measured. The amount of hormone bound to the capture antibody is estimated from the enzyme activity on the bound indicator antibody.

Measurements of hormone–receptor binding

Hormones bind to receptors with high affinity (10^{-9} M) and receptors are normally highly specific for binding a particular hormone. The affinity of hormone binding to its receptor and the effective number of receptors present in a cell can be determined by equilibrium binding experiments and **Scatchard analysis**. Scatchard analysis is similar to a competitive binding assay, but is used to measure the characteristics of receptors instead of the amount of hormone. For this analysis, specific (low capacity, high affinity) binding of hormone to receptors must be determined separately from non-specific (high capacity, low affinity) binding of hormone to other components of the system.

A saturation binding curve is first generated by incubating labelled hormone with a receptor preparation (membrane or nuclear fraction). The bound hormone is separated from the free hormone and the amount of hormone binding is measured. The total binding is the amount of radioactivity bound to the receptor preparation and includes hormone that is bound to receptors as well as non-specific binding of hormone. The non-specific binding is determined by including a large amount of unlabelled hormone (100–1000 times excess) to saturate all the specific receptor sites. When radioactive hormone is added, there are no specific receptors available for binding, so the binding is now due to the large number of non-specific sites available. The specific binding is determined as the difference between non-specific and total binding (Fig. 2.13).

The binding of hormone (H) to its receptor (R) to form a hormone–receptor complex can be described by the equation:

$$R + H \leftrightarrow RH$$

The equilibrium association constant (K_a) and the dissociation constant (K_d) are defined as:

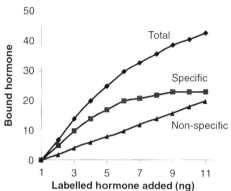

Fig. 2.13. Saturation binding curve for hormone–receptor binding.

$$K_a = \frac{[RH]}{[R][H]} = 1/K_d$$

The ratio of bound to free hormone (B/F) is given by:

$$B/F = \frac{[RH]}{[H]} = K_a[R]$$

The total number of receptors (R_t) is given by:

$$[R_t] = [R] + [RH]$$

Combining these equations:

$$B/F = K_a([R_t] - [RH]) = -K_a[RH] + K_a[R_t]$$

The data from the saturation binding curve is plotted, comparing the ratio of specifically bound to free hormone (y variable) versus hormone specifically bound to the receptor preparation (RH or B) (x variable). The slope of the line is $-K_a$, the x-intercept is R_t and the y intercept is $K_a[Rt]$.

The dissociation constant for the RH complex (K_d) is a measure of the binding affinity of the receptor for the hormone. When the concentration of hormone equals the K_d, one-half of the receptors are occupied. A small K_d indicates a high affinity of the receptor for the hormone, since a low amount of hormone is needed to get a response. The K_d for steroid receptors is in the range of 10^{-8}–10^{-10}. As an approximation, 20 times the K_d is enough to saturate the receptor. At the x intercept, all of the hormone is in the bound form since the bound/free ratio = 0, and this indicates the number of receptors present (Fig. 2.14).

Competition binding

A competition or displacement experiment measures the ability of a test factor to displace the hormone from its receptor. A single concentration of labelled hormone is incubated with increasing concentrations of the test factor, F, and the ability of factor F to compete for binding with labelled hormone is measured. The K_d for factor F is calculated by:

$$K_{d(F)} = IC_{50(F)}/(1 + [H]/K_{d(H)})$$

where $IC_{50(F)}$ is the concentration of F that displaces 50% of bound H. Most factors will not compete with H for binding as the binding of hormone to receptor is very specific and high affinity. This is useful for measuring the binding of hormone analogues to receptors.

Receptor cooperativity is seen when binding of hormone to one receptor affects binding of subsequent hormone molecules.

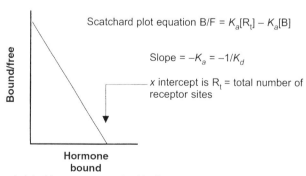

Fig. 2.14. Scatchard plot of hormone–receptor binding.

This is seen as a curve in the Scatchard plot. Positive cooperativity indicates that the initial binding of hormone enhances further binding, and is seen as an upward curve in the Scatchard plot. This may be important in sensitizing a system for a hormone response. Negative cooperativity is seen when the initial binding reduces further binding and produces a downward curve in the Scatchard plot. It may provide a mechanism for desensitizing a tissue to abnormal hormone levels. This is seen with receptors for insulin, TSH and nerve growth factor.

Regulation of receptor number represents another endocrine control mechanism. A number of factors affect the receptor number, including cellular development, differentiation and age. There can be either up or down regulation of receptor numbers. Hormones can induce synthesis (up regulation) of their own receptors; for example, prolactin induces liver receptors. Insulin down regulates its receptors in lymphocytes. Changes in receptor number and affinity can be determined by Scatchard analysis.

When one hormone regulates the receptor number of another hormone, this is heterospecific regulation. This is a mechanism for hormones to act in sequence to amplify or diminish response to other hormones. Homospecific regulation occurs when a hormone regulates levels of its own receptor; this prevents hyperstimulation of cells in pathological states.

2.3 Methods for the Production of Hormones

One of the common methods for manipulating endocrine function is to treat animals with a source of exogenous hormones. Reliable and inexpensive sources of hormones are needed to provide the hormones that can be administered to animals. Very early work used tissue extracts, which were very crude and contained a mixture of hormones and degradation products. In later work, the hormones were isolated and purified and their structure determined, so that usable amounts of the hormone could be synthesized.

Steroids and non-protein hormones

Epinephrine was isolated by Oliver and Shafer from adrenal glands and then synthesized in 1895. In 1919, thyroxine was purified by Kendal from thyroid glands of swine; in 1926, Harrington showed it to be a derivative of tyrosine. Insulin was crystallized by Abel, who demonstrated that it was a protein. Oestrogen was the first steroid to be isolated; Doisy and Butenandt did this from human pregnancy urine. Later, Doisy in 1935 and MacCorquodale in 1936 isolated 12 mg of oestradiol-17β from 4 tonnes of sows' ovaries. In 1931, Butenandt isolated androsterone from male urine and in 1934, progesterone was isolated from sow ovaries by four different groups. In 1935, testosterone was isolated from testes by David and co-workers, who demonstrated that it could be synthesized from cholesterol. In the period of 1936 to 1942, adrenal hormones were isolated by many workers in the US and Europe. From 20,000 slaughter-house cattle, 100 kg of adrenal glands were obtained and used to isolate 300 mg each of 29 steroids. Hench and co-workers demonstrated that cortisone could be partially synthesized and used it to clinically improve rheumatoid arthritis.

Afterwards, it was found that microorganisms such as *Rhizopus* and *Aspergillus* could metabolize steroids to useful compounds. These can be used as bioreactors for the synthesis of steroids. Natural sources of steroids include plant steroids, which are usually conjugated to sugar residues, and bile salts. Hyodeoxycholic acid from bile salts of swine has been used to synthesize progesterone and testosterone.

Steroids all have a cyclopentanoperhydrophenanthrene nucleus consisting of four fused rings, labelled A, B, C and D, and can be members of the pregnane, androstane or estrane families (Fig. 2.15). Different steroids have different functional groups or double bonds on the four rings. The trivial names and systematic names of common steroids are given in Table 2.3.

Synthetic steroids that are more potent steroid derivatives or inhibitors have been chemically synthesized (Fig. 2.16). These can be longer lasting since they are degraded

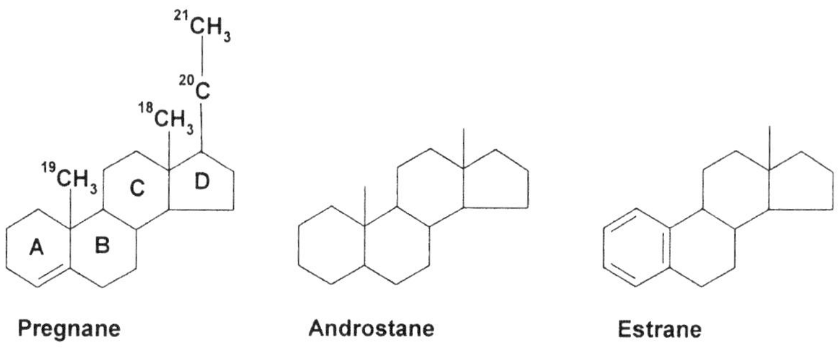

Fig. 2.15. Pregnane, androstane and estrane steroid structures. Pregnane (C21) indicates a steroid structure with methyl groups at C-13 and C-10 and a two-carbon-atom side-chain attached at the C-17 position. Androstane (C19) indicates a steroid structure with methyl groups at both the C-13 and C-10 positions. Estrane (C18) indicates a steroid structure with a methyl group attached at C-13 in addition to the 17 carbon atoms of the ring structure.

more slowly; for example diethylstilbestrol (DES) is a long-lasting oestrogenic compound. Ethinyl oestradiol and norethindrone are used as contraceptive steroids. Dexamethasone is a synthetic glucocorticoid and promegesterone (R5020) is a synthetic progestogen. Other compounds are receptor antagonists. Clomiphene and tamoxifen bind to oestrogen receptors to block them. Tamoxifen has also been shown to have oestrogenic effects. Mifepristone (RU-486) is a progesterone antagonist that is used clinically to induce abortions.

Protein and peptide hormones

Peptide hormones are named based on their endocrine activity rather than their chemical structure. The structure of a particular peptide hormone can vary in different species, while a particular steroid hormone has the same structure regardless of the species in which it is found. For example, growth hormones from pigs and cattle are different, since they have a different amino-acid sequence. However, progesterone isolated from pigs or cattle is the same molecule. Because of this species specificity, peptide hormones must be isolated from the same, or very similar, species for biological activity. They have a higher molecular weight than steroids and this makes structure elucidation more difficult. Proteins are also easily degraded by cellular proteases and this can result in a number of protein fragments being produced during the isolation procedure.

Historically, White and colleagues first crystallized prolactin in 1937. In 1953, the structures of oxytocin and vasopressin were determined, and these peptides were then

Table 2.3. Systematic names of vertebrate steroids.

Class name	Trivial name	Systematic name
Glucocorticoids	Cortisol (hydrocortisone)	11β, 17α, 21-Trihydroxy-pregn-4-ene-3,20-dione
	Corticosterone	11β, 21-Dihydroxy-pregn-4-ene-3,20-dione
Mineralocorticoids	Aldosterone	11β, 21-Dihydroxy-pregn-4-ene-3,20-dione-18-al
	11-Deoxycorticosterone	21-Hydroxy-pregn-4-ene-3,20-dione
Androgenic steroids	Dehydroepiandrosterone	3β-Hydroxy-androst-5-ene-17-one
	Testosterone	17β-Hydroxy-androst-4-ene-3-one
Oestrogenic steroids	Oestradiol	1,3,4(10)-Oestratriene-3,17β-diol
Progestens	Pregnenolone	3β-Hydroxy-pregn-5-ene-20-one
	Progesterone	Pregn-4-ene-3,20-dione

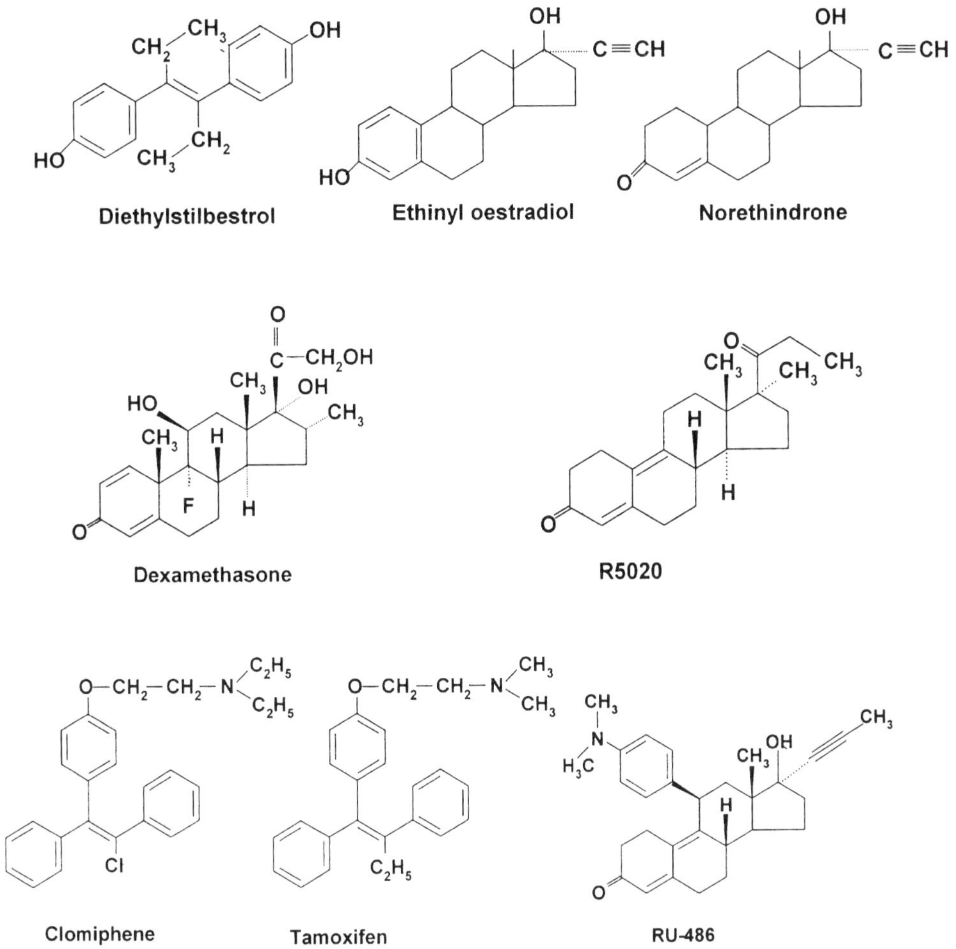

Fig. 2.16. Synthetic steroids and steroid antagonists.

synthesized by du Vigneaud. Sanger determined the structure of insulin, and it was synthesized in the 1960s. Glucagon was crystallized by Staub, and sequenced in 1957 by Bromer and colleagues.

Determination of amino-acid sequence

Peptide and protein hormones consist of a linear chain of amino acids, with the sequence of the amino acids determining the primary structure of the protein. Short peptides can be sequenced directly, but long chains are divided up into smaller chains by enzymatic digestion to obtain overlapping fragments, which are then sequenced.

Peptides can be sequenced from the amino-terminal end by **Edman degradation** (Fig. 2.17). This process is now done by automated methods. The amino-end terminal amino acid of the peptide is first labelled with phenylisothiocyanate and then the labelled amino acid is removed by mild acid hydrolysis. The labelled amino acid that is released by hydrolysis is identified by chromatography. The cycle is then repeated with the shortened peptide.

Peptides and proteins can also be sequenced using matrix-assisted laser desorption ionization–time of flight mass spectrometry (MALDI-TOF). This is used especially in the field of proteomics, to

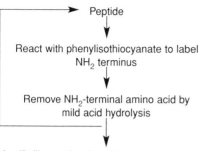

Fig. 2.17. Determination of amino-acid sequence of proteins by Edman degradation.

sequence a large number of proteins that are expressed in a variety of biological samples. This type of mass spectrometry generates overlapping short peptide fragments and determines their molecular weights. The sequence of amino acids is deduced from these molecular weight measurements.

The amino-acid sequence can also be determined from the **DNA sequence**, and this is particularly useful for large peptides and proteins (Fig. 2.18). First, the sequence of 10–20 amino acids is determined from the amino-terminal end of the protein, and then a DNA oligonucleotide primer encoding these amino acids is synthesized. In cases where more than one codon is used for an amino acid, degenerative primers are designed using the codon combinations that are commonly used for this amino acid. The mRNA is isolated from a tissue that expresses the protein, and complementary DNA (cDNA) is produced from the mRNA by reverse transcription. The primers can be used to amplify the cDNA encoding the protein directly, by polymerase chain reaction (PCR). Alternatively, the total cDNA from the tissue can be inserted into a plasmid or phage vector to produce a cDNA library. The library can then be screened using the oligonucleotide primers (or with antibodies if an expression vector is used) to isolate a full-length cDNA encoding the protein hormone. The sequence of the cDNA is determined and the amino-acid sequence is then deduced from it.

Peptide and protein synthesis

The **chemical synthesis of peptides** involves the formation of peptide bonds between amino acids that are arranged in a specific sequence. Since amino acids are at least bifunctional, with both an amino group and a carboxyl group, the groups that are not involved in the formation of peptide bond must be protected (Fig. 2.19). The carboxyl and amino ends, as well as selective side-groups not involved in the desired peptide bond, are first blocked and then the peptide bond is formed. The desired blocking group is then removed from the peptide and the next amino acid is added.

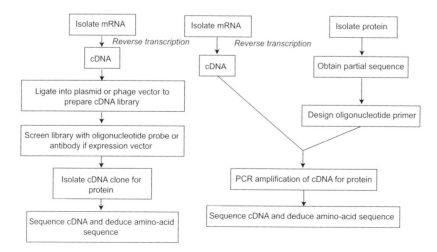

Fig. 2.18. Strategies for deducing amino-acid sequence from cDNA.

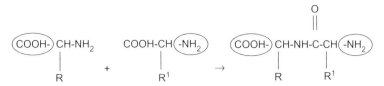

Fig. 2.19. Strategies for the chemical synthesis of peptides.

The chemical synthesis of larger peptides (Merrifield synthesis) uses a solid-phase support to assemble the peptide in a manner that is analogous to the function of a ribosome in the cell (Fig. 2.20). This improves the recovery and efficiency of peptide synthesis. The disadvantages of chemical synthesis are the high cost for equipment and reagents, the limitations of peptide size (~30 amino acids) and the low recovery of the peptide from the synthesis.

Examples of peptides that have been made by solid-phase synthesis include bradykinin (9 residues), vasopressin (9 residues), valinomycin (12 residues), neurotensin (13 residues), β-endorphin (31 residues), corticotrophin (39 residues), ferredoxin (55 residues) and β-lipotrophin (91 residues).

Non-peptide mimics of peptides

A large number of biologically important peptides are known, but small peptides are limited in their usefulness as hormone supplements. Peptides have a poor bioavailability, since they are rapidly degraded by proteases and have a limited ability to penetrate membranes. To overcome these problems, non-peptide compounds, called peptidomimetics or peptoides, have been developed to act as agonists or antagonists for peptide hormones. These non-peptide compounds mimic the three-dimensional structure of the hormone ligand and an unlimited variety of compounds can be made with various biological activities (for example, see Fig. 2.21). Different libraries of chem-

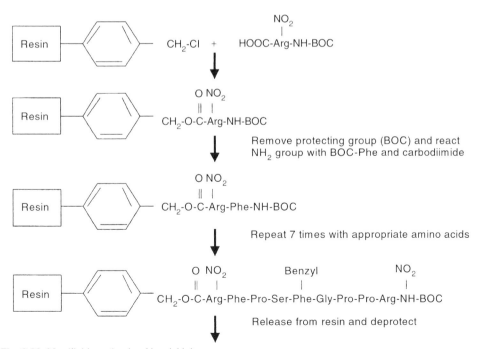

Fig. 2.20. Merrifield synthesis of bradykinin.

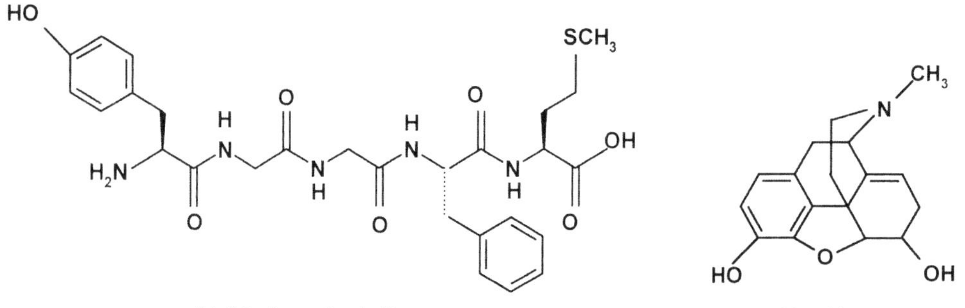

Fig. 2.21. Some examples of non-peptide mimics of hormones.

ical structures can be developed based on a core 'privileged structure' that mimics important parts of peptides (for more information see Patchett and Nargund, 2000). Often small alterations in the molecule, such as alkyl substituents or changes in the stereochemistry, change the biological activity from agonistic to antagonistic.

The earliest examples of these hormone analogues were based on morphine and related opioids, which mimic the biological activity of the enkephalins, natural peptide analgesics. In fact, the structure–activity relationships of the opioids were determined before their receptor was characterized. More recently, non-peptide analogues for the growth hormone secretagogues (GHS) and cholecystokinin (CCK) (see Sections 3.4 and 3.9) have been developed using functional screening assays. GHS analogues were developed based on structural elements in the hexapeptide growth hormone releasing peptide (GHRP) that were thought to be critical for bioactivity. A large number of compounds were then screened for their effectiveness in releasing growth hormone from a rat pituitary cell assay. Similarly, agonists for the CCK_A receptor were screened for their effectiveness in contracting a gall bladder preparation. Once biologically active compounds are found, the pharmacokinetic parameters are measured to determine the *in vivo* effectiveness of the compounds. For further information, see the review by Sugg (1997).

Production of recombinant proteins

An efficient method for producing large peptides or proteins uses recombinant DNA techniques (Fig. 2.22). For the *in vitro* production of proteins, the DNA sequences encoding protein are isolated (as cDNA or genomic DNA) and put into a vector, such as a plasmid, which is transfected into bacteria or a eukaryotic cell line. The cells are then grown to produce the protein, which is then puri-

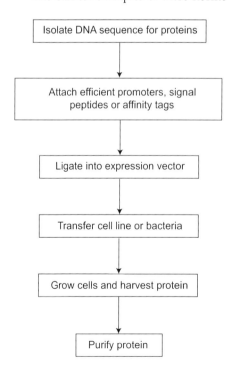

Fig. 2.22. Production of proteins *in vitro* using recombinant DNA techniques.

fied. The use of efficient promoters, and having more than one copy of the gene, can increase the production of the protein.

Purification of recombinant proteins can be simplified by adding a signal peptide to the gene so the protein is exported from the cells. The protein is then harvested from the culture medium, which reduces the number of compounds that have to be removed during the purification.

A common method for expression of recombinant proteins is to produce proteins as fusion products to stabilize the protein, make the protein more soluble, or act as convenient tags for purification by affinity chromatography. The fusion protein can then be cleaved after the purification to produce the recombinant protein. Examples of fusion proteins include β-galactosidase, bacterial alkaline phosphatase, chloramphenicol acetyl transferase, green fluorescent protein (GFP), glutathione-S-transferase, luciferase and polyhistidine. More recently, a small 24-bp FLAG sequence, which contains the rare five-amino-acid recognition sequence for enterokinase (Asp–Asp–Asp–Asp–Lys ↓ X) has been used. The small FLAG peptide is particularly useful as it may not affect the function, secretion or transport of the fusion protein and thus may not need to be removed.

Recombinant protein hormones can also be produced *in vivo* using transgenic animals. This involves creating a transgenic animal that produces the desired protein and deposits it in milk, eggs or semen so it can be easily harvested. This may be a cost-effective method overall, but it requires considerable effort to first generate the appropriate transgenic animal that produces the recombinant product.

The advantages of recombinant DNA technology methods for producing proteins include the lower cost overall, high efficiency and the production of a higher-purity, consistent product. The disadvantages are that an appropriate DNA construct must first be produced and that simple bacterial systems cannot always be used. In some situations, the 'pro' and signal sequences must be removed from the cDNA, and this may affect the folding of the polypeptide chain and reduce the activity. Bacteria use methionine for the initial amino acid, so any protein produced in a bacterial system will begin with methionine. Eukaryotic systems (e.g. yeast) must also be used if post-translational modifications of the protein, such as glycosylation, are necessary. There is also the possibility of degradation of foreign polypeptides when they are expressed in a cell.

2.4 Manipulation of Endocrine Function

The earliest use of hormones to enhance production in farm animals involved feeding iodinated proteins to dairy cows to increase milk production, and using diethylstilbestrol (DES) and dienesterol in broilers for enhanced fat. In the USA, DES was used from 1954 in beef cattle and sheep to increase growth efficiency and lean yield, but its use was banned in 1972 due to potential carcinogenicity in humans. In 1958, oestradiol benzoate/testosterone implants were approved for heifers, and in 1969, zeranol implants were approved. The USA approved bovine somatotrophin (bST) for improving milk production in dairy cattle in 1993.

Hormone delivery methods

Hormones can be administered as **feed additives** if the compounds are not degraded by the digestive system or rapidly metabolized and eliminated. Examples of these orally active compounds are β-agonists (see Section 3.5), or the synthetic progestagen, melengestrol acetate (MGA, Section 3.2). In this case, there is a potential safety concern for feed mill workers from exposure to hormones in the dust. Hormones that are not orally active can be administered as **single or multiple injections**, usually as subcutaneous or intramuscular injections. Multiple injections result in a pulsatile delivery of hormone, while a single dose produces high levels that decrease over a period of time (Fig. 2.23). Hormones can be given as a suspension in oils or waxes to give a more sustained release than hormone in aqueous solution.

Hormones can be delivered with **sustained-release devices** that are implanted in the animal. This gives an initial rapid release of hormone that may be above the optimal

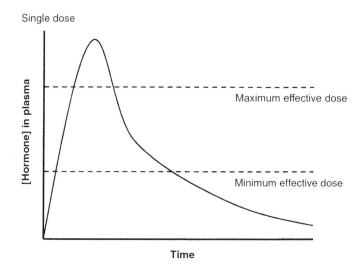

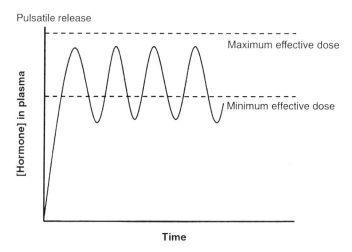

Fig. 2.23. Administration of hormones: pulsatile release versus single injection.

concentration and which decreases gradually (Fig. 2.23). Subsequent implants can be given as the concentration of hormone drops below the effective dose. The hormone composition of the implants could also be varied as the animal develops. Sustained-release devices have the advantage of being less labour intensive than multiple injections.

Types of sustained-release devices

Growth-promoting implants have been used extensively in cattle in North America since the early 1970s. These are implanted in the base of the ear, usually with a large-bore needle. The matrix of implants can be composed of lactose (short acting), cholesterol (long acting), or a large polymer of polyethylene glycol. Silastic tubing impregnated with the hormone can also be used, and in this case the length of the tubing controls the dose. The progesterone-releasing intrauterine device (PRID), which is used for synchronizing oestrus (see Section 5.1), is an example of delivering hormone to the site of action. Osmotic minipumps (Alzet®) can be used for the experimental delivery of hormones. Skin patches could also be used for uptake of the hormone through the skin.

The potential problems of using implants are that protein hormones, such as growth hormone, can form insoluble aggregates over time, so that not all the hormone is released. The hormone can also degrade over time, and may need to be stabilized. The implanting technique can also cause abscesses and expelled or crushed implants.

Pulsatile release of hormone

The pulsatile versus continuous mode of administration can affect the hormone response. Insulin, somatostatin, GH, PTH and LH are secreted in pulses, and pulsatile delivery of these hormones gives the optimum biological effect. This may be due to the suppression of natural hormone secretion from continuous administration of large doses of hormone. Receptor numbers can also be down regulated by negative feedback, or the dimerization of receptor can be decreased by continuous high concentrations of hormone (see Section 1.3).

Pulsatile release can be obtained from implants. Implanted mechanical pumps can be used, but these are expensive and may require surgery. A number of experimental controlled release systems have also been described (Medlicott and Tucker, 1999).

Bulk eroding systems (Fig. 2.24) consist of a mixture of microspheres made of different formulations of polylactic acid/glycolytic acid polymers, which degrade to release hormone at different rates.

Surface eroding systems (Fig. 2.25) comprised of poly(ortho) ester and polyanhydride matrices degrade rapidly from the surface. More than one matrix composition can be used in layers (with hormone and then without hormone) to obtain pulsatile release of hormone. The composition of the surface layer can be varied to produce a mixture of microspheres that degrade at specified rates to produce a series of pulses of hormone release.

In osmotically controlled systems (Fig. 2.26), the outside of the implant is coated with a water-insoluble polymer with a water-soluble pore-former; the interior contains hormone and the osmotic agent. Pores are created to allow water to move inside by osmotic pressure, to rupture the implant and release a pulse of hormone.

Enzymatically activated liposomes can be prepared containing phospholipase A2 with hormone on the inside. The liposome is encapsulated in alginate–poly-lysine. Once the capsule dissolves, the phospholipase degrades the liposomes to release the hormone in a pulse. Triggered pulsatile delivery systems can also be designed to release hormone by exposure to various chemicals, pH

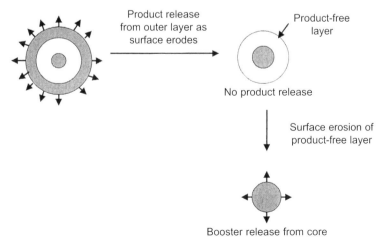

Fig. 2.24. Bulk eroding systems for the pulsatile release of hormones.

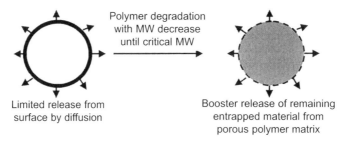

Fig. 2.25. Surface eroding systems for the pulsatile release of hormones.

change, temperature, electric pulses or magnetic fields.

Hormone residues

Hormone residues in edible tissues are a concern when exogenous hormones are used. The acceptable daily human intake (ADI) per unit of body weight is the dose that gives no hormone effect in the most sensitive animal model divided by 100. The potential daily intake (PDI) is the average intake of animal tissue times the residue level. The PDI must be less than the ADI for approval of the hormone treatment protocol.

Usually physiological concentrations of hormone are desired when hormones are given. Higher (pharmacological) doses may be detrimental and act through negative feedback to reduce the effect of the hormone. There are also differences between individuals and time of day effects on hormone activity. The effective dose of hormone depends on how rapidly the hormone is inactivated. Differences in metabolism among different animals may affect the extent of hormone activity.

Potentially, delivery systems can be devised to target particular cells and direct compounds to particular tissues where they are biologically active. The compound could be chemically modified to a derivative that is activated at the target site. Specific carriers can be utilized, such as liposomes, microspheres, nanoparticles, antibodies, cells (e.g. erythrocytes and lymphocytes), and macromolecules. Alternatively, compounds can be coupled to a hormone for delivery to the target tissue of the hormone. For example, *in vivo* treatment of ewes with a GnRH agonist coupled to a cytotoxic agent (pokeweed antiviral antigen) decreased the ability of the pituitary gland to secrete LH (Nett *et al.*, 1999). This could provide a novel approach to sterilizing animals. A conjugate of a lytic

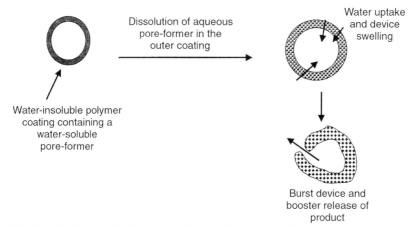

Fig. 2.26. Osmotically controlled system for the pulsatile release of hormones.

peptide (hecate) coupled to a segment of the β-chain of LH selectively killed cultured cells expressing the LH receptor, including prostate cancer cells.

Immunomodulation of hormone action

Antibodies can be generated to enhance or repress hormone action. The **immunoneutralization** of hormone action can be achieved by active or passive immunization. Some applications for the immunoneutralization of hormone action are summarized in Table 2.4. For **active immunization**, antibodies are generated within the animal by immunizing with the hormone. Some animals may not respond to the immunization procedure and generate high antibody titres, so there is reduced effect. For **passive immunization**, purified antibodies against hormone are infused into the animal. This requires a source of purified antibody and is suitable for short-term reversible immunization. Antibodies can also be raised against hormone receptors, to inhibit or stimulate hormone action (anti-idiotypic antibodies). For example, see Erlanger and Cleveland (1992) for a method for producing monoclonal auto-anti-idiotypic antibodies.

Antibodies are immunoglobulins (Igs) that are present in the γ-globulin fraction of plasma proteins. They are comprised of two light polypeptide chains (23 kDa) and two heavy polypeptide chains (50–70 kDa). Disulphide bonds join the chains into a 'Y' configuration, with the variable (V) regions located at the distal ends of the 'Y'. The hypervariable regions in the V regions have binding sites that are specific for particular antigens; these sites are called **idiotypes**. The constant (C) regions next to the variable regions are distinctive for each Ig type (Fig. 2.27).

TYPES OF IMMUNOGLOBULINS. IgM is the first antibody type formed after primary immunization and it acts to protect the intravascular space from disease. IgG is formed after IgM levels decrease and in the secondary immune response. It is the most prevalent serum antibody and is also found in the extravascular space. IgA is found in mucous secretions and is an early antibacterial and antiviral defence. IgE is found in respiratory and gastrointestinal tract mucous secretions, and may be involved in the allergic response.

Anti-idiotypic antibodies naturally exist against idiotypes in another antibody and these are important in the regulation of B-cell function. Secondary antibodies that are directed against primary antibodies to a hormone can also be produced. Some of these secondary antibodies will be specific for the idiotypic region and these are called anti-idiotypic antibodies. Anti-idiotypic antibodies can have hormonal activity, with the

Table 2.4. Applications of immunoneutralization of hormone action.

Production parameter	Immunogen	Mode of action
Increased fertility of sheep	Oestrogen – 'fecundin'	Decreased feedback from oestradiol → increased LH → increased ovulation
	Inhibin	Increased FSH → increased ovulation
Reduced libido – bulls		Decreased GnRH → decreased gonadal function
Contraception – wildlife	GnRH	
Immunocastration – pigs		
Increased growth and lean carcass	Somatostatin	Increased GH due to decreased somatostatin
Manipulation of oestrus in horses	Melatonin	Remove inhibition of reproductive activity in seasonal breeders
Decreased carcass fat	Adipocyte membranes	Decreased development of adipose tissue
Reduced boar taint in uncastrated male pigs	Androstenone	Binding of Ab to androstenone did not decrease accumulation in fat
Wool follicles in sheep	Epidermal growth factor	Increased EGF weakens wool fibre

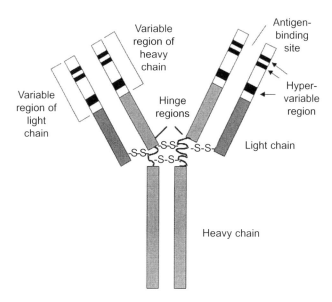

Fig. 2.27. Structure of antibodies.

added advantage of no hormone residues in the edible tissues from treatment. The hormonal activity from anti-idiotypic antibodies made by active immunization will be maintained longer than a hormone implant and may be longer acting (more stable) than hormone treatment. The disadvantage of this procedure is that it is necessary to first generate, test and purify a primary antibody, which is then used to generate the secondary anti-idiotypic antibody. Also, the anti-idiotypic antibody may not have hormonal activity.

Examples of anti-idiotypic antibodies that have been studied include: anti-β-agonist anti-idiotypic antibody, which can act as β-agonists to improve growth and carcass lean; and anti-GH anti-idiotypic antibody, to stimulate milk production in dairy cattle. Anti-idiotypic antibodies can also be tagged with a label (enzyme, chemiluminescent or fluorogenic) and used in non-competitive immunoassays.

Transgenic animals

Uses for transgenic animals

A transgenic animal is an animal that contains exogenous genes or gene modifications. These genes can be from different species and novel gene constructs can be used. Transgenic animals can be produced by inserting new gene(s) or by modifying the expression of existing genes. Interrupting the expression of particular genes can make gene 'knockouts' and these are useful in determining the physiological role of particular gene products. Specific genes can be targeted for knockout by the use of gene constructs that are highly homologous to the target gene, but which contain a mutation that interrupts the function of the gene. The mutation is introduced into the target gene by the process of homologous recombination.

New genes can be used to modify hormone responses and thus affect the performance, growth or health of the animal. Genes for improved growth include GH and the insulin-like growth factors (IGFs). Disease-resistance genes could be used to improve the health and performance of animals. Improved productivity could also come from genes that allow better utilization of feeds or alternative feeds.

Transgenic animals can also be used as bioreactors, to produce a valuable protein product that is deposited in the milk, eggs,

Table 2.5. Projects using transgenic farm animals as bioreactors (from Pinkert and Murray, 1999).

Product	Use	Commercializing firm(s)
α1-Antitrypsin	Hereditary emphysema/cystic fibrosis	PPL
α-Glucosidase	Glycogen storage disease	Pharming
Antibodies	Anti-cancer	CellGenesys, Genzyme, Ligand
Antithrombin III	Emboli/thromboses	Genzyme
Collagen	Rheumatoid arthritis	Pharming
CFTR	Ion transport/cystic fibrosis	Genzyme
Factor IX	Blood coagulation/haemophilia	Genzyme, PPL
Fibrinogen	Tissue sealant development	ARC, PPL
Haemoglobin	Blood substitute development	Baxter
Lactoferrin	Infant formula additive	Pharming
Protein C	Blood coagulation	ARC, PPL
Serum albumin	Blood pressure, trauma/burn treatment	Pharming
Tissue plasminogen activator	Dissolve fibrin clots/heart attacks	Genzyme
Tissues/organs	Engineered for xenotransplantation	Alexion, Baxter, CTI, Novartis

CFTR, cystic fibrosis transmembrane conductance regulator.

semen, etc. This allows the custom synthesis of valuable products such as protein hormones, antibodies and vaccines (Table 2.5).

Transgenic animals can be used as model systems to study hormone function. This includes molecular aspects of hormone/gene interaction and *cis*- and *trans*-acting factors involved in gene regulation *in vivo*. They can also be used to identify genes important in development, to identify factors responsible for tissue-specific gene expression and to study genes involved with disease.

Gene constructs can be made by linking the control region of interest to a marker gene that is easy to measure (Table 2.6). The expression of these test genes can be studied first in tissue culture. A transgenic animal is then made to determine the physiological significance of expression of particular genes and how hormones control the genes.

Production of transgenic animals

Production of a transgenic animal requires a gene to transfer, an efficient method for transferring the gene and an appropriate cell type to receive the transgene. The gene of interest is first identified and isolated. It is then linked to appropriate control elements so that it is expressed in the appropriate cell type in the animal and at the right time in development. These control elements can be found in other genes that are expressed in the tissue of interest. A gene transfer methodology is used to transfect a cell type that gives rise to the germ line. The potential cell types and gene transfer methods are summarized in Table 2.7. Transfection of somatic cells will not produce transgenic offspring, but can be used for transient gene expression in an existing animal.

There are a number of problems and challenges in generating transgenic animals. The number of copies of the transgene that are incorporated is difficult to control. Many promoters that are used do not tightly control the expression of the transgene, so that the transgene may be constitutively expressed. The site of integration can affect transgene

Table 2.6. Marker genes used for gene function studies.

GFP	Green fluorescent protein
β-GAL	β-Galactosidase
CAT	Chloramphenicol acetyltransferase
ALP	Alkaline phosphatase
ADH	Alcohol dehydrogenase from *Drosophila*
LUC	Luciferase
NEO	Resistance to antibiotic G418
HPRT	Resistance to methotrexate

Table 2.7. Methods for gene transfer.

Cell type	Methodology
Single-cell fertilized ovum	Microinjection
Embryonic stem cells	$CaPO_4$, DEAE-dextran, polybrene, PEG, DMSO
Primordial germ cells	Electroporation
Unfertilized ova	Liposomes
Spermatozoa and spermatogonia	Particle bombardment
	Retroviral vectors
	Nuclear transfer

DMSO, dimethylsulphoxide; PEG, polyethylene glycol.

expression, and the integration site is usually random except for targeted gene insertions by homologous recombination. The activities of other genes in the region of the transgene affect the expression of the transgene, and the expression can be inhibited if the transgene inserts in an area of inactive chromatin. Including buffer sequences, such as matrix attachment regions, around the transgene can shield it from these effects. The integration of the transgene can also disrupt other endogenous genes by non-targeted insertional mutagenesis. This can give clues about the function of endogenous genes if the effect is not lethal.

Further details on various applications of transgenic animals can be found in Houdebine (1996, 2000).

Questions for Study and Discussion

Section 2.1 Methods for Studying Endocrine Function

1. Describe a classical endocrine experiment in whole animals to identify endocrine tissues.
2. Discuss the use of various *in vivo* and *in vitro* methods.
3. How are antibodies raised and how can you make a hormone more antigenic?
4. How can you use antibodies to identify endocrine glands or target tissues?
5. Describe the use of various inhibitors as tools to study hormone function.

Section 2.2 Measurement of Hormones and Receptors

1. What are the requirements for assay methods?
2. Describe bioassays. What are the advantages and disadvantages of bioassays?
3. Describe chemical assays. What are the advantages and disadvantages of chemical assays?
4. How can you determine the chemical composition of a hormone (steroid versus protein)? What methods are useful for purifying and measuring protein versus steroid hormones?
5. Describe the different types of chromatography for analytical use.
6. Describe the principle and methods of competitive binding assays for the measurement of hormones.
7. Describe the types of binding proteins and labels that can be used for competitive binding assays.
8. Describe Scatchard analysis; how does it differ from competitive binding assays for hormones?

Section 2.3 Methods for the Production of Hormones

1. What are the sources of steroid hormones? Describe the structure of steroid hormones and their synthetic analogues.
2. How is the sequence of short peptide hormones determined and how can they be synthesized?
3. How are protein hormones produced?
4. What is the relative purity of steroid versus protein hormone preparations?

5. What is the relative purity of hormone preparations from recombinant and natural sources?

Section 2.4 Manipulation of Endocrine Function
1. Discuss the advantages and disadvantages of different methods for delivering hormones.
2. Outline the method and application of immunoneutralization of hormones.
3. Discuss the use of anti-idiotypic antibodies to mimic hormone action.
4. How are transgenic animals made and how can they be used to affect endocrine function?

Further Reading

General

Hadley, M.E. (2000) *Endocrinology*, 5th edn. Prentice Hall, Englewood Cliffs, New Jersey.

Norman, A.W. and Litwack, G. (1997) *Hormones*, 2nd edn. Academic Press, Toronto.

Endocrine methodology

Goding, J.W. (1996) *Monoclonal Antibodies: Principles and Practice*, 3rd edn. Academic Press, New York.

Production of hormones

Erlanger, B.F. and Cleveland, W.L. (1992) Method of producing monoclonal auto-anti-idiotypic antibodies. US Patent #5144010.

Patchett, A.A. and Nargund, R.P. (2000) Privileged structures – an update. *Annual Reports in Medicinal Chemistry* 35, 289–298.

Sugg, E.E. (1997) Nonpeptide agonists for peptide receptors: lessons from ligands. *Annual Reports in Medicinal Chemistry* 32, 277–283.

Manipulation of endocrine function

Houdebine, L.-M. (ed.) (1996) *Transgenic Animals – Generation and Use*. Harwood Academic Publishers, London.

Houdebine, L.-M. (2000) Transgenic animal bioreactors. *Transgenic Research* 9, 305–320.

Medlicott, N.J. and Tucker, I.G. (1999) Pulsatile release from subcutaneous implants. *Advanced Drug Delivery Reviews* 38, 139–149.

Nett, T.M., Ilen, M.C., Wieczorek, M. and Glode, L.M. (1999) A gonadotropin-releasing hormone agonist (GnRH-A) linked to pokeweed antiviral protein (PAP) decreases the ability of the pituitary gland to secrete LH. *Biology of Reproduction* 60(Suppl. 1), 430.

Pinkert, C.A. and Murray, J.D. (1999) Transgenic farm animals. In: Murray, J.B., Anderson, G.B., Oberbauer, A.M. and McGloughlin, M.M. (eds) *Transgenic Animals in Agriculture*. CAB International, Wallingford, UK, pp. 87–96.

3
Manipulation of Growth and Carcass Composition

3.1 Overview

Growth is an increase in body weight and size due to an increase in the number of cells (hyperplasia) or an increase in the size of cells (hypertrophy). Growth can occur from increased deposition of protein, measured as an increase in nitrogen retention in the carcass, as well as increased lipid deposition. Protein deposition is regulated by the balance between protein synthesis and protein degradation; lipid deposition is regulated by the balance between lipogenesis versus lipolysis. Hormones can differentially affect the extent of deposition of muscle and fat, which affects the lean yield of the carcass. Hormones can also affect feed intake, feed conversion efficiency (the ratio of feed:gain), growth performance (the rate of gain) and dressing percentage (the ratio of carcass weight:live weight) (Fig. 3.1). In addition, hormones can affect various aspects of meat quality, including tenderness, juiciness, flavour and water-holding capacity. These hormones are discussed in detail in the following sections of this chapter.

The hormones affecting growth are covered first in this chapter, starting with anabolic steroids and then discussing issues related to castration of pigs and removal of these anabolic hormones. The effects of somatotrophin, β-adrenergic agonists, thyroid hormones and dietary polyunsaturated fatty acids (PUFAs) on growth rate, feed efficiency and carcass composition are then covered. This is followed by a discussion of the many effects of leptin, and then factors such as cholecystokinin that affect appetite are covered. After a short discussion of the use of antimicrobials, the potential effects of dietary

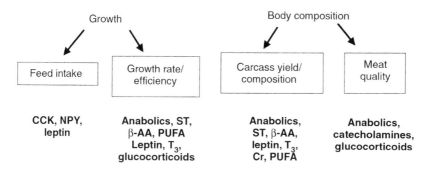

Fig. 3.1. Summary of hormones affecting growth and carcass composition. CCK, Cholecystokinin; NPY, neuropeptide Y; ST, somatotrophin; β-AA, β-adrenergic agonists; PUFA, polyunsaturated fatty acids; T_3, triiodothyronine; Cr, chromium.

chromium and the effects of stress on meat quality are presented.

Many of these factors are interrelated and it is important not to disregard meat quality, for example, when manipulating hormones to improve growth and feed efficiency. Similarly, it is important to stimulate sufficient bone growth to maintain bone integrity and carrying capacity of the increased skeletal mass. For example, treatment of growing pigs with porcine somatotrophin (pST, see Section 3.4) can result in cartilage damage and decreased bone strength. Abnormal skeletal development is a major cause of mortality and downgrading of broiler chickens. This problem is more prevalent in meat-type birds and turkeys than in laying hens and is probably linked to rapid growth.

The increase in knowledge of hormonal effects on various aspects of growth has led to strategies for increasing the lean content in carcasses and improving feed efficiency and growth rate. Improved feed efficiency means that more meat is produced from less feed, which can lower the cost of production and reduce the amount of land necessary for grazing animals. More efficient utilization of feed also means that a lower amount of waste products, such as nitrogen, will be produced in the urine and faeces, thus reducing the impact of intensive production systems. However, public perceptions that exogenously administered hormones are bad, regardless of the available scientific data, can make the commercial application of these compounds difficult.

Effects on growth, feed efficiency and lean yield

Many hormones, including growth hormone, thyroid hormones, catecholamines, insulin, glucocorticoids and sex steroids, affect growth and the metabolic processes involved in the synthesis and degradation of body tissues. The result is a shift in the metabolism to direct more nutrients into muscle with fewer nutrients either deposited into fat or excreted. In contrast to the short-term regulation of metabolism for the maintenance of homeostasis, endocrine regulation of growth involves long-term endocrine changes. The hormonal regulation of metabolism is affected by nutrition and by the normal cycles of endogenous hormone secretion and sensitivity in the animal. It is important to consider these factors to maximize the response of animals to exogenously administered hormones. For further information, see Sejrsen and Vestergaard-Jensen (1990).

The young, growing animal produces mostly muscle mass with little fat, but as the animal continues to grow, the body fat content increases. Thus, the particular endocrine systems affecting growth depend on the species, sex and stage of maturity of the animal. Treatment with growth modifiers to increase protein deposition is less effective in young animals, when the inherent capacity and efficiency of protein deposition is greater, than it is in older animals that are closer to their mature body size.

Protein synthesis and deposition are increased by the intake of dietary protein or limiting amino acids, such as lysine, in the growing pig. When energy is not limiting, there is a linear response of protein synthesis to intake of dietary protein, or the first limiting amino acid, up to a plateau that is determined by the **inherent capacity** of the animal for protein deposition (Fig. 3.2). The slope of this linear response is a measure of the **efficiency** of use of dietary amino acids for protein synthesis. Endocrine manipulations to affect growth could act by increasing the maximal response **plateau** of growth, which

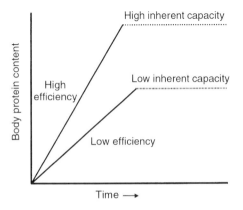

Fig. 3.2. Effects of efficiency versus inherent capacity for protein deposition.

would increase the requirements for dietary amino acids. Alternatively, the efficiency of amino-acid use for protein synthesis could be increased, which would not dramatically increase the dietary requirements of amino acids. Increased efficiency of utilization of dietary amino acids may occur through a decrease in amino-acid catabolism in the liver, which can be measured as a decrease in levels of plasma urea nitrogen.

The effects of endocrine manipulations on increasing protein synthesis versus decreasing protein catabolism have been determined *in vivo* using labelled amino acids. The dilution of isotope and uptake of amino acids can be measured by arteriovenous differences with blood flow measurements across the hind limb of an animal. The release of free amino acids produced from protein catabolism would result in a dilution of the specific activity of the labelled amino acid in the venous blood. Uptake of amino acids for protein synthesis would be measured as a decrease in the amount of labelled amino acids in the venous compared to the arterial blood, with no change in specific activity.

Genetics plays an important role in the hormonal response of an individual animal, and a large proportion of the genetic selection effort to date has been directed to improving muscle development and the partitioning of nutrients away from fat. This has been more successful in poultry and swine, partly because of the shorter generation time, compared to cattle. However, selection for increased muscle mass has resulted in some metabolic problems. For example, selection for increased breast muscle in turkeys has led to problems with pale, soft, exudative (PSE) meat and various cardiovascular problems, such as round heart disease and sudden death syndrome. Selection for decreased fat in swine has reduced some organoleptic qualities and affected flavour and eating quality. Selection for increased muscle development in certain breeds of pigs caused an increase in porcine stress syndrome (PSS). Selection for double muscling in cattle results in pale meat due to lower myoglobin and fat contents, and double-muscled cattle are more susceptible to stress. It is important to include meat quality and health status attributes in selection programmes to reduce these problems.

3.2 Anabolic Steroids and Analogues

Male cattle, pigs and sheep are traditionally castrated, even though intact (uncastrated) males are more efficient at producing lean meat (muscle) than castrates. Cattle are castrated to reduce the undesirable aggressive behaviour of males and to produce a more consistent carcass with improved tenderness and finish. During sexual development at 8–12 months of age, the collagen content increases in the muscles of bulls, which can decrease meat tenderness. Male pigs are castrated to prevent objectionable odours and flavours (boar taint) in the meat (see Section 3.3); male sheep also tend to have undesirable odours when raised to heavier weights. Female animals tend to produce carcasses with less muscle mass and increased fat content than carcasses from intact males. For these reasons, animals have been treated with a variety of steroid hormones to improve lean growth or produce more consistent, high-quality carcasses. Females can be treated with androgens to improve muscle growth and decrease carcass fat. Intact males can be treated with oestrogenic compounds to increase carcass fat and meat tenderness, and to decrease aggressive behaviour. Castrates can be treated with a combination of oestrogens and androgens to produce the most desirable growth rate and carcass composition.

Anabolic steroids have been used in both human and veterinary medicine for over 30 years. They have been used in the treatment of debilitating conditions such as starvation, recovery after surgery and from extreme trauma, such as burns. They have also been used by competing athletes, such as bodybuilders. The use of anabolic steroids in equine athletes began in the 1960s, but by the 1970s it was decided that the presence of anabolic steroids at the time of racing constituted doping.

Anabolic steroids are used therapeutically in horses to increase appetite and protein retention during recovery from injuries and to reduce the catabolic effects of corticosteroid therapy. Anabolic steroids have been reported to stimulate erythropoiesis and have been recommended for the treatment of certain types of anaemia.

Anabolic steroids are sometimes used illegally in horses to increase stamina and to modify behaviour, to increase aggression and 'sharpness' for racing. There are few scientific studies and much anecdotal information on the effects of anabolic steroid use in horses. Racing results are affected by a variety of factors, such as track conditions, weather, other horses and the racer/driver, so determining the effects of doping with anabolic steroids is difficult. For more information, see Snow (1993).

The term '**anabolic**' refers to compounds that increase nitrogen retention. A number of different types of compounds, including insulin, growth hormone, β-agonists and steroids, exert anabolic effects. Steroid hormones currently used as anabolic agents (Fig. 3.3) include testosterone, oestradiol and progesterone, as well as steroid analogues such as zeranol and synthetic steroids such as trenbolone acetate (TBA) and melengestrol acetate (MGA). Zeranol is an oestrogenic compound produced from the mycotoxin zearalenone. TBA is a synthetic androgen, which has 10–50 times the anabolic properties of testosterone with reduced secondary androgenic effects. MGA is an orally active progestagen used for suppressing oestrus and improving the rate of gain and feed efficiency in heifers. Corticoids can also be administered to finishing animals. Oxandrolone is a synthetic analogue of testosterone that has been used in human studies. For further information about the use of anabolic steroids in animal production, see Lone (1997).

Modifications of the structure of testosterone by alkylation at the 17α position produces orally active preparations. However, the most commonly used anabolic steroids in horses have modifications at the C19 or C18 position. These include nandrolone (decadurabolin or 19-nortestosterone), which is also popular for human bodybuilding, trenbolone, which is used in cattle, and methanedieone (methandrostenolone or Dianabol®). Stanazol (Winstrol®) and boldenone are intended primarily for horses (Fig. 3.4).

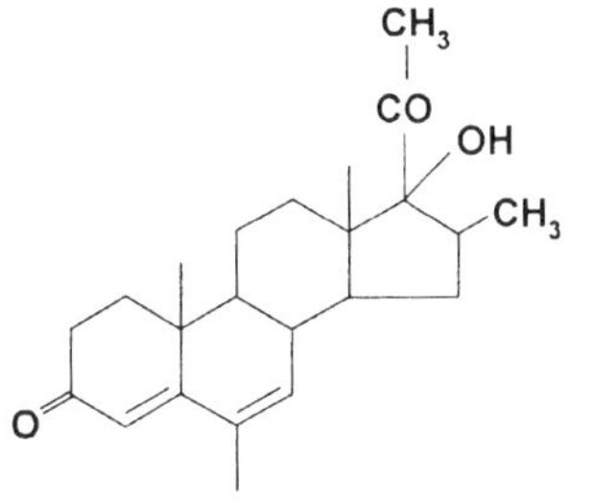

Melengestrol acetate (MGA)

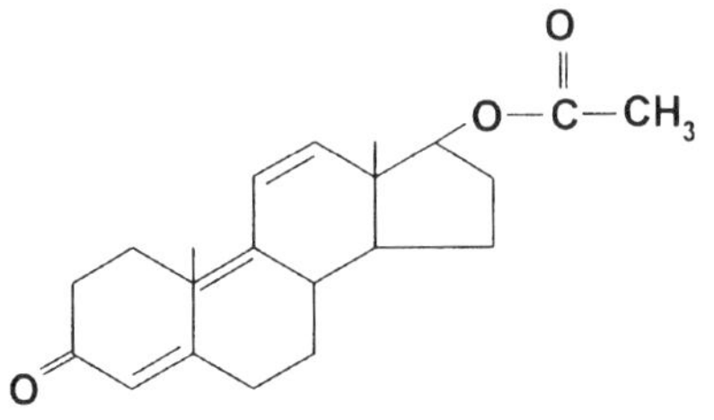

Trenbolone acetate (TBA)

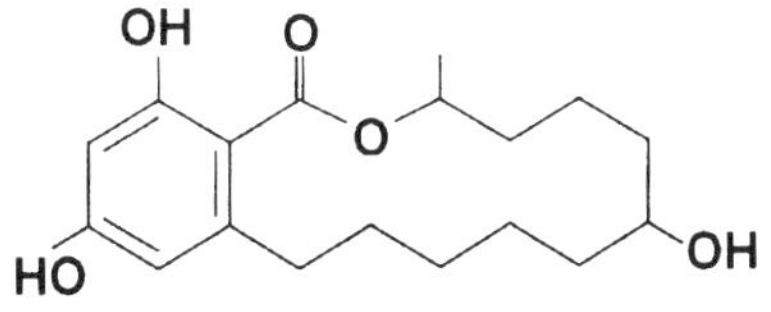

Zeranol

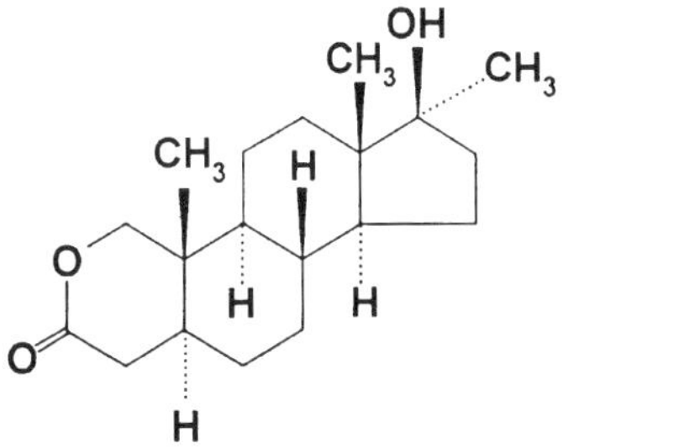

Oxandrolone

Fig. 3.3. Steroids and steroid analogues used as anabolic agents.

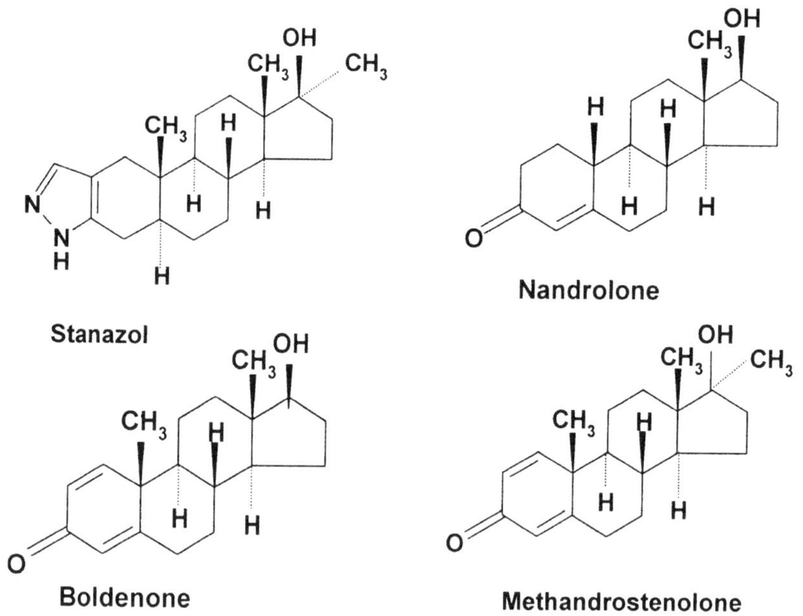

Fig. 3.4. Anabolic steroids used in horses.

These compounds can be esterified at the 17α hydroxyl group and administered intramuscularly in an oil formation for sustained activity. Dosage levels are 0.55–1.1 mg kg^{-1} body weight per injection period. The length of the ester chain affects the rate of absorption and duration of activity. Esters of acetic and propionic acid last for several days, esters of phenylpropionic, cyclopentylpropionic and undecylenic acid last 2–4 weeks and esters of lauric, decanoic and heptanoic acid last several months. Once absorbed, the esters are hydrolysed to produce the free steroids. Stanazol is not esterified, as the pyrazole group on the A ring gives it a 1-week duration of activity.

For use in meat-producing animals, the goal is to maximize the anabolic effects of growth-promoting compounds, while reducing their androgenic effects. The term 'anabolic' refers to the tissue-building properties of the compound, including the increase in muscle mass (nitrogen retention) and decrease in body fat. The term 'androgenic' refers to the increased development of male sexual characteristics by the compound, including behavioural changes and the development of the reproductive tract. Steroid derivatives have been synthesized with specific androgenic and anabolic activities, which are determined by a bioassay using castrated rats. The effects of the compound on the myotrophic activity in levator ani muscle is used as a measure of anabolic activity, while the effects on the development of the prostate or seminal vesicles are used as a measure of androgenic activity. The ratio of anabolic to androgenic activity is the Q value for the compound. These Q values can vary among different laboratories due to non-standardized methods that are used in their determination. The levator ani muscle may also be an androgen-dependent muscle, so true measures of anabolic activity should be based on whole-body nitrogen retention studies.

Mechanism of action

Androgens increase protein synthesis and decrease fat deposition in adipose tissue. Anabolic steroids increase the retention of dietary nitrogen as body protein and increase muscle mass through hypertrophy rather than hyperplasia. Increased body protein can be the result of increased protein synthesis or decreased protein degradation. Anabolic steroids can act by a direct mechanism through the androgen receptors, or by indirect methods. Indirect mechanisms include modulating the production of other hormones, such as growth hormone, thyroid hormone and insulin. For more information, see Sheffield-Moore (2000).

Direct effects

The direct effects of androgens are mediated by interaction with the androgen receptor (AR, see Fig. 3.5). A large number of different androgenic compounds have been developed that are known as selective androgen receptor modulators (SARMs). However, only one androgen receptor has been identified and cloned. The AR gene is located on the X chromosome and codes for a protein of 918 amino acids. As a member of the nuclear receptor family of transcription factors, it contains protein domains that can activate or repress activity. These domains are exposed upon hormone binding to allow interactions with various co-activators or co-repressors. These include cAMP-responsive-element binding protein (CBP/p300), glucocorticoid receptor interacting protein 1 (GRIPI), androgen receptor co-activator (ARA) 54, 55 and 70 and transactivator protein (TAT)-interacting protein (Tip60). The binding of these co-activator proteins is believed to enhance the stability of the pre-initiation complex. Some factors (CBP/p300) have intrinsic histone acetyl transferase activity, which will disrupt nucleosome structure and increase the rate of transcription initiation (see Section 1.3).

The differential effects of androgens in various tissues may be due to differences in the levels of cofactors that affect the AR, or the presence of unique AR-interacting proteins such as ARIP-3, which is specific to testis. In addition, the binding of a particular SARM to the AR may cause a unique conformational change that exposes particular domains that allow the interaction of specific cofactors. This would lead to the selective regulation of individual genes in specific tissues by the AR. Ongoing research will lead to the development of SARMs with selective preferences for individual tissues or activities. For more information on SARMs, see Negro-Vilar (1999), and for oestrogen recep-

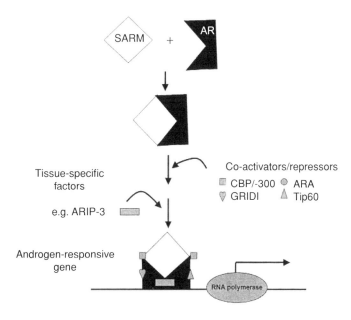

Fig. 3.5. Mechanism of action of the androgen receptor.

tor modulators, see Cosman and Lindsay (1999).

Testosterone can undergo irreversible reduction by the 5α-reductase enzyme in some tissues to form 5α-dihydrotestosterone, which can be further metabolized to 17-ketosteroids and polar metabolites that will have different binding characteristics to the AR. The aromatization of androgens with a Δ-4-3-keto group to oestrogens may also be important in some tissues, since oestrogen-like effects will be produced. The aromatase reaction can be blocked by removal of the C19 methyl function (as in nandrolone) or modification of the A ring (as in oxandrolone and stanazol).

Androgens have a primary effect on skeletal muscle through binding to specific receptors on the muscle. Because the activity of the 5α-reductase enzyme is low in skeletal muscle, testosterone is the major hormone for androgen action in muscle. Increased numbers of androgen receptors are found during the muscle hypertrophy seen with increased exercise or treatment with oxandrolone, which is a synthetic analogue of testosterone that cannot be aromatized to oestrogens.

The number of free androgen receptors is higher in females and castrates compared to intact males. An oestrogen receptor has also been identified, and the free form of this receptor is present in higher amounts in males than in females. Steroids bind to receptors in the cytoplasm, and the hormone–receptor complex enters the nucleus to stimulate the transcription of hormone-specific genes. This increases the synthesis of myofibril and sarcoplasmic proteins to increase the overall muscle mass.

Testicular feminization (Tfm) or androgen insensitivity syndrome (AIS) is an X-linked recessive gene defect in the androgen receptor that results in a male pseudo hermaphrodite. Virtually all androgen-responsive tissues in Tfm individuals are not responsive to testosterone due to the lack of a functional AR. Female carriers have the Tfm gene defect on one X chromosome and a normal AR on the other chromosome. These carriers have an androgen responsiveness that is intermediate between that of normal and Tfm animals.

Indirect effects

Androgens can function indirectly by increasing the production of GH or IGF-I, and prior exposure to androgens can prime cells for the secondary actions of IGF-I. Glucocorticoids may increase the production of IGF-binding proteins to decrease IGF bioactivity. Anabolic steroids also reduce the production of LH and testosterone, delay sexual maturation and reduce testicular growth and spermatogenesis in bulls.

There appears to be a reciprocal relationship between testosterone (an anabolic hormone) and cortisol (a catabolic hormone) in bulls. Anabolic steroids could potentially act as antagonists of the catabolic action of glucocorticoids, by decreasing the concentration of the glucocorticoids, or by displacing the glucocorticoids from their receptors. Studies *in vitro* have shown that testosterone has a high affinity for the glucocorticoid receptor and could thus prevent the normal protein catabolic action of glucocorticoids.

There is some evidence that anabolic agents with oestrogenic activity increase the blood levels of insulin to promote growth. Oestrogens increase growth rate and feed efficiency in ruminants without the loss of desired carcass characteristics. Insulin has anabolic effects on adipose and skeletal muscle, while thyroid hormone (T_3) has a catabolic effect on adipose tissue and an anabolic influence over skeletal muscle.

Delivery systems

Anabolic agents can be given to cattle in three ways:

1. Oral administration by addition to the feed. This requires an orally active hormone (e.g. MGA) and higher levels are needed to stimulate growth performance.
2. Repeated intramuscular injections can be used, but this is labour intensive and results in high levels of hormones at the injection site.
3. Implants containing the steroids can be used for sustained release of the hormone. Implants are normally placed behind the ear, which is a desirable site since it is eliminated

Table 3.1. Steroid anabolic agents used as implants (from Unruh, 1986).

Trade name	Chemical name
Compudose®	Oestradiol 17β
Ralgro®	Zeranol
Synovex -S®	Oestradiol 17β benzoate + progesterone
Synovex-H®	Oestradiol 17β benzoate + testosterone propionate
Finaplix®	Trenbolone acetate
Revalor®	Oestradiol 17β + trenbolone acetate
Forplix®	Zeranol + trenbolone acetate

after slaughter. Steroids can be formulated into a compressed pellet that has a life span of 90–120 days of activity before completely degrading. Silastic rubber implants can also be used to provide a slow and continuous release for 200–400 days. Some commercially available implants are listed in Table 3.1.

The general principle for use of anabolic steroids is to supplement with the hormone that is deficient for the particular animal. In cattle, the expected improvement in growth rate is about 15% and in feed efficiency about 8% from the use of anabolic steroids. Growth responses are likely to be greater in steers than in intact males or females, which already have a source of endogenous hormones. Androgenic compounds are more effective in females, while oestrogens are more effective in males. A combination of androgen and oestrogen or progestagen and oestrogen is given to castrated males.

A combination of androgen and oestrogen can be given to intact males to inhibit testicular function. Implanting zeranol in bulls from birth to slaughter decreases carcass masculinity, improves fat cover to an acceptable finish and improves meat quality and palatability. Steers implanted with trenbolone and oestradiol have weight gains similar to those of intact bulls. Treatment with a combination of androgen and oestrogen or androgen alone increases protein accretion and decreases fat content. Treatment with oestrogens or an oestrogen/progestagen combination increases fat deposition, particularly in uncastrated males. Treatment of bulls with TBA and oestradiol increases tenderness and lowers the connective tissue content, while not affecting the juiciness and flavour of steaks.

A conservative approach using mild oestrogens ensures marbling with a slight growth response. An intermediate approach using a mild combination of androgens and oestrogens gives improved growth with a slight depression on marbling. An aggressive approach using a strong combination of androgens and oestrogens gives maximum improvement in growth with very poor marbling and negative effects on finishing. Mild oestrogens include zeranol, which produces a 5% decrease in marbling, and the combination of oestrogen and progesterone, which gives a 10% decrease in marbling. Treatment with the strong androgen TBA gives a 20–25% decrease in marbling.

The efficacy of growth promoters is also affected by nutrition. Cattle implanted with anabolic steroids require higher levels of dietary protein or nitrogen to sustain the higher growth rates. The level of response may be dependent on the initial nutritional status, level of live weight gain, age, dose and length of treatment. Higher response is found in cattle that are already growing at a rapid rate, since growth in each animal is limited by its genetic potential. In larger-framed cattle, the average daily gain (ADG) should be higher than $1.8–2.0\,\mathrm{kg\,day^{-1}}$ for anabolic treatment to be effective. Anabolic treatment does not affect the digestibility of feed, although feed intake and feed efficiency may increase. Sufficient levels of dietary protein and essential amino acids are needed to sustain the higher growth rates. Anabolic agents are ineffective in the early treatment of intact males, while castrates and females can be treated with anabolic agents at any age. The response to repeat implants is reduced with each sequential implant. Intramuscular injection of synthetic hormones must be repeated every 10–15 days for 8–10 weeks.

Weights of the liver and spleen are increased in steers implanted with anabolic agents. Treatment with corticoids alters thymus, thyroid and adrenal glands and lowers the levels of lymphocytes and neutrophils. These alterations in the immune system can result in health disorders, such as respiratory illness or infection in finishing animals. Treatment of intact males with oestrogenic compounds decreases testicular maturation and spermatogenesis. Growth-promoting implants are not recommended for replacement heifers because of possible detrimental effects on fertility.

Anabolic agents produce similar responses in growth and carcass composition of sheep as in cattle. The results in pigs have been mixed, with few consistent beneficial effects found. Anabolic treatment of fish with methyltestosterone is used to induce sex reversal and promote growth (see Section 5.2). Salmon treated with methyltestosterone had 20–33% higher daily gain, 18–19% improved feed conversion ratio and 19–34% lower ammonia excretion than untreated fish.

Safety issues

Anabolic agents have had a very significant impact on beef production in North America since they were first licensed for use in the early 1950s. In the early years, the synthetic stilbenes, diethylstilbestrol, hexestrol and dienestrol, were used. These were subsequently banned in the EU, the USA and Canada because of their potential carcinogenic activity. Since December 1988, there has been a complete ban on the use of hormonal-type growth promoters in the EU. Consumers generally consider the use of hormones in meat production to be a very high risk. However, in 1999 the Food and Agriculture/World Health Organization (FAO/WHO) Joint Expert Committee on Food Additives (JECFA) reviewed the use of oestradiol, progesterone and testosterone in meat production and declared that they were safe. The no observed effects limits (NOELs) for oestradiol, progesterone and testosterone are $5 \mu g\ kg^{-1}$ BW, $3.3\ mg\ kg^{-1}$ BW and $1.7\ mg\ kg^{-1}$ BW, respectively. Using a safety factor of 100, this limits the acceptable daily intake (ADI) to $50\ ng\ kg^{-1}$ BW, $30\ \mu g\ kg^{-1}$ BW and $2\ \mu g\ kg^{-1}$ BW for oestradiol, progesterone and testosterone, respectively. The theoretical maximum daily intakes for these compounds can be calculated from data on tissue levels of these compounds in treated animals and considering that an individual would consume 300 g muscle, 100 g liver, 50 g kidney and 50 g fat per day. These calculations show that the theoretical daily maximum intakes are substantially less than the ADI from consuming meat from treated animals. For more information, see Waltner-Toews and McEwen (1994).

The ADI values for oestradiol used by the FAO/WHO report have been questioned, based on new data for oestrogen production and clearance in prepubertal boys, with an ADI of 0.04–$0.1\ \mu g\ day^{-1}$. This implies that the additional dose of oestrogen from consuming meat from implanted animals could equal the production rate in prepubertal boys. However, natural levels of oestradiol 17β are significantly higher in meat from animals in late pregnancy than from steers with a hormonal implant. There are also high levels of natural steroids in meat from intact males and in the milk from pregnant cows. Levels of natural oestrogens found in plants and plant products, such as soybean oil, are several orders of magnitude higher than in beef from implanted animals (see Section 6.4).

The maximum residue levels (MRLs) for TBA, zeranol and MGA are $2.0\ \mu g\ kg^{-1}$ muscle, $70\ \mu g\ kg^{-1}$ and $25\ \mu g\ kg^{-1}$ tissue, respectively. A 48 h withdrawal time for MGA is prescribed. The USDA and Agriculture and Agrifood Canada maintain drug testing programmes. There were some instances of high levels of zeranol in 1988 and 1989, but otherwise tests for zeranol, TBA and MGA were negative between 1988 and 1993.

Detection of anabolic steroid use in horses is based on levels in urine or blood, with initial screening by immunoassays and confirmation by GC/MS. However, colts naturally produce testosterone and 19-nortestosterone, so it is more difficult to detect use of these steroids. A ratio of oestrane-3,17-diol to 5(10)-oestrene-3,17-diol levels greater than one indicates that doping with nandrolone has occurred (Houghton et al., 1986).

Anabolic steroid treatment will reduce cycling in mares and sperm production and testis size in stallions. In humans, a number of anabolic steroids have been shown to be hepatotoxic. High doses of anabolic steroids can increase the risk of severe coronary heart disease, decrease high-density lipoprotein (HDL) while increasing low-density lipoprotein (LDL), and reduce the immune response.

3.3 Use of Intact (Uncastrated) Male Pigs

Advantages and problems of intact male pigs

Raising intact male pigs instead of castrates for pork production offers several advantages, due to the natural production of anabolic hormones by intact males. Overall production costs are substantially lower for intact males than for castrates. The labour costs for castration are eliminated, along with death losses and temporary decreases in performance usually seen following castration. Intact males have better feed conversion efficiency than castrates, and may sometimes grow faster than castrates. As a result, the overall output of nitrogen in manure is less with intact males than with castrates. Intact males may also be more resistant to diseases that negatively affect performance.

Intact male pigs, compared to other sexes, have a slightly lower dressing percentage (2–2.5%) but higher lean content (5–7%). Intact males have about 5% less separable fat than castrates, with gilts being intermediate. The total fat content of soft tissues is about 30% lower in intact males than in barrows. Decreased backfat in intact males would result in a higher grading for carcasses, but carcass grade may be underestimated unless special equations are used for carcass grading systems for intact males. The low intramuscular fat content, which can affect the flavour and texture of the meat, may be of some concern in intact males if it is lower than the 2–3% level recommended for optimum sensory quality. Intramuscular fat levels can be higher for gilts than for intact males, but breed differences are more pronounced than sex differences. Meat cuts from intact males are more appealing to the consumer because of the decrease in intermuscular fat.

There are important differences between sexes in the fatty tissue composition, as well as in the composition of the lipid fraction of the fat. The back fat from intact males contains more water, more protein and less lipid than that from castrates and gilts. The lipid fraction of intact male back fat contains more unsaturated fatty acids than that of castrates, with that of gilts being intermediate. The high level of unsaturation of intact male fat is mainly due to the higher levels of linoleic (18:2) and linolenic (18:3) acids, with a lower level of palmitic acid (16:0).

The lower fat content and different fatty acid composition of carcasses from intact males has its advantages as well as disadvantages. Intact male meat, with its lower level of fat and a healthier balance of fatty acids, offers a greater nutritional value and may be more appealing to consumers. The leanness of the meat has been found to be the most important visual criterion in purchasing pork. On the other hand, fat from intact males is softer, due to the high level of unsaturation, and more easily separates from the other tissues. This causes difficulties in handling by meat packers and decreases the quality of the cuts. Also, since unsaturated fatty acids are more susceptible to oxidation, intact male fat turns rancid faster than that of gilts and castrates. However, the degree of saturation of fatty acids in the intact male fat can be manipulated by dietary means. Reducing the level of unsaturated fatty acids or increasing the energy level in swine diets via carbohydrates decreases the level of unsaturation of fatty acids in the fatty tissue. The poorer quality of fat in boars is a result of the leanness of the animals and thus is more important in very lean genotypes.

The processing characteristics of pork from intact males differ very little from that of the other sexes, with the main difference being the slightly lower processing yields.

Intact males are considered to be more aggressive and therefore potentially more susceptible to the development of PSE (pale, soft, exudative) and DFD (dark, firm, dry) conditions in meat. While the frequency of PSE meat is not higher for intact males, they do show a greater tendency for producing DFD meat. Thus, care should be exercised during preslaughter handling of intact males. Avoiding mixing of unfamiliar animals, minimizing lairage time and using good handling practices would reduce the amount of stress and consequently the incidence of DFD. Such measures would also reduce the incidence of skin damage, which is another problem associated with higher aggressiveness of intact males. Sometimes fighting results in only superficial skin blemishes, but in other cases it can cause major carcass bruising and consequently financial losses from increased carcass trim. The incidence of carcass and skin damage varies considerably between slaughter plants, and major problems are likely to occur where good preslaughter practices are not exercised.

A clear disadvantage of intact male carcasses is the higher bone content. Both gilts and barrows have lower bone percentage and consequently lower deboning losses. Due to the much higher fat content in castrates, the lean yield for intact males is higher, despite intact males having more bone. The bone content in pig carcasses is strongly affected by the breed. The dressing percentage is also lower in intact male pigs than in castrates, due to the presence of a developing genital tract. For more information on meat quality in intact male pigs, see the review by Babol and Squires (1995).

Particularly in European countries, animal welfare concerns are putting more and more pressure on the pig production chain to abandon castration. Avoiding surgical castration without anaesthetic undoubtedly improves the welfare of the animals in the short term, although increased aggression may occur with intact males as they approach market weight. As the males approach sexual maturity, there can also be excessive 'riding' or mounting, which results in decreased feed consumption and poorer growth performance.

Avoiding castration would also lower the costs of swine breeding programmes, since the breeding stock could be selected at an older age and the non-selected animals could be sold for meat at a normal market price.

Although the problems with meat quality may be significant, particularly in lean strains of pigs, the main reason that prevents the production of intact males is the presence of an offensive odour, called boar taint, in their fatty tissue. Boar taint is present in only a small portion of carcasses and a high percentage of people are not sensitive to it. Intact males are used to some extent for meat production in several countries including Denmark, Great Britain, Spain and Australia. However, in most countries, including Canada, all male pigs intended for meat production are castrated at a very young age to avoid the potential consumer dissatisfaction from boar taint in the carcasses. Under the current Canadian pork carcass grading system, a carcass from an intact male pig (boar) is not graded and is assigned an index of 67. This compares to a normal grade of 100 for a market hog, making boar carcasses worth about one-third less than a similar carcass from a gilt or castrate.

Effects of sex steroid hormones

The natural sex steroid hormones produced in the testes account for the increase in growth rate and lean meat yield of intact male pigs compared to castrates. The most predominant androgens in the testis of the pig are 5-androstenediol, dehydroepiandrosterone (DHEA) and testosterone. These steroids (and their metabolites 5α-dihydrotestosterone (DHT) and oestrogen) stimulate the development of the reproductive tract and secondary sex characteristics, including actions in the central nervous system. In addition, these steroids have dramatic anabolic effects to stimulate muscle growth, nitrogen retention, phosphorus retention and bone growth, and cause the redistribution of nutrients away from the synthesis of subcutaneous fat. This results in an enhanced growth rate and feed conversion efficiency, along with a reduced backfat thickness and increased lean content of carcasses from intact males compared to females or castrates.

The biological effects of androgens are

mediated by their interaction with specific receptors inside the cells of the target tissues (see Section 3.2). In some tissues DHT is the active androgen that binds to the receptor, while in other tissues it is testosterone. Synthetic anabolic steroids, such as nandrolone decanoate, oxandrolone and stanozolol, have high anabolic activity in skin, muscle and bone but have minimal androgenic activity. These effects may be due to competition between these anabolic steroids and endogenous glucocorticoids for their receptors in non-reproductive tissues. Anti-androgen compounds, such as cyproterone acetate and flutamide, act by binding to the androgen receptor, thus preventing binding of the active androgen. Oestrogen has also been shown to be involved in the development of secondary sex characteristics in the boar.

The testicular synthesis of both the sex steroid hormones and the androst-16-ene steroids (responsible for boar taint) is induced *in vivo* by administration of luteinizing hormone (LH), human chorionic gonadotrophin (hCG) and by sexual stimulation. The production and secretion of testosterone by the testis is controlled by the release of LH from the anterior pituitary gland, which in turn is regulated by gonadotrophin releasing hormone (GnRH) produced by the hypothalamus (Fig. 3.6). As the animal approaches sexual maturity, levels of GnRH rise and stimulate the release of LH into the circulation. The binding of LH to specific receptors on the surface of the Leydig cells stimulates the synthesis of steroidogenic enzymes and thus increases the production of androgens and androst-16-ene steroids. Steroid hormone production can be blocked by the use of GnRH agonists, which cause a down regulation of the number of GnRH receptor sites, and with GnRH antagonists, which block the binding of GnRH to its receptor. Analogues of LH that block LH action also cause decreased production of both androst-16-ene steroids and androgens. Production of antibodies that bind GnRH or LH and thus inhibit their activity, so called 'immunocastration', will also result in decreased androst-16-ene steroid and androgen production.

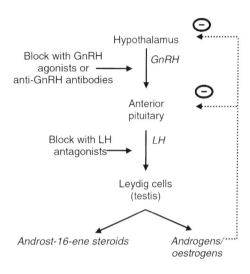

Fig. 3.6. Regulation of androst-16-ene steroid synthesis.

Description of boar taint

Boar taint is the presence of off-odours and off-flavours found predominantly, but not exclusively, in meat of some intact male pigs. The main compounds responsible for boar taint (Fig. 3.7) are 5α-androst-16-en-3-one (androstenone) and 3-methylindole (skatole). There is some evidence that they cannot completely account for the occurrence of boar taint as determined by a trained sensory panel. Intact males near sexual maturity, as judged by the length of the bulbourethral glands, with low levels of androstenone and skatole still had significant boar taint odour and flavour scores. It may be that other factors related to sexual maturity also contribute to boar taint. The presence of 4-phenyl-3-buten-2-one in fat has been shown to increase the perception of androstenone and skatole in fat (Solé and García-Regueiro, 2001).

Androstenone is synthesized in boar testes and is released into the bloodstream (Fig. 3.8). It is removed by a specific binding protein which is a member of the lipocalin family, in the salivary gland (see Section 6.2; Marchese *et al.*, 1998) and released into the saliva, where it acts as a pheromone, inducing a mating response in oestrous sows. Due to its hydrophobicity, androstenone also

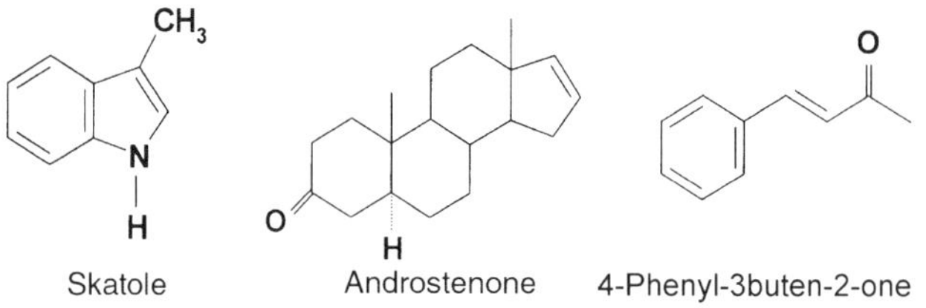

Fig. 3.7. Compounds responsible for boar taint.

accumulates in adipose tissue, causing a urine-like odour when heated. Androstenone is metabolized in the liver, although the details of this metabolism are not known.

Skatole is produced by bacterial degradation of tryptophan in the hindgut. It is absorbed from the gut, metabolized in the liver, partially excreted with the urine and partially deposited in the fatty tissue (Fig. 3.9). It produces a faecal-like odour and a bitter taste.

At usual slaughter weights, the incidence of boar taint is very variable, ranging from 10 to 75% in different studies. Because of the large variation in the incidence of boar taint, and because of the variety of culinary habits between countries, the acceptability of meat from intact males can be quite inconsistent in consumer surveys.

Cut-off levels that define a limit between untainted and tainted samples have been proposed for androstenone and skatole from sensory assessments by trained panels. Cut-off levels for skatole are considered to be 0.20 or 0.25 p.p.m., while cut-off levels for androstenone range between 0.5 and 1 p.p.m. For more information, see the reviews by Bonneau (1998) and Bonneau *et al.* (2000).

Measurement of boar taint

The existence of quick, cheap and reliable methods for the assessment of boar taint on the slaughter line would enable the sorting of carcasses according to levels of boar taint. A quick 'soldering iron' method, or 'sniff test', which involves heating fat directly on the carcass, has poor repeatability and is not realistic in industrial conditions with up to 1000 carcasses processed per line per hour.

A number of different methods have been developed for measuring the levels of skatole and androstenone in carcasses (see Chapter 2 for a discussion of methods for measuring hormones). Immunological methods, including radioimmunoassay (RIA), have been developed for androstenone, and enzyme-linked immunosorbent assay (ELISA) has been developed for skatole and androstenone. The commercial usefulness of

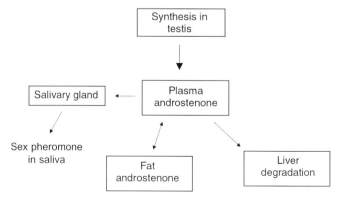

Fig. 3.8. Overview of androstenone metabolism.

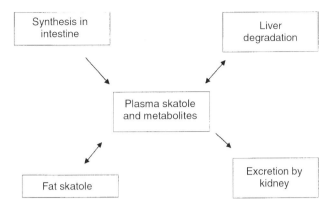

Fig. 3.9. Overview of skatole metabolism.

immunological methods can be limited by the long time required to develop equilibrium binding with the antibody and the requirement for extraction of the fat samples. These factors can also introduce a significant amount of variation in these assays, with coefficients of variation of more than 10% commonly found. Chromatographic methods, such as high-performance liquid chromatography (HPLC) and gas chromatography (GC, see Section 2.2), have been used for measuring skatole and androstenone. Chromatographic methods can have the advantage of measuring a number of related compounds at one time, and are usually quite specific and not affected by interfering compounds. However, they can be time consuming, technically difficult, expensive and prone to equipment failure. Careful extraction of the samples is required for chromatographic analysis, both to ensure good resolution of the samples and to maintain the useful life of the chromatography columns. This makes these methods more suitable for use in experimental analysis rather than for the routine analysis of boar taint compounds on a slaughter line. Colorimetric methods for the analysis of skatole and the 16-androstene steroids have also been described.

None of these methods is practical enough to be used on the slaughter line, as they involve either complicated extraction procedures, or the use of sophisticated and expensive equipment or costly materials, or are too time consuming or unable to handle a large-enough capacity of samples. A colorimetric method for skatole, developed in Denmark, has been used for sorting carcasses in slaughter plants. However, the equipment is quite expensive and has a limited capacity. The detection of compounds causing taint instantaneously on the slaughter line, using probes such as immunosensors, electronic sensors or chemical sensors, would be ideal. Immunosensors couple antibody–antigen reactions to an electronic signal generated by a transducer. Electronic sensors have been used in 'electronic noses' and are composed of semiconductors whose characteristics change when a particular substance is adsorbed on to the surface. Various types of electronic sensors have been investigated for detection of boar taint, and artificial noses have been shown to simulate the response of a laboratory panel to boar taint (Annor-Frempong *et al.*, 1998). Further work is needed in this area to develop commercially useful methods for detecting boar taint.

Use of tainted meat in processed products

Heat processing can reduce the levels of boar taint in processed meat products. Products produced from tainted meat that were not heated during processing, such as dry sausages, dry cured hams and cured bellies, were usually evaluated as having a stronger odour than that of controls made from non-

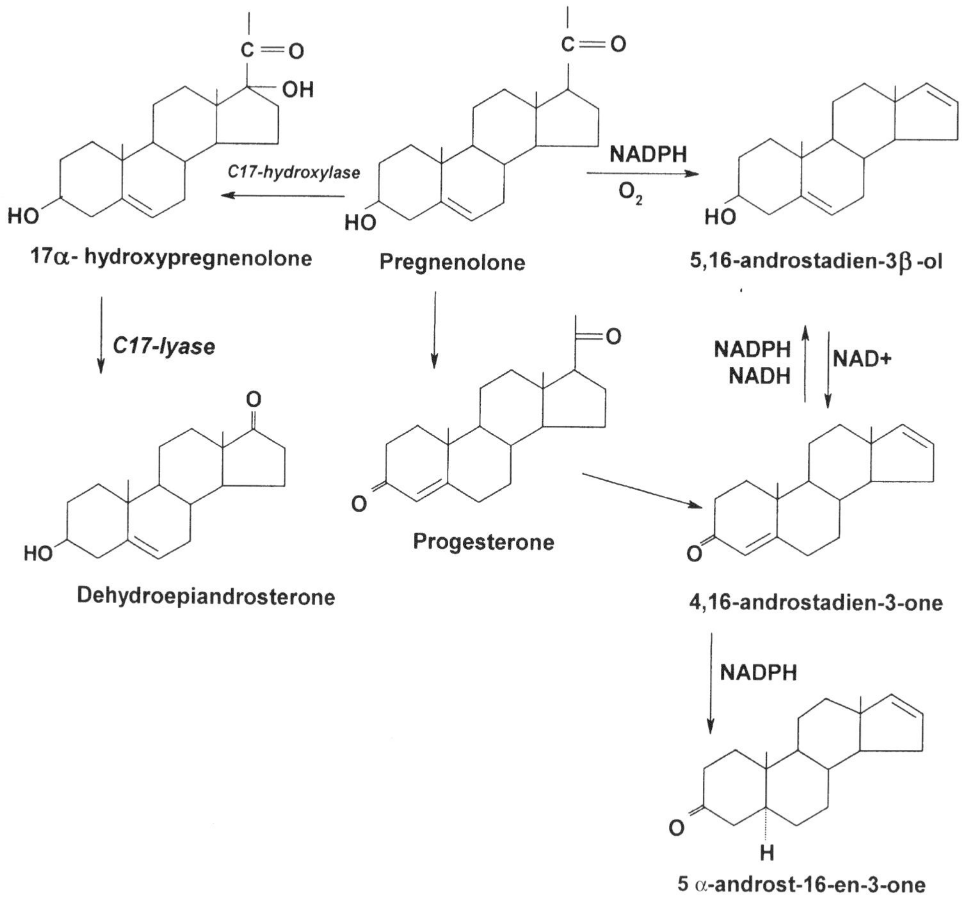

Fig. 3.10. Biosynthesis of androgens and androst-16-ene steroids.

tainted meat. However, cooked products, such as cooked hams, luncheon meat, frankfurters and cooked sausages, were acceptable unless they were prepared from very strongly tainted meat. An important factor influencing the acceptability of processed meat from intact males is the way it is consumed. If it is heated immediately prior to consumption, the perception of the odour is enhanced and the taint is more easily detected. Another way of utilizing tainted meat is to dilute it with non-tainted material and to use flavour-masking substances such as spices.

Metabolism of androstenone and skatole

The extent of skatole and androstenone accumulation in carcasses is controlled by the balance between the processes that produce these compounds and processes that are involved in degrading and removing the compounds.

The 16-androstene steroids (including androstenone) are synthesized in the testis from pregnenolone in a reaction that is catalysed by cytochrome P450C17 along with cytochrome b_5 and the associated reductases. The biosynthetic pathway is given in Fig. 3.10. Cytochrome b_5 has been cloned from a number of species, including pig. Levels of the cytochrome b_5 protein and total cytochrome b_5 mRNA are correlated with the rate of androstenone synthesis. This suggests that levels of cytochrome b_5 in the testis could be used as a marker to reduce boar taint from androstenone.

The 5α-reductase enzyme catalyses the final step in the synthesis of androstenone, which is the reduction of the Δ4 double bond in 4,16-androstadien-3-one. Thus, inhibition of this enzyme could reduce the synthesis of androstenone. However, this enzyme is also responsible for the conversion of testosterone to dihydrotestosterone, which is the ultimate androgen in some tissues. Multiple forms of the 5α-reductase enzyme exist and a specific isoform has been identified in pig testis for the synthesis of androstenone. It may therefore be possible to reduce boar taint from androstenone by inhibition of this isoform without affecting the biological actions of androgens (Cooke et al., 1997).

Skatole is produced by gut bacteria that degrade tryptophan from undigested feed or from the turnover of cells lining the gut of the pig. Skatole is absorbed from the gut and then metabolized primarily in the liver. The major metabolites (Fig. 3.11) that are formed by phase I metabolism are 6-hydroxyskatole, indole-3-carbinol, 3-hydroxy-3-methyloxindole (HMOI), 3-hydroxy-3-methylindolenine (HMI) and 3-methyloxindole (3MOI). The key enzymes in the phase I metabolism of skatole are CYP2E1 and CYP2A6. High levels of activity of these enzymes and the production of 6-hydroxyskatole and HMOI are negatively correlated with skatole accumulation in fat, while production of HMI was positively correlated with skatole levels in fat (Diaz and Squires, 2000). The phase II metabolism of skatole metabolites by phenol sulphotransferase (PST) is also negatively correlated with skatole levels in fat. Thus, pigs with high levels of these liver enzymes will be able to metabolize skatole rapidly and clear it from the body. Pigs with low levels of these enzymes can have high levels of skatole in the fat due to insufficient capacity to metabolize skatole. This becomes a problem if the amount of skatole absorbed from the gut, or the manure in the environment, is high.

Factors affecting boar taint

The extent of boar taint from skatole and androstenone is affected by a number of factors, including diet and management factors and genetic factors. Efficient methods for controlling boar taint are needed before intact males can be used for pork production. If the

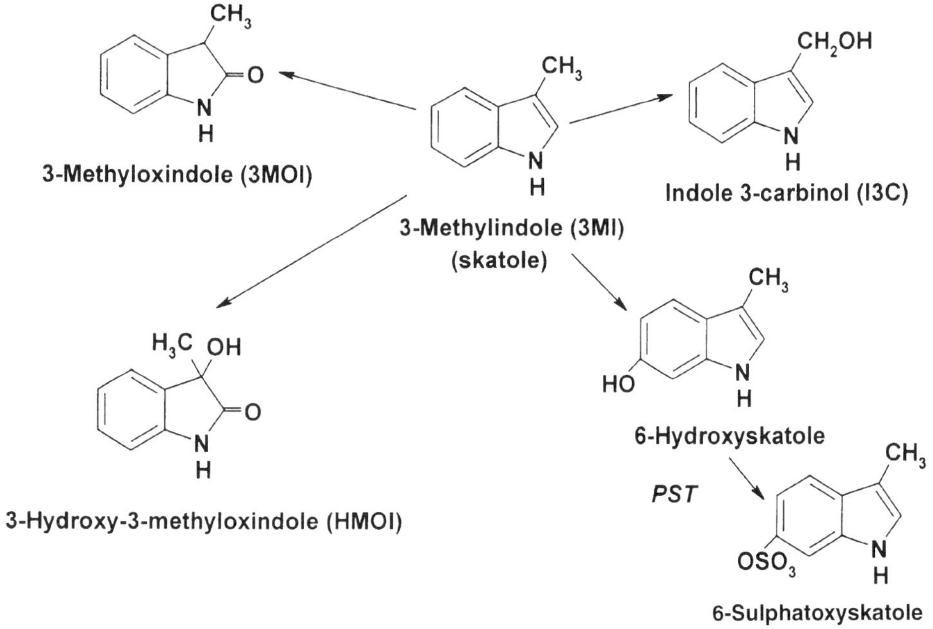

Fig. 3.11. Hepatic metabolism of skatole.

incidence of boar taint in intact male pig populations is sufficiently low and effective detection methods for boar taint were available, tainted carcasses could be sorted out on the slaughter line to be used in processed products.

Diet and management

There is a wide variation among different animals in fat skatole levels at slaughter. Because bacteria in the gut produce skatole, feeding and rearing factors affect skatole production, while genetic factors affect skatole metabolism. Swine diets with a high fibre content or high level of non-digestible carbohydrates provide the energy needed for increased proliferation of bacteria, so that tryptophan can be used for protein synthesis rather than being degraded to skatole. Wet feeding and an unlimited water supply can lower skatole levels in fat. Wet feeding of boars on a low protein diet with virginiamycin as a growth enhancer in fully slatted pens reduced the level of boar taint. Feeding pigs a mixture of inulin and bicarbonate for a few days before slaughter results in a sharp reduction in fat skatole levels. The addition of zinc bacitracin or zeolite to the diet is effective in reducing fat skatole levels. Finally, withholding feed on the evening prior to slaughter has been shown to reduce fat skatole levels. These modifications of the gut microbes can dramatically reduce skatole production. Environmental factors, such as raising pigs on slatted floors, decrease skatole levels compared to animals on concrete, probably because the animals were less dirty. Thus, a proper control of the environment and diet may reduce fat skatole levels substantially.

Diet and management factors do not affect fat androstenone levels, unless they affect the sexual maturity of the intact male at slaughter.

Genetic factors

Androstenone synthesis increases during puberty, along with increased production of the androgens and oestrogens, which are responsible for the better growth performance of intact males. The same regulatory systems appear to control the synthesis of all testicular steroids, so it is very difficult to decrease androstenone without affecting androgens and oestrogens as well. However, there is a wide variation among different intact males in the accumulation of androstenone in fat. Levels of androstenone also vary among breeds of pigs, with Durocs having dramatically higher levels than Yorkshire, Landrace or Hampshire breeds and Large White breeds having higher levels than Landrace breeds.

The sexual maturity and age of the animal at slaughter weight, as well as the genetic potential for androstenone production, affect the synthesis of androstenone. Both mechanisms are genetically determined and this explains the high heritability of fat androstenone content, with heritability estimates ranging from 0.25 to 0.87 (Willeke, 1993). However, selection experiments against androstenone also decreased the production of androgens and oestrogens, and delayed puberty in the gilts of a 'low androstenone' line.

A significant effect of different haplotypes of swine lymphocyte antigen (SLA) and the microsatellite marker S0102 on androstenone level in fat has been proposed. This suggests that a quantitative trait locus (QTL) for androstenone is located on chromosome 7 between these two markers. Other work, based on a selection experiment for low androstenone, proposed a major two-allele gene affecting androstenone levels in fat that was distinct from the SLA system (Fouilloux *et al.*, 1997). Thus it appears that there are at least two major genetic effects on androstenone accumulation in fat. One of these may be related to the expression of cytochrome b_5 and the potential for androstenone production, while the other may be related to other factors, such as the development of sexual maturity. Genetic markers for low androstenone may be developed in the future and used in marker-assisted selection breeding programmes for low androstenone pigs.

Genetic factors also affect the accumulation of skatole in fat. High skatole levels in boars may be related to a higher anabolic potential and increased turnover of intestinal cells, which increases skatole formation in the hindgut. The differences that are seen in the activities of the key enzymes in skatole metabolism (CYP2E1, CYP2A6 and PST) mean that the potential for degrading blood skatole is decreased in some intact male pigs. Genetic markers are needed which can identify those animals expressing high levels of these enzymes. These markers could be used in marker-assisted selection programmes for low-skatole pigs.

Immunological methods to control boar taint

A number of immunological approaches could potentially be used to reduce boar taint from androstenone. Attempts have been made to remove 5α-androstenone by active immunization against the steroid, but it was still deposited in the body tissues. The removal of the androst-16-ene steroids would not adversely affect growth performance, since they show little or no anabolic activity.

Castrating pigs at 2 or 3 weeks before slaughter reduces androstenone content in fat to levels similar to those in castrates and gilts. However, surgical castration of older animals cannot be used in practice. Interfering with testicular function can also reduce boar taint. Antagonists or agonists for GnRH can be used to interfere with gonadotrophin production. Another approach is the immunocastration of male pigs by immunizing against LH or GnRH at an appropriate age. This yields the anabolic effects of the androgens early in life but shuts down all steroid biosynthesis before the synthesis of androst-16-ene steroids at sexual maturity. The performance of immunized animals depends on the response to the vaccine (Turkstra et al., 2002). Immunocastration also results in a reduction of fat skatole content to what is found in castrates. A vaccine for immunocastration (Improvac®) is commercially available in Australia (Dunshea et al., 2001). Two doses of vaccine (2 ml of 200 µg ml^{-1} GnRH conjugate) are given at an interval of at least 4 weeks. The second dose should be given 4–5 weeks before slaughter. The vaccine is administered at the base of the ear with a special vaccinator designed to prevent Whether or not the ly viable and ac remains to be main problem might the technique by the genera. accidentally injected with the also raise antibodies against GnR become sterile.

3.4 Somatotrophin

Somatotrophin or growth hormone (ST, GH) is the most important peptide hormone affecting growth. Experiments in the 1930s demonstrated that rats injected with alkaline extracts of pituitary gland gained more weight and had more muscle and less fat than controls. In 1945, growth hormone was isolated from anterior pituitary and experiments evaluating the effects of crude preparations of porcine growth hormone (pST) began in pigs. However, prior to the 1980s, studies with domestic livestock and commercial application of ST were limited because ST had to be extracted from pituitary glands of slaughtered animals. Thus, there was a very limited supply of ST, and what was available varied greatly in quality. Subsequently, advances in recombinant DNA technology made it possible to produce large quantities of high-purity ST.

Somatotrophins from domestic livestock contain 191 amino acids and there is a high degree of sequence homology between STs from different species (Table 3.2). There are 18 differences in amino-acid sequence between porcine ST (pST) and bovine ST (bST), but only two differences between bST and ovine ST (oST). Chicken ST (cST) has only 77% sequence identity to bST. There are considerable sequence differences between human ST (hST) and bST or pST, so that pituitary preparations from farm animals are not active in humans. Variants of ST are produced by the pituitary gland and four major variants of bST have been identified.

Treatment of pigs with pST increases protein accretion and decreases fat deposition in boars, barrows and gilts in both poor and improved genotypes. The response to pST can vary according to initial body

3.2. Amino-acid sequence of somatotrophins from various species.

	-30	-20	-10	0 10	20	30
Porcine (pST)	MAAGP	RTSALLAFAL	LCLPWTREVG	AFPAMPLSSL	FANAVLRAQH	LHQLAADTYK
Bovine (bST)	MMAAGP	RTSLLLAFAL	LCLPWTQVVG	AFPAMSLSGL	FANAVLRAQH	LHQLAADTFK
Sheep (oST)	MMAAGP	RTSLLLAFTL	LCLPWTQVVG	AFPAMSLSGL	FANAVLRAQH	LHQLAADTFK
Human (hST)	MATGSR	TSLLLAFGLL	CLPWLQEGSA	FPTIPLSRLF	DNAMLRAHRL	HQLAFDTYQE
Chicken (cST)	MAPGS	WFSPLLIAVV	TLGLPQEAAA	TFPAMPLSNL	FANAVLRAQH	LHLLAAETYK
Turkey	MAPGS	WFSPLLIAVV	TLGLPQGAAA	TFPTMPLSNL	FTNAVLRAQH	LHLLAAETYK
Duck	MAPGSW	FSPLFITVIT	LGLQWPQEAA	TFPAMPLSNL	FANAVLRAQH	LHLLAAETYK
Carp	MARV	LVLLSVVLV	SLLVNQGRAS	DNQRLFNNAV	IRVQHLHQLA	AKMINDFEDS

	40	50	60	70	80	90
Porcine (pST)	EFERAYIPEG	QRYSIQNAQA	AFCFSETIPA	PTGKDEAQQR	SDVELLRFSL	LLIQSWLGPV
Bovine (bST)	EFERTYIPEG	QRYSIQNTQV	AFCFSETIPA	PTGKNEAQQK	SDLELLRISL	LLIQSWLGPL
Sheep (oST)	EFERTYIPEG	QRYSIQNTQV	AFCFSETIPA	PTGKNEAQQK	SDLELLRISL	LLIQSWLGPL
Human (hST)	FEEAYIPKEQ	KYSFLQNPQT	SLCFSESIPT	PSNREETQQK	SNLELLRISL	LLIQSWLEPV
Chicken (cST)	EFERTYIPED	QRYTNKNSQA	AFCYSETIPA	PTGKDDAQQK	SDMELLRFSL	VLIQSWLTPV
Turkey	EFERTYIPED	QRYTNKNSQA	AFCYSETIPA	PTGKDDAQQK	SDMELLRFSL	VLIQSWLTPM
Duck	EFERSYIPED	QRHTNKNSQA	FCYSETIPAP	TGKDDAQQKS	DMELLRFSLV	LIQSWLTPVQ
Carp	LLPEERRQLS	KIFPLSFCNS	DYIEAPAGKD	ETQKSSMLKL	LRISFHLIES	WEFPSQSLSG

	100	110	120	130	140	150
Porcine (pST)	QFLSRVFTNS	LVFGTSDRVY	EKLKDLEEGI	QALMRELEDG	SPRAGQILKQ	TYDKFDTNLR
Bovine (bST)	QFLSRVFTNS	LVFGTSDRVY	EKLKDLEEGI	LALMRELEDG	TPRAGQILKQ	TYDKFDTNMR
Sheep (oST)	QFLSRVFTNS	LVFGTSDRVY	EKLKDLEEGI	LALMRELEDV	TPRAGQILKQ	TYDKFDTNMR
Human (hST)	QFLRSVFANS	LVYGASDSNV	YDLLKDLEEG	IQTLMGRLED	GSPRTGQIFK	QTYSKFDTNS
Chicken (cST)	QYLSKVFTNN	LVFGTSDRVF	EKLKDLEEGI	QALMRELEDR	SPRGPQLLRP	TYDKFDIHLR
Turkey	QYLSKVFTNN	LVFGTSDRVF	EKLKDLEEGI	QALMRELEDR	SPRGPQLLRP	TYDRFDIHLR
Duck	YLSKVFTNNL	VFGTSDRVFE	KLKDLEEGIQ	ALMRELEDRS	PRGPQLLKPT	YDKFDIHLRN
Carp	TVSNSLTGNP	NQLTEKLADL	KMGISVLIQA	CLDGQPNMDD	NDSLPLPFED	FYLTMGENNL

	160	170	180	190		
Porcine (pST)	SDDALLKNYG	LLSCFKKDLH	KAETYLRVMK	CRRFVESSCA	F	
Bovine (bST)	SDDALLKNYG	LLSCFRKDLH	KTETYLRVMK	CRRFGEASCA	F	
Sheep (oST)	SDDALLKNYG	LLSCFRKDLH	KTETYLRVMK	CRRFGEASCA	F	
Human (hST)	HNDDALLKNY	GLLYCFRKDM	DKVETFLRIV	QCRSVEGSCG	F	
Chicken (cST)	NEDALLKNYG	LLSCFKKDLH	KVETYLKVMK	CRRFGESNCT	I	
Turkey	SEDALLKNYG	LLSCFKKDLH	KVETYLKVMK	CRRFGESNCN	I	
Duck	EDALLKNYGL	LSCFKKDLHK	VETYLKVMKC	RRFGESNCTI		
Carp	RESFRLLACF	KKDMHKVETY	LRVANCRRSL	DSNCTL		

weight, length of treatment, breed, sex, dose of pST and diet composition. Maximally effective doses can increase average daily gain by 10–20%, improve feed efficiency by 15–30%, decrease feed intake by 10–15%, decrease lipid accretion by 70% and increase protein deposition by 50%. A daily dose of 30–69 µg kg^{-1} body weight given between 50 kg and slaughter improves growth rate, feed conversion efficiency and lean content by 22%, 20% and 10%, respectively, while decreasing carcass fat by 30%. Since pST stimulates protein deposition in all tissues, a slight decrease in dressing percentage may be found due to an increase in weight of the internal organs. There are no detrimental effects on meat tenderness from pST treatment. However, treatment of pigs with pST has been reported to reduce chondrocyte metabolism and compromise cartilage, bone and joint development in growing animals (He et al., 1994).

Bovine ST (bST) is less effective in growing cattle than pST is in pigs. Daily gain, feed efficiency and carcass lean content in growing cattle are improved by 12%, 9% and 5%, respectively, and in sheep by 18%, 14% and 10%, respectively, by bST treatment. Carcass fat content was decreased by 15% in sheep and cattle by bST treatment. The lipolytic effects of ST contribute to improved feed efficiency while maintaining growth rate. As a result, feed costs are lower and carcass grade is improved. Results with chickens have been mixed, but this may be because of the mode of delivery. Episodic administration of cST to older birds markedly improved growth performance, while continuous infusion had no effect. ST is largely ineffective in promoting growth or altering metabolism during the early posthatch period, when the bird is already growing rapidly and the rate of fat deposition is low (Vasilatos-Younken, 1995).

Bovine ST has dramatic effects on milk production and is approved for use in the dairy industry in the USA. This aspect is discussed in Section 4.1.

Control of ST release

ST release from the anterior pituitary gland is regulated by hormones produced in the hypothalamus (Fig. 3.12). ST release is stimulated by growth hormone releasing hormone (GHRH) and decreased by somatostatin (SS). ST release occurs in a pulsatile pattern and this pattern is thought to be regulated by the feedback of ST on periventricular neurons in the hypothalamus to cause the release of SS. The production of SS suppresses the activity of the GHRH neurons in the arcuate nucleus and subsequently decreases the release of ST by the pituitary. The decrease in ST then reduces the negative feedback on the hypothalamus, which increases GHRH production to increase ST release from the pituitary. This cycle of events results in a pulsatile release of ST. For more information, see the review by Müller *et al.* (1999).

Somatostatin exists as both 14 amino-acid (SS-14) and 28 amino-acid (SS-28) forms that interact with different receptors. SS-14 is the predominant form and has the same sequence in different species. Somatostatin is expressed in the brain and in specialized D cells in the gastrointestinal tract and pancreatic islets. It acts to inhibit the secretion of most hormones and, in addition to its endocrine effects, can act locally as a paracrine or autocrine regulator. In the stomach, SS inhibits the release of gastrin and hydrochloric acid, while in the pancreatic islets, SS regulates the secretion of glucagon and insulin. Androgens stimulate while oestrogens decrease the production of SS in the brain; this may, in part, explain the differences in ST secretion patterns between males and females.

Octapeptide analogues of SS are octreotide (Fig. 3.13), lanreotide and vapreotide, and these are used for the treatment of acromegaly and to decrease gastric secretions.

There are five major forms of SS receptors that vary in their tissue distribution. SS receptors are coupled to several populations of K^+ channels and their activation causes hyperpolarization of the cell membranes, leading to decreased action potential and decreased intracellular calcium.

GHRH was first isolated from pancreatic islet tumours. It is a 44 amino-acid peptide that is part of a family of brain–gut peptides which includes glucagon, glucagon-like pep-

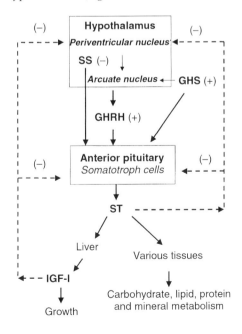

Fig. 3.12. Regulation of growth hormone release.

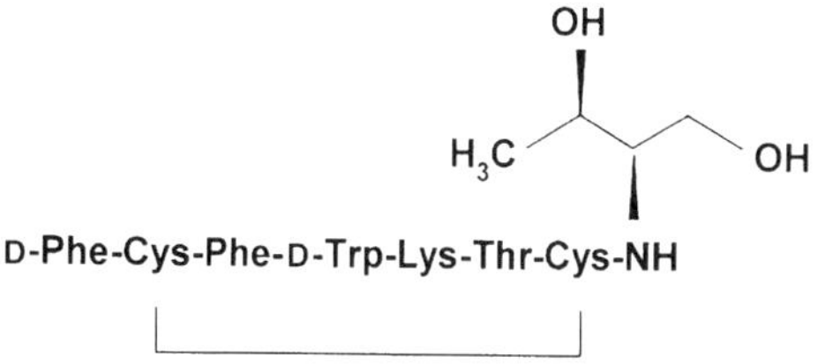
OH
H_3C
OH
D-Phe-Cys-Phe-D-Trp-Lys-Thr-Cys-NH

Fig. 3.13. Structure of octreotide, an analogue of somatostatin.

tide I, vasoactive intestinal polypeptide, secretin and gastric inhibitory peptide. In addition to regulating secretion of ST, GHRH has been implicated in promoting sleep and increasing appetite, particularly for protein. ST secretion also increases during rapid eye movement (REM) sleep.

Pulsatile GHRH release can be stimulated by activation of the growth hormone secretagogue receptor (GHS-R) in the arcuate nucleus of the hypothalamus. The GHS-R is expressed in both the anterior pituitary and central nervous system and is highly conserved across species. The GHS-R is not activated by GHRH or somatostatin, but binds a number of peptide analogues of Leu- and Met-enkephalins known as growth hormone releasing peptides (GHRPs). GHRP-6 (His–D-Trp–Ala–D-Phe–Lys–NH_2) was one of the first analogues developed as a potent stimulator of pulsatile ST secretion, but the GHRPs had low oral bioavailability of about 0.3%. Other non-peptide derivatives were developed (Fig. 3.14) and L-692,429 was the first of these, with 4% bioavailability in humans. Later, the spiropiperidine MK-0677 was developed to have good bioavailability and a longer half-life, being effective for 24 hours. The benzolactam L-739,943 has similar biological and pharmacokinetic properties but different structure from MK-0677. The benzylpiperidine L-163,540 has a shorter half-life

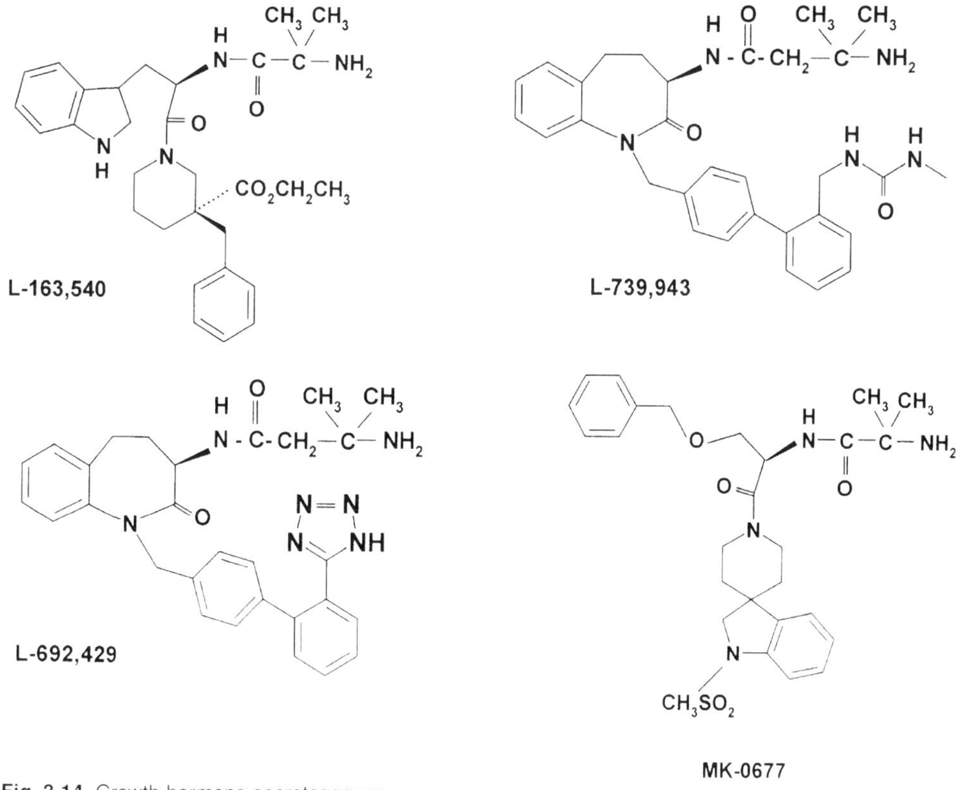
L-163,540
L-739,943
L-692,429
MK-0677

Fig. 3.14. Growth hormone secretagogues.

and is used for finer control over ST and IGF-I levels. For more information of growth hormone secretagogues, see Smith *et al.* (1996, 2001).

Ghrelin is a 28 amino-acid peptide that binds to the GHS-R to stimulate ST release. It has an *n*-octanoly ester on serine 3 that is necessary for biological activity. Ghrelin is produced in the stomach, kidney and placenta, and is the only endogenous GHS-R ligand yet isolated that stimulates GHRH release. Intracerebroventricular injections of ghrelin stimulated feeding and weight gain in rats. Ghrelin blocked the action of leptin (see Section 3.8) in reducing feed intake, suggesting that there is a competitive interaction between ghrelin and leptin in regulating feed intake. Ghrelin also regulates gastric acid secretion by activating the vagus system.

Binding of GHRH to its receptors on somatotrophs increases intracellular levels of cAMP. Somatostatin binds to subtype-2 receptors and hyperpolarizes the somatotroph membranes to inhibit ST release. GHRP-6 acts by stimulating phospholipase C to increase levels of diacylglycerol and IP_3, thereby mobilizing cellular calcium and stimulating protein kinase C (see Section 1.3). The presence of GHRP-6 along with GHRH potentiates the increase in cAMP due to GHRH alone, indicating that there is some interaction between the different signal transduction pathways. Co-administration of GHRH and ghrelin gives a synergistic effect on ST secretion.

Plasma levels of ST and the pulse amplitude for ST are higher in women than in men, but the pulse frequency is the same in males and females (Fig. 3.15). The amplitude and number of oscillations of ST release decline during ageing. Decreased levels of ST are associated with a reduced nitrogen balance and a decline in body condition and muscle tone. The decrease in ST may be due to a decrease in the endogenous ligand(s) of the GHS-R during ageing. Chronic oral administration of MK-0677 once daily to dogs produces a sustained physiological pattern of pulsatile ST release. However, stimulation of GH release by isolated pituitary cells *in vitro* is not sustained, with desensitization occurring within a few minutes. Administration of L-dopa to old rats also restores the amplitude of ST release to what is seen in young animals. It has been suggested that GHS-R ligands may modulate dopamine release from hypothalamic neurons to affect pulsatile release of ST.

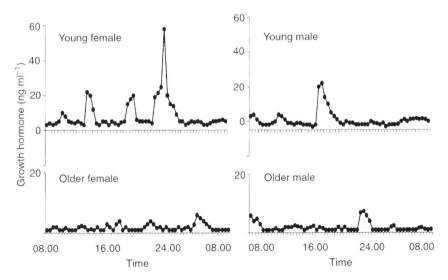

Fig. 3.15. Age and gender effects on the pattern of growth hormone secretion (redrawn from Smith *et al.*, 1996).

Mechanism of action of ST

Somatotrophin affects a wide range of somatogenic and metabolic processes to increase lean tissue growth. It has two distinct types of effects: direct and indirect. The direct effects are due to ST binding to receptors on target cells, and these include effects on carbohydrate, lipid, protein and mineral metabolism. The indirect somatogenic effects of ST are those related to cell proliferation, and these are mediated by insulin-like growth factor-I (IGF-I), which is a potent mitogen with some sequence similarities to insulin.

Direct effects

ST RECEPTORS. The biological actions of ST begin with ST binding to its receptor on the cell surface. The ST receptor is a transmembrane protein of 634–638 amino acids with a molecular mass of about 70 kDa. The free receptor exists in the cell membrane as a single-chain monomer, while the activated receptor complex consists of a receptor dimer that is bound to a single ST molecule. The ST receptor is a member of a superfamily of cytokine receptors, which includes prolactin, erythropoietin, interferons and interleukins (see Section 1.3).

The differential effects of pulsatile versus continuous infusion of ST may be due to effects on the ST receptors. Continuous exposure to high levels of ST may cause down regulation of the number of ST receptors. High levels of ST may also prevent the dimerization of the ST receptor, since the receptor monomers will be bound to different molecules of ST and will not form active dimers. In contrast, at low levels of ST a single ST molecule will be more likely to bind to two ST receptor monomers to form an activated complex (see Fig. 1.28).

A growth hormone binding protein (GHBP), which is essentially the extracellular domain of the ST receptor which is also produced from the ST receptor gene, is present in plasma of many species. GHBP enhances the growth promoting effects of ST, probably by increasing the half-life of ST in the circulation.

METABOLIC EFFECTS. The metabolic effects of ST are summarized in Table 3.3.

There is a dramatic decrease in lipid accretion in ST-treated animals. This allows more nutrients, such as glucose, to be available for increased growth of lean tissue, and improves commercial production efficiency by reducing the proportion of nutrients used for the synthesis of body fat. The effects of ST on lipid metabolism are chronic rather than acute. When animals are in positive energy balance, ST reduces the rate of lipogenesis, while lipolysis is increased in animals under negative energy balance. In growing pigs, ST decreases the uptake of glucose for lipid synthesis by adipocytes through decreased synthesis of the glucose transporter protein, GLUT4. ST also decreases the synthesis of lipogenic enzymes, including fatty acid synthase and acetyl-CoA carboxylase. ST also decreases the stimulatory effects of insulin on glucose uptake and utilization by adipocytes and increases the lipolytic responses to adrenaline. ST treatment of pigs decreases

Table 3.3. Summary of metabolic effects of somatotrophin.

Tissue type	Metabolic effect
Skeletal muscle	Increased protein synthesis/accretion
Adipose tissue	Increased lipolysis, decreased glucose uptake and lipid synthesis
Liver	Increased glucose output
Intestine	Increased calcium and phosphorus uptake
Mammary gland	Increased milk synthesis by increased uptake of nutrients and increased blood flow
Various tissues	Decreased amino-acid and glucose oxidation; increased oxidation of free fatty acids; decreased effects of insulin

glucose utilization by adipose tissue from 40% to 7% of the whole-body glucose turnover.

The protein anabolic effect of ST occurs almost exclusively by increasing protein synthesis, with no measurable effects on protein catabolism. Alterations in tissue response to other endocrine factors, such as insulin, may also be important. Sufficient levels of plasma amino acids are needed to sustain the increased level of protein synthesis. Maintenance energy requirements are also 10–20% higher in pST-treated pigs, because of increased protein synthesis in muscle and in visceral tissues such as liver and gut.

In growing pigs, pST treatment increases the apparent efficiency of use of dietary protein for protein deposition by 25–50%, depending on growth stage, quality of dietary protein and the sex of the pig. Young pigs that are less than about 40 days of age have little response to pST treatment, since they are already growing rapidly with a high efficiency of utilization of dietary protein. The rate of protein deposition is increased by 16%, 25% and 74% in gilts and barrows in the weight ranges of 10–20 kg, 20–55 kg and 55–100 kg, respectively. There is no effect of pST on protein digestibility, and only a marginal increase in the dietary protein requirement for 25–60 kg pigs. Increases in protein deposition from pST treatment are greater in finisher pigs weighing 60–90 kg, and this increases the dietary protein requirement. This is especially important in boars of improved genotypes that already have a high efficiency of use of dietary protein. Thus, pST acts by separately increasing the efficiency of amino-acid utilization and the maximal rate of protein accretion. Decreased levels of plasma urea nitrogen and the activity of lysine α-ketoglutarate reductase are seen following ST treatment, which suggests that oxidation of amino acids is reduced.

Indirect effects

The major role of ST in stimulating growth is mediated via insulin-like growth factor-I (IGF-I, see also Section 4.1). High blood levels of IGF-I decrease the secretion of ST by direct negative feedback action on the pituitary and also by stimulating SS production by the hypothalamus.

IGF-I is synthesized in the liver along with its major binding protein, IGFBP3, and the third member of the ternary binding complex, the acid-labile subunit. Local synthesis of IGF-I in muscle and adipose tissue may also be important. IGF-I is a single polypeptide chain of about 7500 Da and the sequence of human, porcine and bovine IGF-I is identical. There are at least six binding proteins for IGF-I, and almost all of the IGF-I is bound to binding proteins that modulate the interactions of IGF with its target tissues. IGF-I interacts with type I and type II receptors that differ in structure, specificity and signalling mechanism. The type I receptor is similar to the insulin receptor, while the type II receptor is a monomeric protein.

IGF-I stimulates the proliferation of chondrocytes or cartilage cells, to increase bone growth. It also increases the proliferation of satellite cells, which fuse with the myofibre and contribute to muscle fibre growth. It also stimulates amino-acid uptake and protein synthesis. A related polypeptide, IGF-II, may also be important in the control of muscle growth.

Although some of the effects of pST are mediated by IGF-I, treatment of finisher pigs with IGF-I does not affect growth. Treatment of finisher pigs with the potent analogue LR3 IGF-I reduces growth rate, the amplitude of endogenous pST pulsatile release and plasma levels of IGF-I and its major plasma-binding protein. Thus negative feedback limits the effectiveness of IGF-I treatment in finisher pigs. In neonatal pigs, LR3 IGF-I infusion increases growth of the gut and visceral tissues, particularly in pigs under nutritional stress.

Peptide growth factors may be orally active in the neonatal pig and could potentially be added to diets of early weaned or supplemental fed piglets to improve health and growth rate.

Delivery/dose effects

Porcine ST was approved in 1994 for use in pigs in Australia, using daily injections of 3–5 mg per pig for finishing pigs, starting at 35 days before slaughter. So far, it has not been approved for use in other countries. ST is not orally active and most studies have used daily injection as a practical means of delivery of pST. The ADG and rates of protein and ash accretion are maximally stimulated at a daily pST dose of about $100\,\mu g\,kg^{-1}$ BW. However, rates of lipid accretion and feed:gain ratio decrease in a more linear manner up to a dose of $200\,\mu g\,kg^{-1}$ BW. For more information, see Etherton and Smith (1991).

The most investigated formulation for pST delivery has been the implant, with the goal of a one-time treatment that provides pST for the entire finishing period. Sustained-release implants have been designed to deliver approximately 2 mg of pST per day for 42 days. The implants consist of recombinant pST having an N-alanyl residue linked to the natural pST sequence in a solution of 49.5% 1 M sodium phosphate, 49.5% glycerol and 1% Tween 20. Pigs treated with these implants had increased efficiency of gain through decreased feed consumption, but the rate of gain was not increased. It appears that the large peaks in pST caused by daily injection are necessary to activate the physiological processes related to growth.

Administration of ST to rats by constant infusion produces a male pattern of response in liver enzymes, while a pulsatile delivery elicits a female response. In chickens, the pulse amplitude of ST in plasma is high at early ages during rapid growth and decreases at older ages when growth rate is decreased.

Growth hormone secretagogues (GHS) such as MK-0677 have been used in humans as an alternative to injectable ST. They have the advantage of oral dosing and producing a pulsatile, physiological ST profile. However, the GHS-R pathway is subject to negative feedback, so the sustained supraphysiological levels of ST that can be obtained with ST or GHRH injection are not possible with GHS treatment.

A slow-release oil-based formulation of zinc methionyl bST (Sometribove®) is available, which improved growth performance and increased carcass leanness in lambs when administered weekly or once every 2 weeks. Equine ST (EquiGen®) is also used as an aid to improve nitrogen retention in aged horses (at least 15 years of age). The recombinant product is expressed in *Escherichia coli* and has an additional N-terminal methionine. It has been shown to decrease plasma urea nitrogen and urinary nitrogen, indicating that it increased protein synthesis and/or decreased protein breakdown. The recommended dose is $10-20\,\mu g\,kg^{-1}$ BW day^{-1} for 42 days, starting at $10\,\mu g\,kg^{-1}$ BW day^{-1} for the first week and then increasing to $20\,\mu g\,kg^{-1}$ BW day^{-1} for weeks 2–6. ST is given by deep muscular injection, rotating the site between the neck, pectoral muscle and rump. Recombinant bST has also been shown to increase growth rate and improve feed efficiency and protein deposition in both young and older salmonid fish.

Immunization against somatostatin has also been used to increase the secretion of ST. Some work on developing anti-idiotypic antibodies for ST has also been done (see Section 2.4). Transgenic animals with ST genes (pigs, fish) have also been generated.

Safety/quality aspects

ST is a natural protein that is destroyed by cooking and rapidly broken down in the gut, so the potential threat from residues should be small. A slight reduction in meat tenderness in pigs treated with pST may be due to decreased intramuscular fat content.

3.5 β-Adrenergic Agonists

β-Adrenergic agonists (β-AA) are synthetic, orally active **phenethanolamine derivatives** that are related to the naturally occurring catecholamines, dopamine, noradrenaline (norepinephrine) and adrenaline (epinephrine). Noradrenaline functions both as a **neurotransmitter** in the sympathetic nervous system and an endocrine hormone produced by the adrenal medulla. It is biosynthesized from tyrosine in the adrenal medulla and is present in plasma at 2–5 times the levels of adrenaline and dopamine. Adrenaline is synthesized by methylation of noradrenaline.

Catecholamines regulate a variety of physiological functions, including the speed and force of heart contractions, motility and secretions from various parts of the gastrointestinal tract, bronchodilation, salivary gland and pancreas secretions, brown adipose tissue metabolic activity, blood vessel dilation and contraction, uterine contraction and spleen capsule contraction. This wide range of processes is regulated by the presence of distinct groups of α- and β-receptors in different tissues. Adrenaline is more potent than noradrenaline for α-receptors. Noradrenaline is more potent than adrenaline for $β_1$-receptors.

Mechanism of action

In stressful situations, the sympathetic nervous system stimulates the release of the catecholamines, adrenaline and noradrenaline, from the adrenal medulla (see Sections 3.12 and 6.3). These hormones act on α- and β-receptors that are located in many different tissues to produce the 'fight or flight' response. Activation of the β-receptors increases the availability of metabolic fuels (glucose and fatty acids) and the dilation of airways increases the oxygen availability. Catecholamines stimulate the activity of glycogen phosphorylase and inhibit glycogen synthase to increase the conversion of glycogen to glucose in muscle. These effects are also stimulated by the β-AA analogue isoproterenol, and inhibited by the β-receptor antagonist, propanolol. Catecholamines stimulate lipolysis and inhibit lipogenesis in adipose cells and these effects can be blocked with β-adrenergic receptor antagonists.

β-AAs have been used in the treatment of asthma since the 1970s because of their ability to relax smooth muscle in the airways. They are given in small doses directly into the airways using an inhaler, so that little of the drug enters the circulation to affect other tissues. β-AAs are also effective as repartitioning agents that alter nutrient metabolism, to produce dramatic increases in lean and decreases in adipose tissue growth.

β-AA structures

Several synthetic analogues of adrenaline and noradrenaline have been investigated for their effects on increasing skeletal muscle growth and decreasing the fat content of carcasses. β-AAs have the same overall structure, but individual compounds may have quite different chemical and pharmacokinetic properties. In order to be biologically active, β-AAs must have a substituted six-membered aromatic ring, a hydroxyl group on the β carbon and a positively charged nitrogen on the ethylamine side-chain. A bulky R group on the side-chain confers specificity for the β-receptor. The substituents on the aromatic ring at the 3 and 4 carbons are important, since they can form hydrogen bonds to a serine hydroxyl in the β-receptor. Substituting the hydroxyl groups on the aromatic ring for halogen atoms prevents the rapid catabolism and deactivation of β-AA.

Common β-AAs include cimaterol, clenbuterol, ractopamine and L-644,969 (Fig. 3.16). Each of these compounds may have multiple actions on various aspects of nutrient metabolism, which may vary in different species. The positive response to protein accretion is largely confined to skeletal muscle. Clenbuterol and cimaterol act by binding primarily to $β_2$-receptors, while ractopamine is reported to also bind to $β_1$-receptors in muscle. Stimulation of $β_1$-receptors in heart muscle can result in tachycardia. Most β-AA compounds bind to more than one type of β-adrenergic receptor.

β-AA receptors

Most tissues have a mixture of α- or β-adrenergic receptors, with different amounts of each receptor type present in different tissues and in different species. Cloning of β-AA receptors has confirmed the existence of $α_1$-, $α_2$- and $β_1$-, $β_2$- and $β_3$-receptors. β-AA receptors within a species have about 50% amino-acid sequence homology, while individual β-AA subtypes have 75% sequence homology across species. The β-AA receptors are glycosylated with an approximate molecular mass of 65,000 Da.

Some tissues have predominantly one particular receptor subtype. For example, rat heart has primarily $β_1$-receptors, while guinea-pig tracheal muscle has primarily $β_2$-receptors. These tissues were used to classify the specificity of different β agonists and

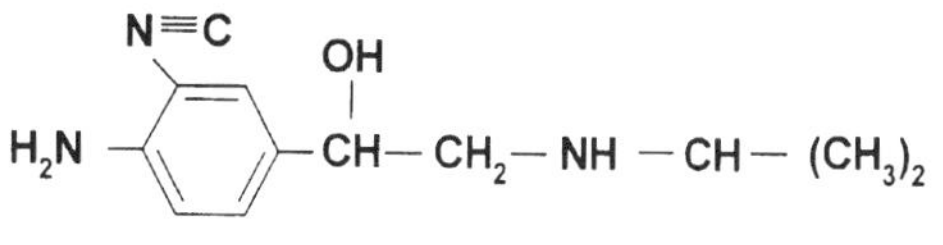

Fig. 3.16. Structures of β-adrenergic agonists.

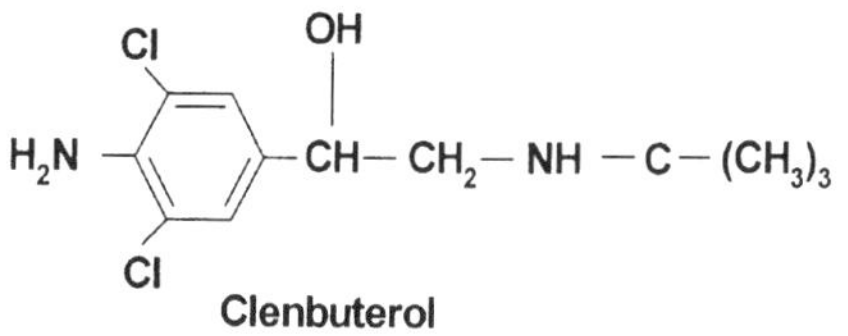

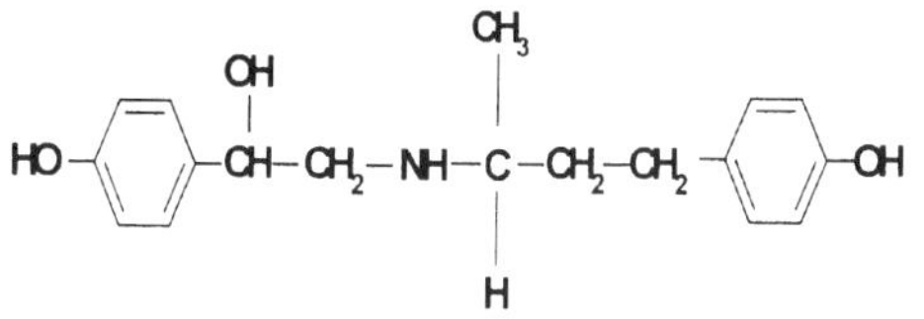

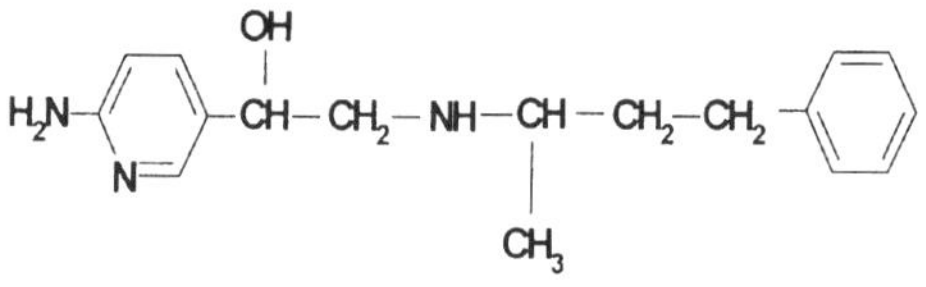

antagonists for different receptor types. For example, the compound CGP 20712A is a specific β_1-receptor antagonist, ICI 118,551 is a specific β_2 antagonist and SR 59,230A is a specific β_3 antagonist. Isoproterenol is a universal β receptor agonist. Compounds with a high degree of specificity for a particular receptor subtype were then used to classify different tissues for the receptor subtypes present. However, there are species differences in the effectiveness of different β-AAs for subtypes of β-receptors. There may also be differences in the cellular distribution of the different β-receptor subtypes.

Heart contractility is stimulated primarily by β_1-receptors, although β_2-receptors are also present in heart muscle in some species. In cattle and sheep, skeletal muscle and adipose tissue have β_2-AA receptors. The β_3-receptor subtype is the predominant receptor in brown and white adipose tissue in rats and in brown adipose tissue in the bovine fetus. It is pharmacologically different from β_1 and β_2 and is not as readily inactivated by phosphorylation. Adipocytes in cattle and sheep have predominantly β_2-AA receptors.

Binding of β-AA to its receptor activates the G_s-protein, and the α-subunit of the G_s-protein then activates adenylate cyclase. The cAMP produced by adenylate cyclase activates protein kinase A to release its catalytic subunit to phosphorylate and activate a number of intracellular proteins (see Section 1.3).

These include hormone-sensitive lipase, the rate-limiting enzyme responsible for triacylglycerol degradation in adipocytes. The CREB (cAMP-response-element binding protein) is also phosphorylated by protein kinase A and subsequently binds to the cAMP response element in the regulatory region of genes to stimulate gene transcription. Acetyl-CoA carboxylase, which is the rate-limiting enzyme for fatty acid biosynthesis, is inhibited by phosphorylation (Fig. 3.17).

The β-AA receptor is eventually inactivated by uptake and metabolism of the β-AA. Noradrenaline and adrenaline are catabolized by catechol-o-methyl transferase, which is an enzyme that methylates the hydroxyl groups on the aromatic ring, and by deamination by monoamine oxidase. The β-AA

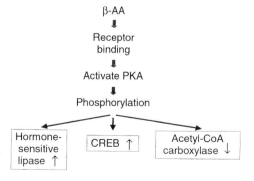

Fig. 3.17. Mechanism of action of β-adrenergic agonists.

Table 3.4. Effects of oral β-AA in different species (% change) (from Mersmann, 1998).

Animal	Weight gain	Feed consumption	Gain/feed	Muscle	Fat
Cattle	+10	−5	+15	+10	−30
Sheep	+15	+2	+15	+25	−25
Pigs	+4	−5	−5	+4	−8
Chicken	+2		+2	+2	−7

receptor can also be inactivated by phosphorylation of the receptor after β-AA binding by protein kinase A. Chronic stimulation with β-AA also reduces the number of β-AA receptors.

Physiological responses to β-AA

The effects of β-AA are generally greatest in cattle and sheep and least in chickens, with pigs being intermediate; but the results are quite variable (Table 3.4). This may be due to the fact that some species, such as broiler chickens, have been intensively selected for growth and are closer to the biological maximum growth rate. There may also be differences in the ability of different β-AAs to activate receptors in different species.

β-AAs increase muscle growth in both obese and lean genotypes of swine, but the genotype differences are not eliminated by β-AA treatment. The effects of β-AA on growth performance and carcass composition are much smaller in young pigs and ruminants that are not laying down fat at a fast rate compared to market-weight animals. It is not known whether this is due to low receptor numbers or increased desensitization of receptors in younger animals.

Carcass fat is reduced by 20–30% in cattle and sheep, with less dramatic improvements of about 10% in pigs. Small but significant improvements in dressing percentage and feed conversion are found. β-AAs are also effective in poultry, but the effects are less than those seen in mammals. Cimaterol reduces carcass fat by 10% in female broilers and by 5% in male broilers. Dose-dependent improvements in growth and carcass composition are also found in ducks and turkeys.

The effects of β-AA on metabolism are summarized in Fig. 3.18. Insulin regulates the rate of lipogenesis, and β-AAs reduce the rate of lipogenesis through a reduction in the number of insulin receptors and in the binding of insulin to adipocytes. β-AAs also act directly on adipose tissue via the β-receptor to stimulate lipolysis or decrease lipogenesis. Increased lipolysis by β-AA results in acutely increased levels of non-esterified fatty acids in the plasma of pigs and cattle.

β-AA treatment causes muscle hypertrophy rather than hyperplasia. The response is greater in type II fibres (fast contracting, mixed glycolytic–oxidative) than in type I fibers (slow contracting, oxidative). Long-term treatment with β-AAs may increase the proportion of type II fibres in a muscle.

β-AA treatment can also increase blood flow to the skeletal muscle and thereby increase the availability of energy and amino acids required for protein synthesis. Increased blood flow to the adipose tissue may help to remove the non-esterified fatty acids produced from lipolysis.

The effects of β-AA on muscle metabolism are probably mediated by direct effects, rather than indirect effects on other

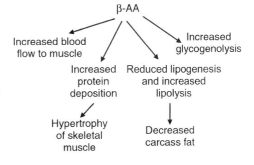

Fig. 3.18. Metabolic effects of β-adrenergic agonists.

endocrine systems, since the effects of β-AA are also seen in hypophysectomized or severely diabetic rats. Thus, the mode of action of β-AA is by direct receptor-mediated stimulation of muscle growth. Treatment with β-AA can both increase protein synthesis and decrease the degradation of muscle protein. Increased levels of mRNAs for muscle-specific proteins, such as myosin light chain and α-actin, as well as reduced activity of calpains and other specific proteolytic systems, have been reported.

Dry matter intake is commonly reduced after initial exposure to β-AAs, but this returns to normal after a short period. The repartitioning effects of β-AA occur in both feed-restricted and adequately fed animals, but the increase in growth rate is only seen in well-fed animals.

The increased protein deposition with β-AA is accompanied by a proportional increase in dietary protein requirements, and no response to ractopamine was found in pigs when dietary crude protein levels were less than $140\,g\,kg^{-1}$. At higher dietary protein levels, the maximal protein deposition rate is 23% higher in ractopamine-treated gilts compared to controls. This suggests that β-AAs increase the maximum rate of protein deposition. Sufficient levels of dietary energy are required to sustain the increased rate of protein deposition with β-AA treatment. However, maintenance energy requirements are not increased by β-AA. This may be because the energy requirements of increased muscle mass are offset by decreased energy requirements of a smaller visceral mass.

An important advantage of β-AAs is that they stimulate protein deposition primarily in skeletal muscle, while pST stimulates protein deposition and growth in all tissues. The increased energy and protein required for muscle synthesis comes from both increased utilization of dietary energy and protein and from turnover of other body tissues, such as liver, kidneys and gut. The weight of the liver, heart and gut are often reduced by β-AA treatment. There is little or no effect of β-AA on bone weight in ruminants, but β-AA treatment results in decreased bone weight in pigs. Thus the dressing percentage (carcass:offal ratio) is increased by β-AA treatment.

The effects of β-AA on meat quality can vary with the different β-AAs used, the species, dose and length of treatment. Overall, the anabolic effects of β-AA tend to increase meat toughness, as measured by shear force values. Trained sensory panels found decreased tenderness in beef, but not in pork. Other objective measures of meat quality, including pH, colour, drip loss and protein solubility, are not affected. There is some risk of developing DFD meat, due to the decreased muscle glycogen and plasma glucose at slaughter (see Section 3.12).

Delivery/dose

The effects of β-AA are dose dependent, with a tendency for reduced effects at very high doses. Doses of $5\,\mu g\,kg^{-1}$ BW resulted in peak plasma levels of $0.5\,ng\,ml^{-1}$ of clenbuterol after 3–5 h in cattle, with plasma levels increasing to $2\,ng\,ml^{-1}$ as the treatment was continued. Doses of 50–$80\,\mu g\,kg^{-1}$ BW of salbutamol resulted in plasma levels of about $5\,ng\,ml^{-1}$. For pasture-fed cattle, use of β-AA is impractical, since β-AAs are not available as implants.

The response to β-AA decreases over time if they are administered at a constant rate on a daily basis. The initial marked responses to β-AA peaks at 14 days after treatment and then is greatly decreased by 21 days of treatment. This occurs through densensitization of receptors and uncoupling of receptors with the adenyl cyclase system and down regulation of receptor numbers in the cell membrane. Thus, a relatively short treatment time towards the end of the finishing period will maximize the response to β-AA.

A withdrawal period can reduce the overall effectiveness of β-AA treatment. The withdrawal period should be kept relatively short (about 5 days), since compensatory growth of adipose tissue is seen after withdrawal of β-AA, while muscle mass is maintained.

Safety aspects

The oral potencies of different β-AAs can differ by as much as three orders of magnitude, due to differences in the structures, pharmacokinetics and metabolism of the different β-AAs. β-AAs that have halogenated aromatic ring systems, such as clenbuterol, are metabolized by both oxidative and conjugative pathways and have a long half-life. β-AAs that have hydroxylated aromatic rings are metabolized solely by conjugation and have a short half-life. The metabolism occurs in the liver and in the intestine during absorption.

The β-AAs are all orally active, in contrast to ST and most anabolic steroids. β-AAs are also heat stable and are not destroyed by cooking. This increases concern about the potential effects of residues of β-AA in meat. Half-life estimates for plasma cimaterol in steers was 54 min, and for plasma clenbuterol in veal calves it was 18 h for the initial distribution phase and 55 h for the terminal half-life. These estimates are much larger than for adrenaline. The very long half-life for clenbuterol means that high concentrations can accumulate in liver and adipose tissue. The effective therapeutic dose for clenbuterol is 10–20 µg, with a no observed effects limit (NOEL) of 2.5 µg day^{-1}. In contrast, the effective dose for other common β-AAs is 2000–10,000 µg. These less potent β-AAs suggest that safe and effective use of β-AAs may be possible. For more information on tissue residues for β-AAs, see the review by Smith (1998).

Ractopamine was approved for use in swine in the USA in December 1999; otherwise β-AAs are not approved for use in modifying the growth of animals raised for meat. In Canada and Europe, clenbuterol is approved only for therapeutic use as a bronchodilator in horses and calves, and for tocolysis to prevent premature birth in cows. Because of its high potency, clenbuterol has been used illegally in cattle, and this led to the hospitalization of 135 people in Spain in 1990 after clenbuterol poisoning caused by calf liver contaminated with 160 and 291 µg kg^{-1} of clenbuterol. The symptoms include muscle tremors, cardiac palpitations, nervousness, headache, muscular pain, dizziness, nausea, vomiting, fever and chills. The highest accumulation of clenbuterol occurs in pigmented tissues, such as hair and the retina in the eye, due to binding with melanin. This finding can be used as a highly sensitive test for detecting residues of clenbuterol in these tissues up to 60 days after treatment. Intensive monitoring programmes have shown a decrease in the illegal use of clenbuterol in The Netherlands, Germany, Northern Ireland and Spain by the mid 1990s. In the USA, clenbuterol has been used illegally for some show-ring animals, to accentuate muscle definition and reduce subcutaneous fat. For more information on the illegal use of β-AAs in Europe and the USA, see the reviews by Kuiper *et al.* (1998) and Mitchell and Dunnavan (1998).

Alternative approaches for using growth promoters

The effect of combined use of growth promoters depends on the mode of action of the different compounds. An additive effect can be expected when two growth promoters act via different mechanisms. Anabolic steroids and β-adrenergic agonists increase meat yield by reducing protein degradation, while ST increases protein synthesis more than protein degradation. Studies in pigs and veal calves combining ST treatment with β-AAs showed additive effects on feed efficiency and loin eye area from the two treatments.

Vaccines can potentially be used to mimic the effects of β-AA and ST, without the need to administer natural or synthetic hormones. This might alleviate concerns about the use of hormonal growth promoters and also make it feasible to improve growth in animals on extensive grazing systems.

Antibodies can be raised against natural hormones to block their effects. Targets for this research have been somatostatin, which blocks ST release, and ACTH, which stimulates release of the catabolic hormone, cortisol. Antibodies can also be generated which bind to a hormone to increase the activity of the hormone. This might occur by protecting the hormone from degradation and increasing the biological half-life or improving the delivery to the target tissue. This has been demonstrated with ST and IGF-I. Antibodies can also be directed to the target tissues to block the receptors of catabolic hormones, or stimulate the receptors of anabolic hormones (anti-idiotypic antibodies, see Section 2.4), as has been demonstrated for β-AA. Antibodies can also be developed which destroy cells. An example would be removing fat cells to prevent the accumulation of lipid. So far, these approaches are still under development. Since antibodies are proteins that will be inactivated by cooking, there should be significantly less concern about residues in meat from this immunological approach.

3.6 Thyroid Hormones

Thyroid hormones regulate two main types of processes. First, they affect metabolic pathways to modulate oxygen consumption, basal metabolic rate and lipid, carbohydrate and protein metabolism. There are also indirect effects due to altering the levels of other hormones. Secondly, thyroid hormones trigger cell differentiation and maturation in a number of tissues and indirectly affect growth. These effects on metabolism and cell differentiation are interrelated.

Synthesis and metabolism

Thyroid hormone is synthesized in the follicles of the thyroid gland. The precursor of thyroid hormones is a large glycoprotein called thyroglobulin, which contains iodinated tyrosine residues. These residues are coupled to form iodinated thyronine, mainly in the form of tetraiodothyronine or thyroxine (T_4, see Section 1.2). Hydrolysis of thyroglobulin causes the release of T_4. Thyroid hormones are highly hydrophobic and are bound to transthyretin, as well as other proteins, for transport and stabilization in the blood. T_4 is converted to the most biologically active form of thyroid hormone, triiodothyronine (T_3), outside the thyroid gland by the action of the type I and type II 5'-deiodinases (Fig. 3.19). T_4 and T_3 are metabolized to the inactive diiodotyrosine (3,3'-T_2) by the type III 5'-deiodinase enzyme. The deiodinase enzymes have been cloned and shown to be selenium-dependent enzymes with active selenocysteine residues. A potent inhibitor of the type I enzymes is 6-propyl-2-

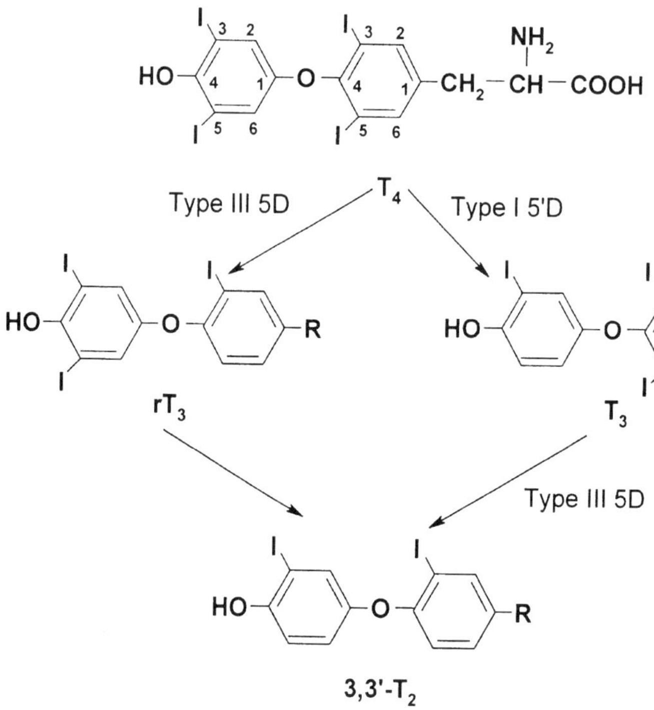

Fig. 3.19. Metabolism of thyroid hormones.

thiouracil (PTU), while all of the deiodinase enzymes are inhibited by iopanoate. The tissue-specific expression patterns of these enzymes suggest that they play an important role in the local and systemic availability of active T_3. Thyroxine can also be deaminated to form inactive tetraiodothyroacetate, or conjugated to glucuronides by UDP-glucuronyl transferase and then excreted. For more information on deiodinase enzymes, see the review by Köhrle (2000).

Thyroid hormone release and synthesis is stimulated by thyrotrophin (TSH), which is synthesized in the basophilic thyrotrophs of the pars distalis in the adenohypophysis. The amino-acid sequence of TSH is highly conserved and TSH from mammalian sources stimulates thyroid activity in most vertebrates. TSH interacts with receptors in the thyroid follicular cells to activate adenylate cyclase and increase cAMP production, resulting in increased synthesis and release of thyroid hormones. Thyroid hormones exert a negative feedback on TSH secretion.

Thyroid hormones act via two different mechanisms. There is a direct action of T_4, reverse T_3 (rT_3) or T_2 on the plasma membrane or on subcellular organelles. T_3 also acts by binding to nuclear receptors and affecting gene expression. There are two main classes of receptors (c-erb Aα and c-erb Aβ) encoded by different genes. Different forms exist within these classes. Most are stimulatory, while others, such as the c-erb Aβ2 in the pituitary gland, which is responsible for the negative feedback of T_3 on TSH synthesis, are inhibitory. The expression of some forms of thyroid hormone receptors (e.g. β_1) is developmentally regulated. For more information on the mechanism of action of thyroid hormone receptors, see Zhang and Lazar (2000).

Metabolic effects

The effects of thyroid hormones on metabolism and growth and development are summarized in Fig. 3.20. Thyroid hormones increase heat production and oxygen utilization by heart, liver, kidney and pancreas. In chickens, a precocial species, there is an increase in thyroid activity in the perihatch period in response to cooling, while doves, an altricial species, show little thermoregulatory development until 1–2 weeks after hatching. At older ages in chickens, the conversion of T_4 to T_3 is increased by short-term cold exposure and decreased by warm temperatures. Noradrenaline also plays an important regulatory role on non-shivering

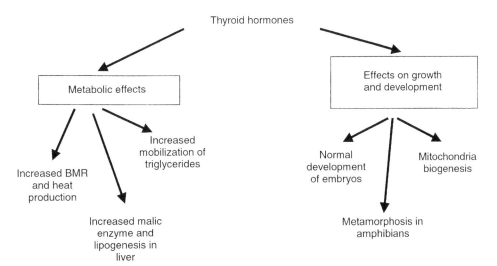

Fig. 3.20. Effects of thyroid hormones on metabolism, growth and development. BMR, basal metabolic rate.

thermogenesis, and thyroid hormones play a permissive role for this action of catecholamines by increasing the number of β-adrenergic receptors.

The maximal rate of stimulation of the basal metabolic rate (BMR) by thyroid hormone, as measured by oxygen consumption, is 100–150% in humans. Heat production involves stimulation of the Na^+/K^+ ATPase ion pump. The thermogenic effects of thyroid hormones are mediated through both short-term (minutes) and long-term (hours) effects on mitochondria. The short-term effects are due to T_3 binding to c-erb Aα1 isoform p28 on the inner mitochondrial membrane. Inhibitors of protein synthesis do not affect these short-term effects, which can also be demonstrated in isolated mitochondria, so they do not involve gene transcription. The long-term effects occur at the nuclear level, and include changes in phospholipid turnover and increased synthesis of uncoupling protein, which increase proton leakage at the inner membrane. T_3 binds to c-erb Aα1 isoform p43 in the inner mitochondrial matrix to stimulate mitochondrial genome transcription. T_3 is also involved in the production of new mitochondria, by activating both nuclear and mitochondrial genome expression.

In the liver, T_3 stimulates the synthesis of malic enzyme, which converts malate to pyruvate and provides NADPH for lipogenesis. Thyroid hormones also increase glucose transport and storage. Insulin increases the effects of T_3, but has little effect by itself, while glucagon is inhibitory. The effects are not seen on malic enzyme in non-hepatic tissues, indicating that tissue-specific factors are important. In chickens, malic enzyme does not increases in response to T_3 during the perihatch period when the lipid supply from the yolk is high, but only after the birds begin to feed in the early posthatch period, when carbohydrate from the diet becomes the primary energy source.

Thyroid hormones affect not only growth, but also carcass composition. Thyroid hormones increase the activity of lipoprotein lipase to increase the mobilization of triacylglycerol stored in adipose tissue and increase the levels of non-esterified fatty acids. Low levels of thyroid hormones in poultry are associated with increased fatness, while hyperthyroidism reduces abdominal fat. Similarly, pigs exposed to high temperatures have decreased levels of thyroid hormones and increased uptake and storage of triacylglycerols in adipose tissue.

Levels of T_3 decrease during fasting and increase again during refeeding. This is due to changes in the conversion of T_4 to T_3, as well as changes in TSH release from the pituitary. The levels of dietary carbohydrate affect both of these factors.

Effects on growth and development

Thyroid hormones are necessary for proper growth and development, and a deficiency in thyroid hormones results in severe growth retardation. Hypothyroidism during early stages of development results in deficiencies in somatic, neural and sexual development and decreased metabolic rate (termed cretinism in humans). Inadequate thyroid hormone production leads to high levels of TSH and hypertrophy of the thyroid gland, to cause goitre. This was traditionally observed in humans when dietary iodine was inadequate. Thyroid deficiency also results in decreased wool growth in sheep and hair growth in a number of different species.

It has been known for a long time that thyroid hormones are necessary for metamorphosis in tadpoles, and this is the basis for a bioassay for endocrine disruptor chemicals that affect thyroid hormones (see Section 6.4). In the rat, plasma T_3 and T_4 increase during the first 3 weeks after birth and then decrease to adult levels. Supplementation of broiler chicken diets with T_3 at $0.1\,mg\,kg^{-1}$ diet stimulated growth, while higher levels ($0.3\,mg\,kg^{-1}$ diet) depressed growth. Levels of T_3 can also be affected by altering the selenium status of broilers, since this affects the activity of the deiodinase enzymes. It is likely that increases in heat production and basal metabolic rate from increased levels of T_3 lower feed efficiency and growth (Jianhua et al., 2000).

There is an interaction between ST and thyroid hormones. Thyroid hormones are required for secretion of ST and for its systemic actions. Both thyroxine and ST increase the production of muscle proteins and there is a synergistic action of both hormones on muscle and whole-body growth. In birds, conversion of T_4 to T_3 is increased by pulsatile administration of ST but not by continuous ST administration.

Mitochondria have been implicated in the control of cellular proliferation and apoptosis. T_3 affects mitochondrial biogenesis and activity, and thus is involved in the differentiation and maturation of various cell types. These include neurones and glial cells of the central nervous system. Levels of T_3 are maintained in the brain during periods of hypothyroidism by increases in the activities of the type I 5'-deiodinase enzymes that convert T_4 to T_3 and decreases in the activity of the type III 5'-deiodinase enzymes that convert T_3 to diiodotyrosine. T_3 is also a major regulator of myoblast differentiation. For more information, see the review by Wrutniak-Cabello *et al.* (2001).

3.7 Dietary Polyunsaturated Fatty Acids

Polyunsaturated fatty acids (PUFAs) are important components of cell membranes that influence membrane function, signal transduction, eicosanoid metabolism and gene expression. The number and position of the double bonds in the fatty acid (for example 18:2, *n*-3 means an 18-carbon fatty acid with 2 double bonds starting at carbon 3 counting from the methyl end of the molecule) has dramatic effects on its biological activity. Animals cannot insert double bonds into fatty acids beyond carbon 9 from the carboxyl end of a fatty acid. Thus, both linoleic acid (LA, 18:2, *n*-6) and α-linolenic acid (ALA, 18:3, *n*-3) are essential PUFAs that must be obtained from the diet. These fatty acids are converted through desaturation and elongation reactions to longer-chain *n*-3 and *n*-6 fatty acids.

Mechanism of action

PUFAs and their metabolites are involved in regulating the expression of enzymes involved in lipid and carbohydrate metabolism (Fig. 3.21). PUFAs cause decreased lipid synthesis in the liver, increased fatty acid oxidation in the liver and skeletal muscle and increased synthesis of glycogen. Thus, PUFAs act as repartitioning agents, increasing the oxidation of fatty acids rather than storage as triacylglycerols and increasing glucose storage as glycogen.

PUFAs are also involved in the control of adipogenesis. Activation of adipocyte genes

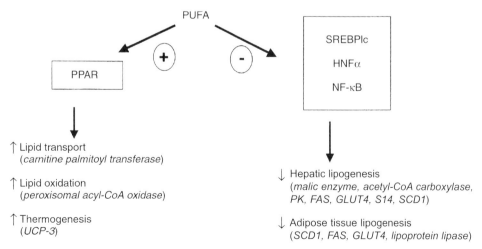

Fig. 3.21. Metabolic effects of dietary PUFA. PPAR, Peroxisome proliferator-activated receptor; SREBP, sterol regulatory elements binding protein; HNF, hepatic nuclear factor; PK, pyruvate kinase; FAS, fatty acid synthase; GLUT4, glucose transporter 4; SCD, stearoyl-CoA desaturase; UCP, uncoupling protein.

leading to adipocyte differentiation is stimulated by a metabolite of PGD_2 (15-deoxyprostaglandin J2) interacting with the peroxisome proliferator-activated receptor γ (PPARγ). Adipocyte differentiation is inhibited by $PGF_{2\alpha}$ interacting with a cell-surface receptor and altering intracellular calcium levels. The balance between these two pathways determines the overall effect of PUFA metabolites on adipocyte differentiation.

Fatty acids can regulate gene expression by activating nuclear transcription factors, including PPARα, β and γ, hepatic nuclear factor (HNF4α), NFκB, and sterol regulatory elements binding protein (SREBP1c). The PPAR bind responsive elements (PPRE) on DNA as a heterodimer with the retinoid X receptor (RXR) to increase gene expression (see Chapter 1). PPARα is activated by lipid catabolic fibrate drugs and PUFA to increase expression of genes involved in lipid transport, oxidation and thermogenesis, including carnitine palmitoyltransferase, peroxisomal acyl-CoA oxidase and uncoupling protein 3. For further details, see the reviews by Sessler and Ntambi (1998) and Clarke (2001).

PUFAs inhibit hepatic lipogenesis by decreasing the expression of enzymes involved in glucose metabolism and fatty acid biosynthesis. Malic enzyme, acetyl-CoA carboxylase, L-type pyruvate kinase, fatty acid synthase (FAS), glucose transporter 4 (GLUT4), S14 protein and stearoyl-CoA desaturase (SCD1) in the liver are decreased by PUFAs. The expression of SCD1, FAS, GLUT4 and lipoprotein lipase in adipose tissue are also decreased by PUFAs. The suppression of lipogenesis is not mediated by PPAR, but through decreased levels of transcription factor SREBP1c and by reducing the DNA-binding activity of other transcription factors. PUFAs can affect both the level of gene transcription and mRNA stability.

Nuclear transcription factors (TF) can be activated by direct binding of fatty acids, fatty acyl-CoA or oxidized fatty acids, and responsive elements have been identified in the promoter regions of several PUFA-regulated genes. Alternatively, cell-surface receptors linked to G-proteins can be activated by eicosanoids that are produced by the metabolism of PUFA, or intracellular calcium levels can be regulated by oxidized fatty acids. This can result in intracellular signalling cascades and subsequent activation of transcription factors (Fig. 3.22).

Linoleic acid, linolenic acid and γ-linolenic acid

Linoleic acid (LA, 18:2, *n*-6) and α-linolenic acid (ALA, 18:3, *n*-3) are essential dietary

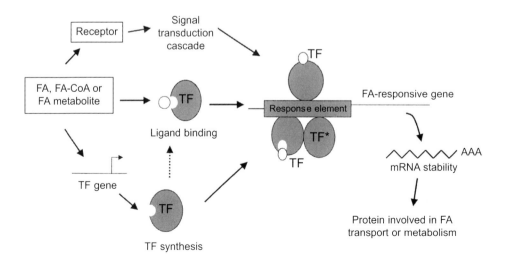

Fig. 3.22. Mechanism of action of dietary PUFA on transcription factors (adapted from Duplus *et al.*, 2000).

polyunsaturated fatty acids (PUFAs) that are converted by elongation and desaturation reactions to other n-6 and n-3 fatty acids, respectively. LA is ultimately metabolized to arachidonic acid (AA), which is the precursor of 2-series prostaglandins and 4-series leukotrienes (see Section 1.2). Dihomo-γ-linolenic acid (DGLA), an intermediate in the formation of AA, is converted into 1-series prostaglandins and 3-series leukotrienes. ALA is metabolized to eicosapentaenoic acid (EPA) and docosahexaenoic acid (DHA). EPA is converted into the 3-series prostaglandins and 5-series leukotrienes. This is summarized in Fig. 3.23.

Changes in the metabolism of the n-3 and n-6 fatty acids can alter the structural properties of cell membranes, such as fluidity and permeability. This can affect the activity of ion channels and membrane proteins. In addition, the availability of 20-carbon precursors for the synthesis of prostaglandins and leukotrienes can have dramatic endocrine effects. The Δ6 desaturase enzyme catalyses the conversion of LA to γ-linolenic acid (GLA), which is the rate-limiting step in the n-6 reaction cascade. The activity of the Δ6 desaturase is increased by insulin, and inhibited by epinephrine, cortisol, thyroxine, glucagon, saturated fat and ageing. Thus, dietary supplementation with GLA may be necessary to provide sufficient precursors for the synthesis of DGLA and the series-1 prostaglandins. GLA is found in natural plant oils, including evening primrose oil (9%), borage oil (23%) and blackcurrant oil (18%).

The n-3 and n-6 PUFAs have opposing physiological functions and their balance is important for normal growth and development. The n-6 and n-3 pathways are believed to share the same enzymes, so an increase in LA results in an increase in AA and a decrease in the formation of the n-3 fatty acids, EPA and DHA. Along with the total amount of fatty acids available, the n-6/n-3 ratio affects the formation of the long-chain PUFAs. The affinity of the enzymes is higher for the n-3 fatty acids, and efficient conversion of ALA to long-chain n-3 PUFAs can occur at a 4:1 ratio of LA to ALA in the diet. Since the longer chain n-3 PUFAs are the most biologically active, dietary supplementation with DHA and EPA is necessary if the metabolism of ALA is impaired. Good dietary sources of DHA and EPA are various marine oils, while ALA is found in flax, canola (rapeseed), perilla and soybean oils.

The n-3 fatty acids have been implicated in the prevention and management of coronary heart disease and hypertension. n-3 fatty acids stabilize myocardial membranes electrically, resulting in decreased susceptibility to ventricular arrhythmia and reducing the risk of sudden death. DHA is essential for growth and functional development of the brain in infants, and deficiencies of DHA are

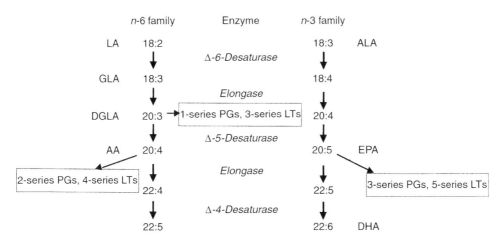

Fig. 3.23. Metabolism of n-3 and n-6 fatty acids.

associated with many types of mental disease. The n-3 fatty acids have potent anti-inflammatory effects, and high doses of n-3 fatty acids can reduce plasma triacylglycerol levels.

As the levels of DHA and EPA increase, the levels of AA decrease in cell membranes of platelets, erythrocytes, monocytes, neutrophils and liver cells. This decreases the levels of the eicosanoids produced from AA that enhance platelet aggregation, immunosuppression and the development of inflammation and allergic reactions. These include PGE_2, thromboxane A_2 (a potent platelet aggregator and vasoconstrictor), leukotriene B_4 (an inducer of inflammation and leucocyte adherence and chemotaxis), and thromboxane A_3 (a weak platelet aggregator and vasoconstrictor). There is also increased production of prostacyclin PGI_3 and maintenance of PGI_2 (both active vasodilators and inhibitors of platelet aggregation) and increased levels of leukotriene B_5 (a weak inducer of inflammation and chemotaxis). Thus, an increase in the formation of eicosanoids from AA increases the formation of blood clots, increases blood viscosity and vasoconstriction and increases inflammation.

The appropriate balance of n-6/n-3 fatty acids in the diet is important. Deficiencies of the n-6 fatty acids have been linked with depressed growth, impaired immune function, cardiovascular disease, diabetes and cancer. However, the diets of Western countries have high levels of LA, which has been promoted for its cholesterol-lowering effects, so deficiencies are unlikely, except under conditions where there is decreased activity of the Δ6 desaturase enzyme.

Supplementation with GLA, an n-6 PUFA that has the same number of double bonds as ALA, has similar physiological effects to supplementing with n-3 PUFA. GLA supplementation increases tissue levels of DGLA and the 1-series prostaglandins, particularly PGE_1, which has anti-inflammatory and immunoregulating properties. Levels of AA are not increased since the levels of the Δ5 desaturase are low. DGLA competes with AA for the cyclooxygenase enzyme and thus further lowers the synthesis of the pro-inflammatory series-2 prostaglandins from AA. GLA supplementation also lowers blood pressure and plasma cholesterol. GLA supplementation has also been shown to decrease the accumulation of body fat in rats. For more information on GLA, see the review by Wu and Meydani (1996).

Applications

Supplementing diets with n-3 fatty acids may be effective in treating metabolic diseases of poultry, including ascites and sudden death syndrome (heart attack) in broilers and turkeys. Ascites is caused by an imbalance between the oxygen requirements of the bird for growth and metabolism and the delivery of oxygen to the tissues, resulting in tissue hypoxia. This increases the haematocrit and blood viscosity, which leads to pulmonary hypertension and hypertrophy of the right ventricle of the heart. Eventually, the atrioventricular valve fails, causes backpressure and ascites fluid leakage from the liver. Supplementing broiler diets with 5% flax oil decreased blood viscosity and reduced the incidence of ascites. The addition of flaxseed, flax oil or other sources of n-3 fatty acids to layer rations reduces the accumulation of fat in the liver of laying hens. In addition, supplementing animal diets with n-3 fatty acids not only improves the health of the animals but also provides a good dietary source of n-3 fatty acids for human consumption.

A major limitation to supplementing animal diets with PUFA is the potential for increased lipid peroxidation and rancidity; these problems have been at least partly addressed by increasing levels of vitamin E and other antioxidants. n-3 Fatty acids can also lead to 'fishy' odours and taste. n-3-Enriched eggs are now produced commercially by including flax in layer rations, and it has been demonstrated that cows' milk can be enriched with n-3 fatty acids. Research on enriching the n-3 content of meat from poultry, pigs and cattle without adversely affecting the stability and organoleptic properties is continuing.

Conjugated linoleic acid

Conjugated linoleic acid (CLA) refers to a group of isomers of linoleic acid (*cis*-9, *cis*-12 octadecadienoic acid), the most biologically important being *cis*-9, *trans*-11 CLA (also known as rumenic acid) and *trans*-10, *cis*-12 CLA (Fig. 3.24). CLA has a broad range of biological activities, including anti-cancer effects, regulation of energy partitioning and nutrient metabolism, reduction of the catabolic effects of immune challenge, and reduction in blood lipids and prevention of atherosclerosis. Many of these effects have been demonstrated in animal models and have not yet been shown conclusively in humans. From an animal production point of view, CLA acts as an effective repartitioning agent in pigs, to increase feed efficiency, decrease back fat and increase lean muscle in the carcass.

CLA was first isolated from grilled beef and can be formed by heating linoleic acid in the presence of a base. CLA is produced biologically by the biohydrogenation of linoleic and linolenic acid by rumen microbes, and various CLA isomers are found in milk, cheese and beef. The *cis*-9, *trans*-11 isomer is the predominant form of CLA produced by rumen microbes. Most feeding trials with animals use a mixture of *cis*-9, *trans*-11 CLA and *trans*-10, *cis*-12 CLA in approximately equal amounts, with other isomers at considerably lower levels. Some commercial preparations may contain isomers with conjugated double bonds at carbon 8, 10, 11 or 13, and this may affect the biological activity. *Trans*-vaccenic acid (*trans*-11-C18:1) is the major *trans* fatty acid of rumen fermentation. It can be converted in mammalian tissues to *cis*-9, *trans*-11 CLA by the Δ9 desaturase enzyme.

Metabolic effects of CLA isomers

Dietary CLA increases feed efficiency and decreases body fat with a lesser increase in lean content. CLA at 0.5% in the diet of mice decreased body fat by approximately 60% and increased whole-body protein. Results in pigs and poultry have been less dramatic.

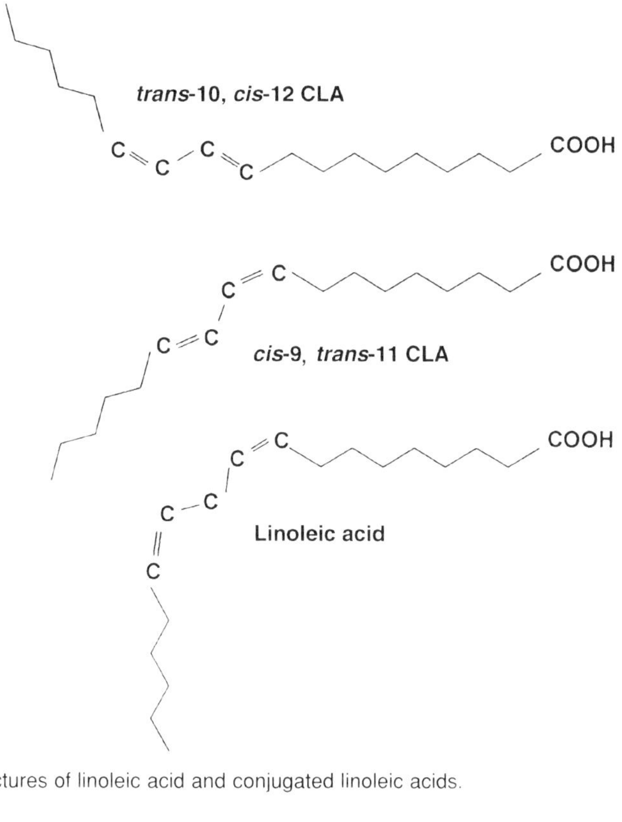

Fig. 3.24. Structures of linoleic acid and conjugated linoleic acids.

Pigs fed diets containing 1% CLA tended to have about a 6% improvement in feed efficiency, 7% less subcutaneous fat and 2.5% more lean, with no difference in rate of gain compared to controls. In chickens, CLA supplementation reduced abdominal fat content, but also reduced feed intake and growth rate.

In mice, the decrease in body fat has been attributed to the *trans*-10, *cis*-12 CLA isomer. This CLA isomer reduced lipoprotein lipase and decreased intracellular levels of triacylglycerol and glycerol in adipocytes. It also reduces the expression of stearoyl-CoA desaturase in adipocytes and depresses milk fat synthesis in cows. CLA also increases carnitine palmitoyltransferase, which is a rate-limiting enzyme in the β-oxidation of fatty acids. CLA also enhanced noradrenaline-induced lipolysis and hormone-sensitive lipase activity.

Trans-10, *cis*-12 CLA also reduces the secretion of apolipoprotein B and triacylglycerol in the human hepatoma HepG2 cell line, which would explain the role of CLA in the prevention of atherosclerosis. CLA supplementation in rabbits reduced LDL cholesterol and the incidence of aortic plaques, and did not affect HDL cholesterol.

The *trans*-10, *cis*-12 CLA isomer is also likely to be responsible for many of the effects of CLA on the immune system, including enhancing the immune system, reducing the catabolic effects of immune stimulation, and reduced production of PGE_2 and leukotriene B4. CLA appears to shift the immune response from a T_H2-type response (allergic reactions) to a T_H1-type response (cell-mediated functions; see Section 6.3).

CLA is an effective inhibitor of carcinogenesis, and both CLA isomers may be equally effective in this regard. Fish oil is usually required at about 10% of the diet to exert a beneficial effect, while CLA is effective at levels of 1%. The anticarcinogenic effect of CLA may be due to a decrease in the production of PGE_2 and an increase in tissue levels of retinol.

The CLA-induced increase in growth in young rodents may be due to *cis*-9, *trans*-11 CLA and this may be blocked by the *trans*-10, *cis*-12 CLA isomer. The *cis*-9, *trans*-11 CLA is a potent and high-affinity ligand for the peroxisome proliferator-activated receptor α (PPAR-α, see above). PPAR-α is found in liver, skeletal muscle, spleen, kidney and brown adipose tissue, and is involved in increasing lipid catabolism and reducing fatty acid synthesis. CLA supplementation produces antidiabetic effects, including lower plasma glucose, insulin and free fatty acids.

Mechanism of action of CLA

The effects of CLA may be due to alterations in eicosanoid metabolism and reduced synthesis of PGE_2. The isomers of CLA are elongated and desaturated in the same manner as linoleic acid to produce CLA-derived eicosanoids. This can reduce the conversion of arachidonic acid into PGE_2 and thereby alter eicosanoid signalling pathways. Reduced levels of prostaglandins, such as PGE_2, will decrease the synthesis and action of tumour necrosis factor-α (TNF-α) and interleukin 1 (IL-1). These cytokines induce the inflammatory response in immune cells, catabolism in skeletal muscle, changes in cell-surface proteins and increase the production of the intestinal peptide cholecystokinin, which induces anorexia (see Section 3.9). TNF-α is a key mediator in many chronic pathological conditions, including cachexia (physical wasting), atherosclerosis, carcinogenesis and obesity. Some of the effects of CLA may be due to reduced activity of TNF-α. For more information on the mechanism of action of CLA, see Pariza *et al.* (2000).

CLA preparations

It is apparent that the different isomers of CLA have different biological properties. It is thus essential that the isomer composition of commercially available preparations of CLA be known. Early preparations of CLA, produced by alkali isomerization during the first half of the 1990s, consisted of approximately 20–40% of *cis*-9, *trans*-11 and *trans*-10, *cis*-12 CLA, with the remainder consisting of other positional isomers. The CLA products available starting in 2000 are almost entirely the two desired isomers. The different CLA isomers can be measured as fatty acid methyl esters by gas chromatography or HPLC using silver-impregnated columns.

Enhancement of cows' milk with CLA is hampered by the dramatic decrease in milk fat caused by supplementation of diets with CLA. This is due to a decrease in fatty acid synthesis and desaturation of fatty acids by CLA. For more information on CLA in milk, see the review by Parodi (1999).

Further studies are needed with single isomers of CLA to determine their effects in commercial animals and humans.

3.8 Leptin

Leptin is a 16 kDa protein hormone produced by white adipose tissue, and is the product of the *obese* (*ob*) gene. Leptin secretion is higher in subcutaneous than omental adipocytes. It is thought to function as an 'adipostat' and long-term regulator of energy reserves in the form of adipose tissue in the animal. When there is a lot of adipose tissue, production of leptin increases to activate the satiety centres in the hypothalamus and reduces food intake. Conversely, when adipose tissue reserves decrease due to limited availability of food, leptin levels decrease and appetite increases. Leptin is thus an important regulator of appetite, whole-body energy balance and body composition. Leptin is also important in metabolic adaptation to starvation, by regulating metabolic rate and inhibiting reproductive and thyroid function (Fig. 3.25).

Plasma leptin varies in a pulsatile manner, with approximately 30 pulses in 24 h in humans. There is also a diurnal rhythm in leptin levels, with highest levels at 01.00 hours and lowest levels at 11.00 hours, but varying the timing of meals can alter this rhythm.

Leptin is mutated and inactive in the homozygous obese (*ob/ob*) mouse; leptin treatment of these mice reduces feed intake and body weight by about 40% after 33 days of treatment. Pair-fed (*ob/ob*) mice that are not treated with leptin lose significantly less weight than those treated with leptin. Body weight loss from leptin treatment is due almost exclusively to loss of adipose tissue, with lean mass unaffected. Leptin treatment of normal wild-type mice also reduces feed intake and body fat.

Up to six different serum-binding proteins for leptin have been identified. The amount of the binding proteins is reduced and the percentage of free and unbound leptin is increased in obesity.

Leptin receptors

Leptin receptors are located in the hypothalamus as well as on peripheral target tissues. Spontaneous mutations in the leptin receptors in the *db/db* mice and *fa/fa* rats result in inactive leptin receptors and severe obesity that is resistant to leptin treatment. The receptor has been cloned from a variety of tissues and is similar to the class I cytokine receptor family, which includes growth

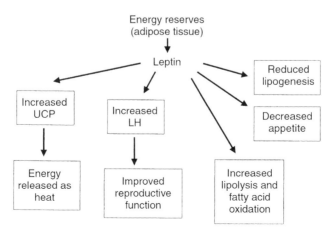

Fig. 3.25. Metabolic effects of leptin.

hormone, prolactin, interleukin-6 and leukaemia inhibitory factor. This receptor does not have an intrinsic tyrosine kinase activity, but acts by association with cytoplasmic Janus kinases (JAK, see Section 1.3). The receptor exists in multiple forms with a common extracellular sequence and a variable-length cytoplasmic domain. Alternate splicing of the mRNA from a single gene produces the six isoforms of the leptin receptor.

The short form of the receptor (OB-R_a) in the choroid plexus is thought to act as a transport protein to allow leptin to cross the blood–brain barrier. The kidney is the major site of leptin metabolism and clearance. A short form of the leptin receptors in the adrenal medulla may be involved in adrenaline secretion. The long form of the leptin receptor (OB-R_b) is found in areas of the hypothalamus that have been implicated in the regulation of appetite and body weight, namely the arcuate, dorsomedial, ventromedial and paraventricular nuclei. The effects of leptin on food intake may be mediated in part by inhibiting the actions of neuropeptide Y (NPY), which is a potent stimulator of food intake. For more information, see the reviews by Ahima and Flier (2000) and Fried et al. (2000).

Involvement in energy metabolism and reproduction

Leptin treatment of *ob/ob* mice decreases serum insulin and glucose and increases oxygen consumption, locomotor activity and body temperature to that found in normal wild-type mice. Leptin is more effective in the *ob/ob* mice than in normal animals, and even the highest dose did not cause metabolic indices to exceed normal levels. These effects may be mediated by the sympathetic nervous system, through the effects of noradrenaline on brown adipose tissue. Catecholamines and β-agonists decrease leptin levels in the long term. Leptin levels are also increased by high-fat diets. Insulin and glucocorticoids act synergistically to increase leptin levels over the long term.

Leptin also causes local tissue effects that are mediated by mitochondrial uncoupling proteins (UCPs). It stimulates synthesis of UCP mRNA and protein in muscle and adipose tissue. The UCPs increase proton leakage through the inner mitochondrial membrane and thus uncouple oxidative phosphorylation and ATP synthesis from mitochondrial electron transport, with the energy released as heat.

Females with poor body condition do not reproduce well, and leptin may act to signal the reproductive system that sufficient body fat is present to support a pregnancy. Leptin treatment increases sexual development in both male and female *ob/ob* mice, resulting in increased serum LH and ovary and uterine weights in females, and increased serum FSH, testes and seminal vesicle weights and sperm counts in males. Levels of leptin are 2–4 times higher in women than in men, and leptin levels increase near puberty in girls but not in boys. Leptin production is decreased by androgens (testosterone and dihydrotestosterone) but not by oestrogens.

Leptin levels vary during fetal development and are correlated with fetal size. Leptin may be involved in signalling nutrient availability and in regulating growth and development of the fetal–placental unit during embryonic development.

Leptin stimulates secretion of ST and restores pulsatile ST secretion after fasting. It decreases GHRH receptors but increases growth hormone secretagogue receptors (GHS-R) in isolated pituitary somatotrophs. Thus leptin decreases GHRH stimulation while increasing GHS stimulation of ST secretion.

Direct effects

Leptin has direct effects on a variety of tissues, and this has been shown in isolated cells and tissues treated with physiological levels of leptin *in vitro*. It reduces lipid synthesis and increases lipolysis in isolated adipocytes and fat pads. Leptin reduces the effect of insulin on stimulating gluconeogenesis in HepG2 liver cells and reduces insulin secretion by pancreatic islets. It stimulates fatty acid oxidation and decreases triacylglycerol synthesis in muscle. Glucose uptake and glycogen synthesis is stimulated in myotubes by leptin.

Applications

Leptin can affect nutrient intake and carcass composition, including the level and rate of fattening. This, together with the effects on reproductive efficiency, makes leptin of interest to animal producers. Leptin and its receptors are potential candidate genes for the development of genetic markers for improvements in animals. Since leptin decreases the efficiency of energy utilization by increasing thermogenesis, the use of leptin antagonists may improve feed utilization and increase the efficiency of animal production. Leptin has broad effects on reproductive processes and may be useful for inducing early puberty in young, thinner animals and to reduce the interval from parturition to oestrus. For more information, see the reviews by Hossner (1998) and Houseknecht *et al.* (1998).

3.9 Cholecystokinin and Appetite

Mechanism of action

The capacity of pigs for growth is limited to some extent by the amount of feed they consume, so methods for increasing appetite should increase weight gain and shorten the time taken to reach slaughter weight. Feed intake is increased by stimulation of the lateral hypothalamic nuclei, while stimulation of the ventral medial hypothalamus decreases feed intake. Cells within these regions have glucoreceptors that monitor blood glucose levels to control feed intake. Other hormones or metabolites may also affect the feeding and satiety centres in the hypothalamus. One such hormone is cholecystokinin (CCK), which is a peptide hormone produced by the duodenal mucosa. The transfer of digesta (particularly amino acids, HCl and certain fatty acids) from the stomach to the duodenum stimulates its release and causes emptying of the gall bladder and release of pancreatic enzymes. CCK or a related peptide also acts as an important satiety signal to reduce food intake, so interfering with CCK function can be used to increase food intake.

CCK is structurally related to gastrin (Fig. 3.26) and the common C-terminal pentapeptide is present in a wide range of species. The tyrosine at residue 7 from the carboxy-terminal of CCK is sulphated, and the sulphated form of the hormone binds 160 times more strongly to the CCK receptor (CCK_A) than the non-sulphated form. The gastrin receptor (CCK_B) does not require sulphated hormone for activation and subsequent stimulation of acid secretion in the stomach. CCK exists in a number of forms, from a 33-amino-acid polypeptide in the gastrointestinal (GI) tract to the 8-amino-acid C-terminal peptide (CCK-8) that is found in the central nervous system (CNS). CCK-8 is inhibitory to food intake as well as stimulating pancreatic secretions and gall-bladder contraction.

Synthetic analogues of CCK inhibit food intake as well as induce gall-bladder contraction and release of pancreatic enzymes, suggesting that these functions are regulated by a common mechanism. Satiety signals are transferred to the hypothalamus via the vagus nerve, but there is also evidence that CCK acts locally within the ventral medial hypothalamus. This is another example, along with somatostatin and ghrelin, of a hormone that has actions in both the gut and brain.

Other hormones involved in satiety are leptin and neuropeptide Y (NPY). NPY is a

$$SO_3H$$
$$|$$
Gastrin$_{(17)}$ (pyro)Glu-Gly-Pro-Trp-Leu-Glu-Glu-Glu-Glu-Glu-Ala-Tyr-Gly-Trp-Met-Asp-Phe-NH_2

$$SO_3H$$
$$|$$
CCK$_{(17-33)}$ -Asp-Pro-Ser-His-Arg-Ile-Ser-Asp-Arg-Asp-Tyr-Met-Gly-Trp-Met-Asp-Phe-NH_2

Fig. 3.26. Structures of gastrin and cholecystokinin.

36-amino-acid peptide found in the peripheral nervous system and in the brain. Feed restriction increases transcription and translation of NPY and intracerebroventricular administration of NPY increases appetite. A related peptide YY (PYY) is localized in endocrine cells of the intestinal mucosa. NPY and PYY are also involved in regulating blood flow, motility and electrolyte secretions in the intestine. Leptin administration decreases feed intake, and this occurs at least in part by decreased activity of NPY. Leptin receptors have been co-localized with NPY neurons in the arcuate nucleus of the hypothalamus. Animals that do not produce functional leptin (such as *ob/ob* mice) or functional leptin receptors overexpress NPY and are hyperphagic. This has led to the proposal that NPY is the major mediator of the actions of leptin.

Other peptides, such as galanin and orexins, also influence feed intake. For more information on endocrine signals regulating food intake, see the review by Bray (2000).

Applications

Differences in plasma levels of CCK have been found in pigs with different genetic potential for feed intake. This suggests that satiety effects of CCK are involved in the genetic differences between lines of pigs for feed intake. Decreased production of CCK has been found in humans with bulimia nervosa. Increased levels of CCK, or sensitivity to CCK, have been implicated in age-related anorexia. CCK may also prove useful in the control of obesity in humans.

Immunizing pigs, but not sheep, against CCK-8 increased growth rate by about 10% and feed intake by 8%, with no change in carcass composition. Antibody titres were highly variable and were correlated with weight gain (McCauley *et al.*, 1995). Treatment of pigs and immature rainbow trout with a CCK_A receptor antagonist (MK-329, see Fig. 3.27) increased feeding behaviour, feed intake and weight gain. For an example of work on developing non-peptide ligands for the CCK receptor, see Bernad *et al.* (2000).

3.10 Antibiotics, Antimicrobials and Other Factors

Antibiotic use for enhancing meat production began in the 1940s, when it was discovered that they improved growth rate by 4–20% and feed conversion efficiency by 4–10% if used in small amounts in the diet of pigs and poultry. These effects may be greater in younger than in older animals. There are no significant effects on carcass quality, such as length and depth of the carcass, fat content and dressing percentage.

Since the effects of antibiotics may not be seen in germ-free animals, it is thought that they may improve growth and feed efficiency by changing the intestinal microflora. This may, in turn, reduce the production of toxins, decrease the destruction of nutrients and increase the absorptive capacity of the intestinal mucosa. There may also be direct effects of antibiotics on the metabolism of the animal. Antibiotics may reduce low-level and sub-chronic infections and thus cause increased growth performance.

In ruminants, antibiotics can affect rumen fermentation and improve digestibility of feeds. Many of these antibiotics are ionophores, which are also effective coccidiostats in poultry. Improvements in daily gain of 15–25% and feed conversion efficiency of 3–16% are seen with beef cattle fed these antibiotics. The extent of the effects are dependent on diet, with greater improvements in grazing animals eating low-quality forages compared to those fed high-energy-concentrate diets.

A number of factors can affect the eco-

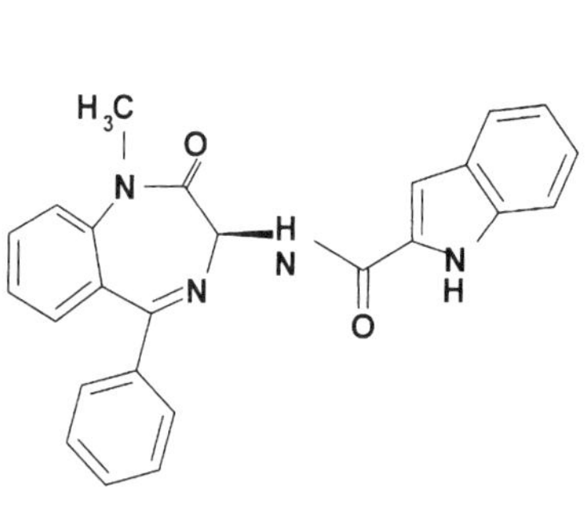

Fig. 3.27. Structure of a CCK_A receptor antagonist (MK-329).

nomic advantages of using growth enhancers. Responses in feed conversion and growth rate obtained in controlled experiments may not be obtained under field conditions. One exception would be the use of antimicrobial growth enhancers in the feed, which can give greater responses in commercial use due to the higher bacterial challenge and feed variability that is likely under production conditions. The economic benefits can also be affected by the duration of treatment that is necessary. Changes in health status, either positive or negative, may result in effects that continue long after treatment is completed. In this regard, the potential carry-over of antibiotic residues to humans is a concern. This has led to the banning of some feed-delivered antibiotics, especially in the EU, with consequently increased emphasis on the use of 'natural' ingredients, such as herbal extracts, in animal feeds to replace them. Changes in the volume and composition of manure may be important in regions that have strict environmental controls, such as The Netherlands. Increasing the lean content of carcasses makes economic sense only if the premiums offered to the producers for lean are greater than the cost of producing the extra lean. In some situations, such as for fresh pork cuts, a low fat content is desirable, while for some processed products a higher fat content is desirable.

3.11 Dietary Chromium and Insulin

Chromium is an essential trace mineral that is active in biological systems as the trivalent form (Cr^{3+}). It has been shown to have antioxidant properties and to stabilize proteins and nucleic acids. Chromium, as part of a 70 kDa protein, has been shown to enhance RNA synthesis in mice. However, the major role of dietary chromium is to increase the sensitivity of cells to insulin, perhaps by facilitating the binding of insulin to its receptor on the cell surface. Chromium is an essential component of an organometallic complex called glucose tolerance factor (GTF), which also contains nicotinic acid, glutamic acid, glycine and cysteine. The exact structure of the GTF insulin-potentiating complex has not been determined. The biologically active *in vivo* form of Cr, known as the low molecular weight chromium-binding substance (LMWCr), may be a Cr-containing oligopolypeptide.

Insulin

Insulin is a polypeptide hormone produced by the β cells in the islets of Langerhans in the pancreas. Its primary role is to regulate blood glucose levels by increasing the uptake of glucose into tissues and storage as glycogen or lipid. It also plays a key role in promoting amino-acid uptake in species where little fluctuation in blood glucose is seen, such as in functional ruminants and carnivores. Insulin consists of two polypeptides, an A chain of 21 amino acids and a B chain of 30 amino acids, that are linked by disulphide bridges (see Section 1.2). The amino-acid sequence of insulin is very similar among vertebrates, with minor changes mainly at positions 8, 9 and 10 of chain A and position 30 of the B chain. As a result, insulin isolated from one species is active in another. Insulin is synthesized as proinsulin, with the active molecule formed from cleavage of the 23-amino-acid C peptide. There is a high degree of species variability in the C-peptide sequence.

Insulin acts on the liver to increase glucose uptake and formation of glucose-6-phosphate and to activate glycogen synthetase. In adipose tissue, glucose is converted to glycerol and combined with free fatty acids to form triacylglycerol. Lipid synthesis is increased by stimulation of citrate lipase, acetyl-CoA carboxylase, fatty acid synthase and glycerol-3-phosphate dehydrogenase. In muscle, insulin stimulates the uptake of glucose and amino acids and stimulates glycogen and protein synthesis. Protein catabolism is also decreased. Insulin also has a direct vasodilatory effect to increase blood flow and nutrient supply to muscles.

Rapid growth and leanness in domestic animals is related to enhanced sensitivity of muscles to insulin and enhanced glycolytic metabolism of muscles. Decreased insulin sensitivity results in increased basal levels of insulin and increased carcass fat with decreased growth rate. For a review on the insulin signalling pathways, see Nystrom and Quon (1999).

Glucagon

Glucagon is a peptide hormone that is produced in the α cells of the pancreas islets and acts to increase blood glucose by increasing glycogenolysis or gluconeogenesis. Glucagon is a single-chain polypeptide of 29 amino acids that is identical in sequence across all mammals and highly conserved across all species.

The effects of glucagon are opposite to those of insulin. In the liver, it stimulates glycogenolysis and gluconeogenesis from amino acids and glycerol, to maintain sufficient levels of blood glucose. This is particularly important during prolonged fasting, exercise or during neonatal life. Glucagon stimulates lipolysis in adipocytes.

Chromium improves the function of insulin to reverse the effects of clinical hyperglycaemia by increasing the uptake of glucose into tissues for lipogenesis and glycogenesis. Chromium also increases the effects of insulin in stimulating uptake of amino acids and protein synthesis in muscle.

The signs of chromium deficiency are related to impaired insulin function and include impaired glucose tolerance, elevated levels of insulin, cholesterol and triacylglycerols in plasma, glucosuria, impaired growth and decreased longevity and fertility. Chromium supplementation is effective in humans receiving parenteral nutrition or with type 2 (insulin independent) diabetes. Chromium supplementation is also effective in animals under various forms of stress, such as newly arrived feedlot cattle and first lactation dairy cows.

Mechanisms of action

The insulin receptor (see Section 1.3) consists of extracellular α-subunits and intracellular β-subunits held together by disulphide bonds. Insulin binding to the α-subunit causes autophosphorylation of the β-subunit and activates its tyrosine kinase. Insulin activates a phosphatase enzyme that dephosphorylates and activates glycogen synthetase.

Detection of glucose levels in plasma by the pancreatic islet cells, as well as uptake and release of glucose by tissues, involves glucose transport proteins (GLUTs) in the cell membrane. These proteins catalyse the transport of glucose down concentration gradients and into target cells. There are five glucose transporter isoforms, GLUT1–4 and GLUTX1 (which is involved in early blastocyst development), along with a fructose transporter, GLUT5.

The GLUTs differ in tissue distribution and kinetic properties and play a key role in glucose homeostasis. GLUT1 is present at high levels in erythrocytes and in endothelial cells lining the blood vessels in the brain. GLUT3 is present in neurones, and together with GLUT1, allows the brain to take up adequate glucose. GLUT2 is a low-affinity transporter that is part of the glucose sensor system in pancreatic cells and is involved in absorbing dietary glucose across the basolateral membranes of intestinal epithelial cells. GLUT2 is also important in the transport of glucose out of the liver and kidney and into the blood. GLUT4 is the major insulin-responsive isoform found in adipose tissue and striated muscle. It is normally located in intracellular storage compartments and is translocated to the plasma membrane in response to insulin. When insulin levels decline, GLUT4 is recycled back into intracellular storage.

The mode of action of chromium in glucose metabolism has not been firmly established. Chromium may be involved in stabilizing insulin, or affecting the interaction of insulin with its receptor. The LMWCr complex has also been shown to increase the activation of the insulin receptor phosphotyrosine kinase (PTK) in the presence of insulin. The increased PTK activity increases the translocation of the GLUT4 from the Golgi apparatus to the cell membrane. This improves glucose uptake in adipose tissue and skeletal muscle.

Chromium also has been reported to be involved in stabilizing nucleic acids and stimulating RNA and protein synthesis, but the mechanism behind these effects has not been established.

Physiological effects

The potential physiological responses to chromium are summarized in Fig. 3.28. Supplementing newly arrived feedlot cattle with dietary chromium is reported to decrease morbidity and lower plasma cortisol, but the response is variable. The immune response is also improved, as measured by the blastogenic response of peripheral blood mononuclear cells cultured with T-lymphocyte mitogens. The beneficial effects of chromium were not seen when calves were given a long-acting antibiotic shortly after arriving at the feedlot. In growing cattle, chromium supplementation improves glucose metabolism by enhancing tissue responses to insulin as well as improving immune status. For a review on the effects of chromium on health and performance, see Borgs and Mallard (1998).

Chromium supplementation is effective in first-lactation dairy cows in the transition period, but not in multiparous cows. Supplemental chromium given through late gestation and into the first weeks of lactation tends to improve milk yield in the first 4–6 weeks of lactation. The metabolic stress during transition is also lowered, with decreases in levels of blood ketone bodies. There is also improved cell-mediated immune response in lactating cows via changes in production of cytokines. The specific role of chromium in cattle undergoing stress has not been established.

The effectiveness of chromium in pigs is quite variable. However, when given as a dietary supplement ranging from 0.5 to 5.0 mg kg^{-1} diet, chromium has been shown, under some circumstances, to alter metabolism in pigs to improve growth rate, increase lean muscle mass and decrease fat content in carcasses and improve reproductive performance.

Supplemental chromium chloride at 20 mg kg^{-1} increases the rate of glucose utilization by livers of chicks and poults. There is some evidence of improved growth rate and feed efficiency and decreased mortality and serum and yolk cholesterol from chromium supplementation in poultry. Research with rabbits has shown that the cholesterol and plaque contents in the aorta are decreased with supplemental chromium. Chromium supplementation has also been reported to alter glucose metabolism in fish, and to increase weight gain, energy disposition and liver glycogen.

For a comprehensive review of the role of chromium in animal nutrition, see NRC (1997).

Dose

The response of animals to chromium supplementation of practical diets is variable, since some diets may already have adequate levels of chromium. The chromium status of the animals can also vary, so that in some situations the animal's requirement for chromium is increased. Thus, diets must be deficient in chromium or the animal must require additional dietary chromium before supplementation of diets with chromium can be effective. A chromium intake of between 50

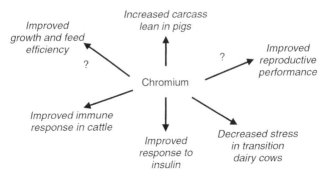

Fig. 3.28. Potential physiological responses to chromium.

and 200 µg day^{-1} is recommended for adult humans, but because there is no accurate measure of chromium status, daily chromium requirements for animals are difficult to define and no recommended dietary intake has been made.

Trivalent chromium has been used as a dietary supplement in the form of chromium picolinate (CrPic), chromium nicotinate (CrNic) and high-chromium yeast, which are absorbed more efficiently than inorganic chromium chloride. The efficacy of dietary chromium may be affected by inefficient absorption of chromium, particularly if it is not complexed with an organic molecule such as picolinate. The absorption of chromium is inversely related with dietary intake, ranging in humans from 2% absorption at 10 µg day^{-1} in the diet to 0.5% absorption with >40 µg day^{-1} in the diet. Absorption is increased in diabetes, possibly because diabetics are deficient in chromium.

Factors affecting the bioavailability of dietary chromium in feeds and the relative bioavailability of chromium from different supplements are not well understood. The low bioavailability of inorganic chromium may be due to the formation of insoluble chromic oxide, binding to chelating agents in feeds, and slow conversion to the biologically active form. Measurements of the total chromium content of feeds are therefore poorly related to the amount of biologically active chromium.

Plasma levels of chromium range from 0.01 to 0.3 µg l^{-1} and are lower during infections and after glucose loading. Chromium is transported in the blood bound to β-globulins and is transported into tissues bound to transferrin or as GTF. Turnover of chromium follows a three-compartment model with half-lives of 0.5, 6 and 83 days. Plasma levels of chromium are not a good indicator of tissue chromium levels, since these different chromium pools do not seem to be in equilibrium. Uptake of chromium by tissues decreases with age.

Chromium is excreted mostly in the urine, with 24 h excretion rates of 0.22 µg day^{-1} in humans. Excretion increases after glucose loading and is higher in diabetics. Stress and exercise also increase chromium excretion, with several times more chromium excreted after extreme trauma.

Safety issues

Chromium toxicity is primarily associated with exposure to hexavalent Cr^{6+} salts, which are strong oxidizing agents used in alloying and tanning and in the production of corrosion-resistant paints. Workers exposed to these compounds suffer from dermatitis, cancer, gastroenteritis, nephritis and hepatitis. Hexavalent Cr^{6+} is more soluble, more readily absorbed and is at least five times more toxic than Cr^{3+}. The LD$_{50}$ for Cr^{3+}–nicotinic acid complex injected intravenously is 60 mg kg^{-1} BW in rats. The lethal single dose of Cr^{6+} in young rats is 130 mg kg^{-1} BW, while as much as 650 mg kg^{-1} BW of Cr^{3+} produced no toxicosis. There is no information on chromium toxicity in swine. Feeding a diet containing 2000 mg kg^{-1} CrCl$_3$ decreased growth rate of chicks, while levels of 1000 mg kg^{-1} had no effect on growth. The concentrations of trivalent chromium that are typically added as dietary supplements for most food-producing animals are assumed to be safe and non-toxic. However, aside from CrPic, chromium has not been approved as a feed additive for livestock, due to the lack of information on the dietary requirements for chromium and its role as an essential nutrient. The NRC (1980) set maximum levels of chromium oxide at 3000 p.p.m. and chromium chloride at 1000 p.p.m. for livestock feeds.

3.12 Effects of Stress on Meat Quality

Pale, soft, exudative and dark, firm, dry meat

Pale, soft, exudative (PSE) meat is a potential problem with poultry (chicken and turkey) and with pork. The PSE condition has been most extensively studied in pigs and can be caused by pre-slaughter stress, which induces a rapid post-mortem breakdown of muscle glycogen and subsequent increase in anaerobic glycolysis to form lactate. This increases the temperature of the muscle and dramatically decreases muscle pH, which denatures the sarcoplasmic reticulum proteins and decreases water-binding capacity of the tissue. An increased level of metabolites in the cells also increases the osmotic pressure, which causes water to move into the cells from the extracellular space. The muscle of pigs becomes pale, watery (exudative), sour-smelling and loose-textured. There is increased water (drip) loss and decreased water binding capacity during storage, leading to decreased shelf-life and poor consumer appeal.

The water-binding capacity of meat is affected by pH. It is lowest at the isoelectric point of meat between 5.0 and 5.1, since the protein has no net charge and the solubility of the proteins is at a minimum. Thus, as the pH of meat declines after slaughter, the water-binding capacity also decreases. In PSE pork, the exudate fills the spaces between the muscle fibre bundles and contributes to the soft texture of the pork. The paleness of meat is caused by the increased scattering of light within the meat due to protein denaturation. The pale colour of PSE pork takes at least 1.5 h to develop, and the development of pale colour may continue up to 4 days post-slaughter.

Stress can also cause dark, firm and dry (DFD) meat when levels of muscle glycogen are drastically depleted before slaughter. This limits the amount of lactate that is formed post-mortem and results in a high pH in the meat. The high pH results in increased water-holding capacity.

The initial pH taken at 45 min after slaughter and the ultimate pH taken at 24 h after slaughter can be used to predict PSE and DFD meat (Fig. 3.29). A pork carcass with a pH of 6.0 at 45 min would have a rapid rate of glycolysis and a strong possibility of becoming PSE. At 24 h, normal pork has a pH of between 5.4 and 5.7, PSE pork can have a pH below 5.0 and DFD pork has a pH of greater than 6.0. For further details, see Swatland (1994).

PSE poultry meat is a problem since it is unsuitable for further processing due to excessive colour variation, defective water-holding capacity and poor binding ability, which reduce product yield and texture. PSE pork is also less suitable for the production of cured and processed products.

PSE pork was first described as a post-slaughter quality defect related to stunning technique, temperature of scalding water, duration of scalding and time until the carcass was chilled. A survey of US packing plants in 1992 showed that only 16% of hams were the desirable reddish-pink, firm, non-exudative (RFN) quality. Fifty-eight per cent of hams were reddish pink, soft, exudative (RSE), 16% were pale, soft and exudative (PSE) and 10% were classified as dark, firm and dry (DFD). The method of restraint and stunning dramatically affects the incidence of PSE meat. Stunning with carbon dioxide produces the lowest incidence of PSE, followed by electrical stunning without restraint, with the highest incidence when electrical stunning with restraint was used. High-voltage

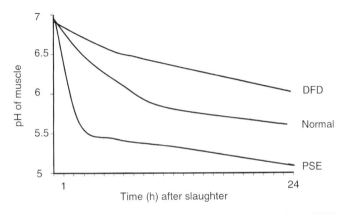

Fig. 3.29. Effects of muscle pH on dark, firm, dry (DFD) and pale, soft, exudative (PSE) meat.

(330–700 V) stunning increased PSE and blood splash compared to low-voltage (70 V) stunning. Carbon dioxide stunning is considered to be both humane and efficient and is increasingly used in Europe. Handling of hogs in holding pens, truck-loading techniques, mixing of animals, crowding, temperature, travel time and access to feed and water are also important.

Drip loss in PSE pork is about 1.70% of trimmed carcass weight compared to 0.77% for normal pork and drip loss increases with storage. For an estimated incidence of 18% of PSE carcasses, with an extra shrink loss of 5–6%, this amounted to US$95 million loss to the US pork industry in 1972.

DFD meat is a problem with beef, pork and probably also with poultry. In beef, DFD meat is referred to as dark-cutting meat and is most severe in young bulls, due to their aggressive behaviour, which reduces muscle glycogen stores. In poultry there is some evidence that cyanosis is due to DFD meat. Cyanosis is the fourth leading cause of condemnation of chickens in Canada and cyanotic meat is condemned because little is known about its safety and quality. It is thought to be caused by chronic hypoxia from overcrowding, but may be related to general shipping stress.

Porcine stress syndrome

Porcine stress syndrome (PSS) is a genetic defect that leads to malignant hyperthermia (MH) in the live pig and the formation of pale, soft, exudative (PSE) pork. The syndrome can be traced to early 20th century Germany and is particularly prevalent in the Piétrain breed. This breed may have begun by inadvertently selecting animals with the PSS mutation due to their exceptional muscle development.

PSS pigs that are stressed develop hyperthermia (body temperature >41°C) and take rapid, shallow breaths (dyspnoea). The skin becomes blotchy and the muscles tremble and become rigid or weak. The animal may die from acidosis, hyperkalaemia, vasoconstriction and cardiac arrest. Survival depends on cooling the animal and providing adequate oxygen. These effects are due to rapid muscle metabolism, which increases oxygen consumption up to threefold and produces heat. The endocrine stress response is also stimulated (see below), which exacerbates the metabolic response.

In PSS pigs, the muscle pH can drop to 6.0 in less than an hour due to excessive lactic acid accumulation, and the muscle temperature can rise to above 40°C, resulting in PSE meat. The DFD condition could be produced in PSS pigs that survived the initial stress response, but were slaughtered with very low levels of muscle glycogen. This would result in low levels of lactic acid in the meat and an ultimate pH greater than 6.0.

PSS and PSE are not the same thing. PSS is a genetic defect of pigs while PSE is a nonspecific meat condition that may occur in meat from any animal source with accelerated post-mortem glycolysis. However, PSE meat is more prevalent in carcasses of pigs that suffer from PSS, with 60–70% of PSS pigs developing PSE meat, depending on post-mortem handling. Stress factors that can trigger PSS include breeding, high temperatures, crowding/mixing, transportation to the abattoir, vaccination, castration, moving within the abattoir, and oestrus. Although PSS pigs are more susceptible to developing PSE meat, pigs that are free from the genetic defect can produce PSE meat as well. An examination of PSE loins for the PSS genotype showed that 3.6% were homozygous and 29% were heterozygous for the recessive stress gene, while the remainder had a normal genotype. This indicates that, while a high proportion of PSS pigs produce PSE meat, PSE meat can also be produced in normal pigs.

PSS is a genetic susceptibility to stress that is due to changes in calcium metabolism in the muscle (Fig. 3.30). Intracellular Ca^{2+} levels in relaxed muscle are 10^4 times lower than in the extracellular fluid. During muscle contraction, Ca^{2+} that is stored in the sarcoplasmic reticulum in the muscle cell is released through a calcium release channel (CRC, also known as the ryanodine receptor for its binding of the plant alkaloid, ryanodine) and intracellular Ca^{2+} is increased tenfold. Relaxation occurs as the Ca^{2+} is pumped back into the sarcoplasmic reticulum. The CRC is a large protein (molecular mass

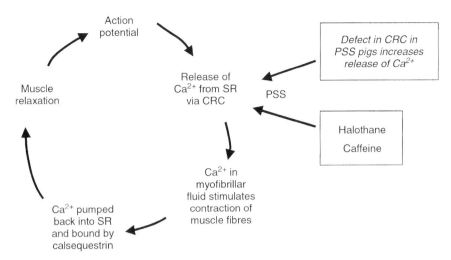

Fig. 3.30. Defect in calcium metabolism leading to porcine stress syndrome (PSS). CRC, Calcium release channel; SR, sarcoplasmic reticulum.

564,743 Da). It is activated by phosphorylation and is under negative feedback control from intracellular Ca^{2+}. It is also inactivated by calsequestrin (a protein that stores Ca^{2+} within the sarcoplasmic reticulum) and the action potential, which causes membrane depolarization and is measured by a voltage sensor that is physically connected to the CRC. Mg^{2+}, calmodulin and repolarization of the membrane that is detected by the voltage sensor inactivate the CRC. Non-physiological activators of the CRC include halothane, ryanodine and caffeine.

The PSS defect is due to a substitution of a single nucleotide (position 1843 of the *ryr-1* gene) that causes a single amino-acid replacement (cysteine instead of arginine) in amino acid 615 in the CRC. The mutation causes the CRC to be more sensitive to biochemical activators, making it easier to open and more difficult to close. This in turn makes skeletal muscle contraction easier to activate and more difficult to relax (Fujii *et al.*, 1991).

Testing for PSS

PSS pigs were first reliably identified by the halothane challenge test, which was developed in 1974. In this test, 2–3-month-old pigs are given 3–5% halothane in oxygen through a face mask. PSS-susceptible pigs develop extensor muscle rigidity when exposed to halothane gas. This is analogous to malignant hyperthermia in humans. PSS is inherited as an autosomal recessive trait and the halothane challenge test is highly sensitive for homozygotes, but only 25% of heterozygotes are detected. The response in the test is increased if the pigs are stressed by fighting or rough handling. Skeletal muscle from PSS-susceptible pigs, as well as from malignant hyperthermia-sensitive people, is more sensitive to drug-induced contraction. This forms the basis for caffeine and halothane contraction tests using muscle biopsy specimens.

Following the development of the halothane challenge test, it was found that PSS was rarely found in Hampshire, Duroc, Large White and Yorkshire breeds, while it was prevalent in Piétrain, Landrace and Poland China pigs. The incidence of PSS found in Landrace pigs at different locations in the 1970s suggested that the defect spread from Germany and Belgium. Haplotype analysis of the *ryr-1* gene suggests that the PSS mutation arose from a single founder animal. It is now found in virtually all other domestic breeds used for intensive pork production worldwide and in more that 25% of the breeding stock in Europe and North America. In the early 1990s, the PSS mutation was found in 97% of Piétrain, 80% of Poland China, 37% of Landrace, 22% of Large White, Duroc and Hampshire and 17% of Yorkshire pigs. The spread of the mutation is due to the relatively small number of breeding pigs that are providing the genetics for the slaughter pigs, the introduction of new breeding stock in genetic improvement programmes and the fact that there is increased carcass lean in animals with the PSS mutation, which resulted in it being selected for in 'improved' pig genotypes. In North America, there has also not been any direct penalty to producers for carcasses that have PSE.

Canadian researchers have developed a PCR-based test that identifies the gene mutation in pigs (Fig. 3.31). DNA is first isolated from blood, muscle, semen or other tissues, with the preferred procedure being a dried blood sample on filter paper. A 659-bp portion of the *ryr-1* gene that includes the mutation is then amplified by PCR and the amplified DNA is subjected to restriction fragment length polymorphism (RFLP) analysis. This involves digesting the DNA with the restriction endonuclease *Bsi*HKAI, which cuts the amplified DNA at a site common to both normal and mutant genes as well as at the mutation site itself. The number and position of DNA fragments that are found after agarose gel electrophoresis is used to identify the PSS genotype (O'Brien *et al.*, 1993).

However, PSS has not been eliminated. The DNA test is costly, therefore some producers do not test their herds. There are no economical incentives for producers to test their herds. PSS pigs that survive to slaughter have higher lean yield, and a higher lean yield is also found in pigs that are heterozygous for the mutation. Thus, it may be advantageous to control the frequency of the mutation, and to raise heterozygous animals rather than to completely eliminate it.

The PSS mutation results in improved carcass characteristics and feed conversion in heterozygous animals. Landrace pigs that are heterozygous for the mutation have 5% less back fat and 2.8% increased lean yield, while heterozygous Yorkshire pigs have 5.7% improved feed conversion, 6.3% decreased back fat and 4.1% increased lean yield compared to non-mutant pigs. The increased lean and decreased fat may be a direct consequence of the increased sensitivity of the muscle to contraction and increased metabolic activity of the muscle in heterozygous pigs. This may act directly as a stimulus for muscle hypertrophy and mobilization of fat in PSS pigs.

The *ryr-1* gene is also expressed in brain, while two other CRC genes, *ryr-2* and *ryr-3*, are expressed in smooth muscle and brain. The PSS mutation does not, therefore, affect cardiac or smooth muscle. The presence of the PSS mutation in the brain may mean that the central nervous system is more sensitive to stress in mutant animals. Regional differences in levels of neurotransmitters in the brain have been reported in affected pigs, but these may be secondary effects from the altered muscle metabolism.

PSS is analogous to malignant hyperthermia (MH) found in humans, dogs and horses. Although MH is due to a single mutation in pigs, there are many different mutations that can cause MH in humans. MH is also 5000- to 25,000-fold less common in humans than in pigs. This is because modern domestic pigs are highly inbred and the PSS mutation has likely been selected for due to the improvements in carcass composition and feed conversion associated with it.

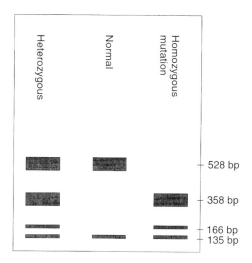

Fig. 3.31. DNA test for porcine stress syndrome (PSS).

Endocrine factors that affect PSS pigs and PSE pork

During stress, secondary responses of the endocrine system occur with adrenaline and cortisol (see also Section 6.3 for a discussion of stress). The adrenal catecholamine, adrenaline is synthesized from noradrenaline in the adrenal medulla. The catecholamines act as short-term response hormones to cause vasoconstriction and other changes for the 'fight or flight' response. There are both α-adrenergic receptors, which lead to vasoconstriction and β-adrenergic receptors, which stimulate vasodilation (see Section 3.5). Arterial and venous systems contain mostly α-receptors, while skeletal and cardiac muscle contain both α- and β-receptors. Noradrenaline stimulates α-receptors and causes vasoconstriction, while adrenaline stimulates both α- and β-receptors. The overall vascular response of an organ to catecholamines thus depends on the relative amounts of adrenaline and noradrenaline and the ratio of α- to β-adrenergic receptors present. In skeletal muscle, high levels of catecholamines result in vasoconstriction.

During the stress response (Fig. 3.32), adrenaline levels in PSS pigs rise quickly and adrenaline acts to increase the activity of glucagon. Glucagon and adrenaline work via a cAMP-dependent pathway to stimulate glycogenolysis, which is the breakdown of glycogen stores. Glycogen is converted to glucose-6-phosphate during glycogenolysis, and during glycolysis, glucose-6-phosphate is converted to pyruvate. After the pig is slaughtered, pyruvate is converted to lactate under anaerobic conditions. Lactate builds up in muscle to cause a dramatic decrease in muscle pH, leading to PSE pork.

Adrenaline causes vasoconstriction and a hypoxic condition results in the muscle of PSS pigs, due the lack of oxygenated blood. This causes lactate to accumulate from anaerobic glycolysis. In a live pig under anaerobic conditions, the excess lactate would be shunted in the blood to the liver. This lactate would then be converted back to glucose; this is known as the Cori cycle. Under aerobic conditions, pyruvate from glycolysis is converted to acetyl-CoA which enters the TCA cycle.

Cortisol is a glucocorticoid steroid hormone produced in the adrenal cortex and is a

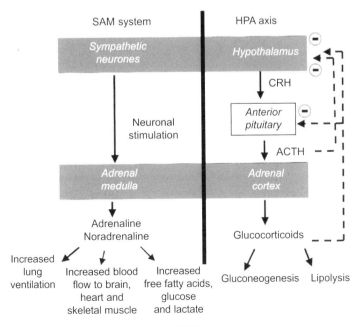

Fig. 3.32. Hormonal and metabolic responses to stress.

long-term stress hormone. Its major target tissue is the liver, and other important target tissues are the lymphoid cells, thymus gland and kidney. Cortisol affects carbohydrate metabolism by promoting gluconeogenesis from body proteins, enhancing fatty acid mobilization and oxidation, increasing plasma cholesterol and triacylglycerol levels and glycogen deposition in the liver. This affects water and electrolyte distribution in tissues by opposing water shift into the cells.

Gonadotrophin secretion is also affected, as cortisol inhibits GnRH stimulation of LH release. In the gastrointestinal tract, prolonged high levels of cortisol encourage the development of ulcers. The inhibition of prostaglandin synthesis is possibly involved in this process. Cortisol also has effects on immunity and inflammation (see Section 6.3). Cortisol is involved in placental endocrine function. Maternal cortisol levels are higher and the placenta contains cortisol receptors. Cortisol promotes the secretion of chorionic gonadotrophin, which is a hormone required for the maintenance of pregnancy. However, increased production of cortisol by the fetus acts as a trigger for parturition (see Section 5.1).

The effects of cortisol are more pronounced in PSS pigs. When the pig is stressed, cortisol levels rise. Because of the problems with calcium sequestration in PSS pigs, the animal's muscles continue to contract. Cortisol levels remain high for extended periods and this causes the proliferation of gluconeogenesis from body proteins. When a pig is stressed during transportation or upon entry into the abattoir, cortisol levels will be high. Cortisol has permissive effects on catecholamines; therefore adrenaline will more readily convert glycogen to glucose. Upon death, glucose will be utilized by anaerobic glycolysis to produce lactate in the muscles, leading to PSE meat.

Thyroid hormones may also play a role in the development of PSE meat. Thyroid hormones normally stimulate heat production by increasing aerobic metabolism. Removing the thyroid glands from pigs decreases the rate of post-mortem glycolysis, while supplementing pigs with thyroid hormones increases the rate of pH decline in the muscles.

Manipulations to reduce the incidence of PSE

Reducing stress prior to slaughter will reduce the incidence of PSE meat. Adequate pen space should be provided in the holding pens at the slaughter plant (0.6 m^2 per 115 kg pig). All pigs must have room to lie down, and during hot weather the animals can be wet down with sprinklers. Two to 4 hours of rest should be allotted prior to stunning. Animals should be handled and driven quietly, with minimal electric prod usage. Overcrowding and exposure to unfamiliar animals should be avoided. The animals should have room to turn so they can enter the race more easily. Pigs must always have access to water. Supplementing animals with electrolytes and feeding a high-quality protein such as casein may also reduce the stress response.

Rapid cooling of the carcass would reduce the formation of PSE conditions, since it would reduce the denaturation of sarcoplasmic proteins caused by the excessive heat from the rapid metabolism. Injection of sodium bicarbonate has been shown to reduce the rate or extent of post-mortem pH decline. Injection of 10% by weight of 0.3 M sodium bicarbonate at 15 min after death improved colour, reduced drip loss and increased the ultimate pH of longissimus and biceps femoris muscles. Including 0.7% NaCl along with the sodium bicarbonate improved juiciness and flavour. Injection at 24 h after death improved drip loss but not the colour defect (Kauffman *et al.*, 1998).

Summary and Conclusions

In this chapter, a number of methodologies have been described for manipulation of the growth rate, carcass composition and meat quality of meat animals. Some of these methods, such as the use of anabolic steroids and antimicrobials, have been used for many years to improve growth rate, performance and lean yield. The use of uncastrated male pigs has also been used for a long time in a few countries, but has yet to be a universally accepted procedure for pig production. Other technologies, such as the use of somatotrophin or β-agonists and reducing the incidence of PSS, have been used for a shorter period of time. Other endocrine systems, such as leptin, control of appetite with CCK antagonists, and use of various dietary PUFAs are only now being considered for use. A thorough understanding of the underlying biology is necessary for development of effective methods by which to manipulate these endocrine systems. In a number of cases, this has allowed the development of specific receptor agonists or antagonists that affect only the appropriate components of the endocrine system, to produce the desired results without adverse side-effects.

Questions for Study and Discussion

Section 3.2 Anabolic Steroids

1. Describe the use of anabolic steroids in meat animal production and in horse racing.
2. Describe methods used to distinguish between the anabolic and androgenic effects of various steroids. Give examples of structural changes that have been used to optimize these effects.
3. Describe the direct and indirect mechanisms of action of anabolic steroids.
4. Discuss potential safety issues with the use of anabolic steroids.

Section 3.3 Use of Intact Male Pigs

1. Discuss the advantages and disadvantages of raising intact male pigs for pork production.
2. Describe the metabolism of the boar taint compounds, androstenone and skatole.
3. Discuss potential methods to control boar taint.

Section 3.4 Somatotrophin

1. Outline the effects of ST on carcass composition.
2. Discuss the control of pulsatile ST release.
3. Describe the direct metabolic effects of ST.
4. Describe the indirect effects of ST via IGF-I.
5. Outline potential methods to manipulate ST.

Section 3.5 β-Adrenergic Agonists

1. Describe the role of the sympathetic nervous system and catecholamines.
2. Describe the types of β-AA receptors and their interaction with various β-AA compounds.
3. Outline the metabolic effects of β-AA that result in changes in carcass composition.
4. Comment on the safety concerns for β-AA use.

Section 3.6 Thyroid Hormones

1. Describe the synthesis and metabolism of thyroid hormones.
2. Discuss the metabolic effects of thyroid hormones.
3. Describe the role of thyroid hormones in growth and development.

Section 3.7 Dietary PUFA

1. Discuss the effects of PUFA and their metabolites on gene expression.
2. Describe the metabolic pathways of LA, ALA and GLA leading to eicosanoid production.
3. Outline the metabolic effects of CLA isomers.
4. Describe the potential applications of dietary PUFA.

Section 3.8 Leptin

1. Describe the mechanism of action of leptin and the model systems used.
2. Outline the role of leptin as an adipostat.
3. Describe the effect of leptin on uncoupling proteins and heat generation.
4. Describe the role of leptin in regulation of reproduction.

Section 3.9 CCK and Appetite

1. Describe the structure of CCK, gastrin and their receptors.
2. Describe the role of CCK and NPY in regulating appetite.
3. Outline potential applications of CCK in controlling feed intake.

Section 3.11 Dietary Chromium

1. Describe the role of chromium in the glucose tolerance factor and its effects on insulin activity.
2. Outline the potential physiological roles of dietary chromium.
3. Discuss the effective dose and potential safety concerns of dietary chromium.

Section 3.12 Effects of Stress on Meat Quality

1. Discuss the post-mortem factors that lead to the development of PSE and DFD meat.
2. Describe the genetic defect leading to PSS and testing procedures for PSS-susceptible pigs.
3. Discuss the endocrine responses that lead to PSE.

Further Reading

General

Hocquette, J.F., Ortigues-Marty, I., Pethick, D., Herpin, P. and Fernandez, X. (1998) Nutritional and hormonal regulation of energy metabolism in skeletal muscles of meat-producing animals. *Livestock Production Science* 56, 115–143.

Hornick, J.L., Van Eenaeme, C., Gerard, O., Dufrasne, I. and Istasse, L. (2000) Mechanisms of reduced and compensatory growth. *Domestic Animal Endocrinology* 19, 121–132.

Mersmann, H.J. (1987) Nutritional and endocrinological influences on the composition of animal growth. *Progress in Food and Nutrition Science* 11, 175–201.

National Research Council (1994) Mechanisms of action of metabolic modifiers. In: *Metabolic Modifiers: Effects on the Nutrient Requirements of Food-producing Animals.* NRC, Washington, DC, pp. 5–22. Available at: www.nap.edu/openbook/0309049970/html

Preston, R.L. (1999) Hormone containing growth promoting implants in farmed live stock. *Advanced Drug Delivery Reviews* 38, 123–138.

Sejrsen, K. and Vestergaard-Jensen, M. (1990) *An Overview of Endocrine Mechanisms Important to Regulation of Metabolism and Growth in Farm Animals.* Cornell Nutrition Conference, Cornell University, Ithaca.

Anabolic steroids

Cosman, F. and Lindsay, R. (1999) Selective estrogen receptor modulators: clinical spectrum. *Endocrine Reviews* 20, 418–434.

Houghton, E., Ginn, A., Teale, P., Dumasia, M.C. and Moss, M.S. (1986) Detection of the administration of anabolic preparations of nandrolone to the entire male horse. *Equine Veterinary Journal* 18, 491–493.

Lone, K.P. (1997) Natural sex steroids and their xenobiotic analogs in animal production: growth, carcass quality, pharmacokinetics, metabolism, mode of action, residues, methods, and epidemiology. *Critical Reviews in Food Science and Nutrition* 37, 93–209.

Negro-Vilar, A. (1999) Selective androgen receptor modulators (SARMs): a novel approach to androgen therapy for the new millennium. *Journal of Clinical Endocrinology and Metabolism* 84, 3459–3462.

Sheffield-Moore, M. (2000) Androgens and the control of skeletal muscle protein synthesis. *Annals of Medicine* 32, 181–186.

Snow, D.H. (1993) Drug use in performance horses – anabolic steroids. *Veterinary Clinics of North America: Equine Practice* 9, 563–576.

Unruh, J.A. (1986) Effects of endogenous and exogenous growth-promoting compounds on carcass composition, meat quality and meat nutritional value. *Journal of Animal Science* 62, 1441–1448.

Waltner-Toews, D. and McEwen, S.A. (1994) Residues of hormonal substances in foods of animal origin; a risk assessment. *Preventive Veterinary Medicine* 20, 235–247.

Intact male pigs

Annor-Frempong, I.E., Nute, G.R., Wood, J.D., Whittington, F.W. and West, A. (1998) The measurement of the responses to different odour intensities of 'boar taint' using a sensory panel and an electronic nose. *Meat Science* 50, 139–151.

Babol, J. and Squires, E.J. (1995) Quality of meat from entire male pigs. *Food Research International* 28, 210–212.

Bonneau, M. (1998) Use of entire males for pig meat in the European Union. *Meat Science* 49, S257–S272.

Bonneau, M., Walstra, P., Claudi-Magnussen, C., Kempster, A.J., Tornberg, E., Fischer, K., Diestre, A., Siret, F., Chevillon, P., Claus, R., Dijksterhuis, G.B., Punter, P., Matthews, K.R., Agerhem, H., Béauge, M.P., Oliver, M.A., Gisper, M., Weiler, U., von Seth, G., Leask, H., Font i Furnols, M., Homer, D.B. and Cook, G.L. (2000) An international study on the importance of androstenone and skatole for boar taint. IV. Simulation studies on consumer dissatisfaction with entire male pork and the effect of sorting out carcasses on the slaughter line, main conclusions and recommendations. *Meat Science* 54, 285–295.

Cooke, G.M., Pothier, F. and Murphy, B.G. (1997) The effects of progesterone, 4,16-androstadien-3-one and MK-434 on the kinetics of pig testis microsomal testosterone-4-ene-5α-reductase activity. *Journal of Steroid Biochemistry and Molecular Biology* 60, 353–359.

Diaz, G.J. and Squires, E.J. (2000) Metabolism of 3-methylindole by porcine liver microsomes: responsible cytochrome P450 enzymes. *Toxicological Sciences* 55, 284–292.

Dunshea, F.R., Colantoni, C., Howard, K., McCauley, I., Jackson, P., Long, K.A., Lopaticki, S., Nugent, E.A., Simons, J.A., Walker, J. and Hennessy, D.P. (2001) Vaccination of boars with a GnRH vaccine (Improvac) eliminates boar taint and increases growth performance. *Journal of Animal Science* 79, 2524–2535.

Fouilloux, M.N., Le Roy, P., Gruand, J., Renard, C., Sellier, P. and Bonneau, M. (1997) Support for single major genes influencing fat androstenone level and development of bulbo-urethral glands in young boars. *Genetics, Selection, Evolution* 29, 357–366.

Marchese, S., Pes, D., Scaloni, A., Carbone, V. and Pelosi, P. (1998) Lipocalins of the boar salivary glands binding odours and pheromones. *European Journal of Biochemistry* 252, 563–568.

Solé, M.A.R. and García-Regueiro, J.A. (2001) Role of 4-phenyl-3-buten-2-one in boar taint: identification of new compounds related to sensorial descriptors in pig fat. *Journal of Agriculture and Food Chemistry* 49, 5303–5309.

Turkstra, J.A., Zeng, X.Y., van Diepen, J.T.M., Jongbloed, A.W., Oonk, H.B., van de Wiel, D.F.M. and Meloen, R.H. (2002) Performance of male pigs immunized against GnRH is related to the time of onset of the biological response. *Journal of Animal Science* 80, 2953–2959.

Willeke, H. (1993) Possibilities of breeding for low 5α-androstenone content in pigs. *Pig News and Information* 14, 31N–33N.

Somatotrophin

Etherton, T.D. and Smith, S.B. (1991) Somatotropin and β-adrenergic agonists; their efficacy and mechanisms of action. *Journal of Animal Science* 69(Suppl. 2), 2–26.

He, P., Aherne, F.X., Nam, D.S., Schaefer, A.L., Thompson, J.R. and Nakano, T. (1994) Effects of recombinant porcine somatotropin (rpST) on joint cartilage and axial bones in growing and finishing pigs. *Canadian Journal of Animal Science* 74, 257–263.

Müller, E.E., Locatrelli, V. and Cocchi, D. (1999) Neuroendocrine control of growth hormone secretion. *Physiological Reviews* 79, 511–607.

Smith, R.G., Pong, S., Hickey, G., Jacks, T., Cgeng, K., Leonard, R., Cohen, C.J., Arena, J.P., Chang, C.H., Drisko, J., Wyvratt, M., Fisher, M., Nargund, R. and Patchett, A. (1996) Modulation of pulsatile GH release through a novel receptor in hypothalamus and pituitary gland. *Recent Progress in Hormone Research* 51, 261–286.

Smith, R.G., Leonard, R., Bailey, A.R.T., Palyha, O., Feighner, S., Tan, C., McKee, K.K., Pong, S.-S., Griffin, P. and Howard, A. (2001) Growth hormone secretagogue receptor family members and ligands. *Endocrine* 14, 9–14.

Vasilatos-Younken, R. (1995) Proposed mechanisms for the regulation of growth hormone action in poultry: metabolic effects. *Journal of Nutrition* 125, 1783S–1789S.

β-Adrenergic agonists

Kuiper, H.A., Noordam, M.Y., van Dooren-Flipsen, M.M.H., Schilt, R. and Roos, A.H. (1998) Illegal use of β-adrenergic agonists: European Community. *Journal of Animal Science* 76, 195–207.

Mersmann, H.J. (1998) Overview of the effects of β-adrenergic receptor agonists on animal growth including mechanisms of action. *Journal of Animal Science* 76, 160–172.

Mitchell, G.A. and Dunnavan, G. (1998) Illegal use of β-adrenergic agonists in the United States. *Journal of Animal Science* 76, 208–211.

Smith, D.J. (1998) The pharmacokinetics, metabolism, and tissue residues of β-adrenergic agonists in livestock. *Journal of Animal Science* 76, 173–194.

Thyroid hormones

Jianhua, H., Ohtsuka, A. and Hayashi, K. (2000) Selenium influences growth via thyroid hormone status in broiler chickens. *British Journal of Nutrition* 84, 727–732.

Köhrle, J. (2000) The deiodinase family: selenoenzymes regulating thyroid hormone availability and action. *Cellular and Molecular Life Sciences* 57, 1853–1863.

Wrutniak-Cabello, C., Casas, F. and Cabello, G. (2001) Thyroid hormone action in mitochondria. *Journal of Molecular Endocrinology* 26, 67–77.

Zhang, J. and Lazar, M.A. (2000) The mechanism of action of thyroid hormones. *Annual Review of Physiology* 62, 439–466.

Dietary polyunsaturated fatty acids

Clarke, S.D. (2001) Polyunsaturated fatty acid regulation of gene transcription: a molecular mechanism to improve the metabolic syndrome. *Journal of Nutrition* 131, 1129-1132.

Duplus, E., Glorian, M. and Forest, C. (2000) Fatty acid regulation of gene transcription. *Journal of Biological Chemistry* 275, 30749-30752.

Pariza, M.W., Park, Y. and Cook, M.E. (2000) Mechanisms of action of conjugated linoleic acid: evidence and speculation. *Proceedings of the Society for Experimental Biology and Medicine* 223, 8-13.

Parodi, P.W. (1999) Conjugated linoleic and other anticarcinogenic agents of bovine milk fat. *Journal of Dairy Science* 82, 1339-1349.

Sessler, A.M. and Ntambi, J.M. (1998) Polyunsaturated fatty acid regulation of gene expression. *Journal of Nutrition* 128, 923-926.

Wu, D. and Meydani, S.N. (1996) γ-linolenic acid and immune function. In: Huang, Y.-S. and Mills, D.E. (eds) *γ-Linolenic Acid Metabolism and its Roles in Nutrition and Medicine*. AOCS Press, Champaign, Illinois, pp. 106-117.

Leptin

Ahima, R.S. and Flier, J.S. (2000) Leptin. *Annual Review of Physiology* 62, 413–437.

Fried, S.K., Ricci, M.R., Russell, C.D. and Laferrere, B. (2000) Regulation of leptin production in humans. *Journal of Nutrition* 130, 3127S–3131S.

Hossner, K.L. (1998) Cellular, molecular and physiological aspects of leptin: potential application in animal production. *Canadian Journal of Animal Science* 78, 463–472.

Houseknecht, K.L., Baile, C.A., Matteri, R.L. and Spurlock, M.E. (1998) The biology of leptin: a review. *Journal of Animal Science* 76, 1405–1420.

Cholecystokinin and appetite

Bernad, N., Burgaud, B.G.M., Horwell, D.C., Lewthwaite, R.A., Martinez, J. and Pritchard, M.C. (2000) The design and synthesis of the high efficacy, non-peptide CCK1 receptor agonist PD 170292. *Bioorganic and Medicinal Chemistry Letters* 10, 1245–1248.

Bray, G.A. (2000) Afferent signals regulating food intake. *Proceedings of the Nutrition Society* 59, 373–384.

McCauley, I., Billinghurst, A., Morgan, P.O. and Westbrook, S.L. (1995) Manipulation of endogenous hormones to increase growth of pigs. In: *Manipulating Pig Production V. Proceedings of the Fifth Biennial Conference of the Australasian Pig Science Association*, pp. 52–61.

Dietary chromium and insulin

Borgs, P. and Mallard, B.A. (1998) Immune–endocrine interactions in agricultural species: chromium and its effect on health and performance. *Domestic Animal Endocrinology* 15, 431–438.

NRC (1997) *The Role of Chromium in Animal Nutrition*. National Academy Press, Washington, DC. Available at: www.nap.edu/openbook/030906354x/html

Nystrom, F.H. and Quon, M.J. (1999) Insulin signalling: metabolic pathways and mechanisms for specificity. *Cellular Signalling* 11, 563–574.

Effects of stress on meat quality

Fujii, J., Otsu, K., Zorzato, F., De Leon, S., Khanna, V.K., Weiler, J.E., O'Brien, P.J. and MacLennan, D.H. (1991) Identification of a mutation in porcine ryanodine receptor associated with malignant hyperthermia. *Science* 253, 448–451.

Kauffman, R.G., van Laack, R.L.J.M., Russell, R.L., Pospiech, E., Cornelium, C.A., Suckow, C.E. and Greaser, M.L. (1998) Can pale, soft, exudative pork be prevented by postmortem sodium bicarbonate injection? *Journal of Animal Science* 76, 3010–3015.

O'Brien, P.J., Shen, H., Cory, C.R. and Zhang, X. (1993) Use of DNA-based test for the mutation associated with porcine stress syndrome (malignant hyperthermia) in 10,000 breeding swine. *Journal of the American Veterinary Medical Association* 203, 842–851.

Swatland, H.J. (1994) The conversion of muscles to meat. In: *Structure and Development of Meat Animals and Poultry*. Technomic Publishing Co., Lancaster, pp. 495–599.

4
Endocrine Effects on Animal Products*

4.1 Mammary Gland Development and Milk Production

Introduction

The endocrine effects on milk production are important in the growth and development of the mammary gland (mammogenesis), the initiation of lactation (lactogenesis) and the maintenance of lactation (galactopoiesis). Metabolic diseases related to lactation, such as ketosis and milk fever, are also affected by endocrine factors.

The mammary gland or udder is a skin gland located outside the abdominal cavity. In cows, the right and left sides of the udder are entirely separate, with no common blood supply. There is also no internal crossover of the milk duct system among the individual udder quarters, so that substances injected into the teat and duct system of one quarter will only be found in that quarter. Sheep and goats do not have the four quarters of the udder that cows have.

Milk is synthesized in small, sack-like structures called alveoli (Fig. 4.1). Groups of alveoli are organized into lobules, and groups of lobules are organized into lobes. The epithelial lining of the alveolus is surrounded by myoepithelial cells, which contract in response to the hormone oxytocin to squeeze the milk out into small ducts. The small ducts connect with larger ducts, which deliver the milk to the cistern within the gland for storage. There is no direct innervation of the secretory system to affect milk delivery.

Mammary gland development

The number of milk-synthesizing cells is one of the major factors that determines the milk yield capacity of the mammary gland. Factors that affect mammary development during rearing can affect the number of milk-producing cells and thus affect the future milk yield capacity of the mature cow.

Mammary gland development is related to reproductive development, which occurs in distinct phases during fetal life, puberty, pregnancy and lactation (Fig. 4.2). At birth, the non-epithelial tissues and basic structures of the mammary gland, including the connective tissue and blood and lymph vessels, have already formed. However, the secretory and glandular tissue is still rudimentary. At 2–3 months of age until shortly after puberty is completed in cattle, the mammary gland undergoes an **allometric** growth phase in which it grows faster than other parts of the body. During this period, the fat pad and mammary ducts grow rapidly but no alveoli are formed. At puberty, the mammary glands of heifers weigh about 2–3 kg, with 0.5–1 kg of parenchymal tissue that consists of 10–20% epithelial cells, 40–50% connective tissue and 30–40% fat cells. Increased growth rate due to feeding high-energy diets near puberty can

---

*This chapter deals with hormonal systems that can affect the production of animal products other than meat, namely milk, eggs and wool.

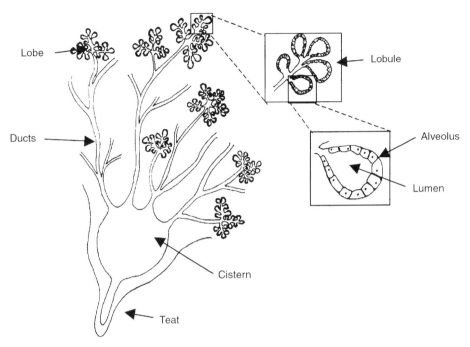

Fig. 4.1. Diagram of mammary gland structure.

lead to permanently reduced milk yield potential. This is due to excess fat deposition in the mammary gland and decreased growth of parenchyma. This may not occur when a high-energy and high-protein diet is fed, so that dietary protein does not limit mammary gland development. Higher body-weight gain after puberty and during pregnancy has no effect on mammary growth and potential milk yield. For more information, see Sejrsen et al. (2000).

Very little mammary growth occurs after puberty until pregnancy. In early pregnancy, the mammary ducts grow and there is extensive lobulo-alveolar development from mid-pregnancy. The growth rate is exponential, with the majority of the growth occurring in the last trimester of pregnancy. The growth and development during this period determine the number of milk-secreting cells and thus the extent of milk production. Cell numbers continue to increase even after parturition. Increased nursing intensity increases mammary growth during lactation. The lactating mammary gland in dairy cattle weighs 15–25 kg, with the parenchyma consisting of

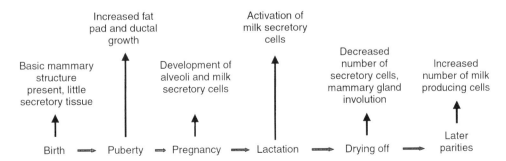

Fig. 4.2. Mammary gland changes during development.

40–50% epithelial cells as ducts and alveoli, 15–20% lumen, 40% connective tissue and almost no fat cells.

During lactogenesis, the alveolar secretory cells are activated. The amount of rough endoplasmic reticulum, Golgi and mitochondria increases in alveolar cells. The activities of enzymes associated with lactation, such as acetyl-CoA carboxylase and fatty acid synthase, are increased, along with the transport systems of the substrates for milk synthesis, including amino acids and glucose. Casein micelles and fat droplets accumulate in the cytoplasm and are released into the lumen of the alveolus.

There is a two- to sixfold increase in blood flow to the mammary gland, starting 2–3 days postpartum. However, the decrease in production that normally occurs later in lactation is not due to decreased blood flow.

Mammary duct development increases by around 8% in each oestrous cycle and then regresses during metoestrus and dioestrus. After the first pregnancy, cell numbers can be higher for subsequent lactations. Milk yield increases until the cow is 8 years old, with mature cows typically producing 25% more milk than 2-year-old heifers. For more information on mammary cell changes during pregnancy and lactation, see Knight and Wilde (1993).

Involution and the dry period

As lactation progresses in dairy cattle, the number of mammary cells decreases due to apoptosis, resulting in decreased milk yield. Pregnancy also decreases milk yield, possibly due to the effects of oestrogen produced by the placenta on mammary function. The decrease in milk yield eventually results in an end to lactation and mammary gland involution. In contrast with other milk proteins, the synthesis of lactoferrin is increased during involution; lactoferrin may act as a non-specific disease-resistance factor. The level of the lysosomal enzyme N-acetyl-β-D-glucosaminidase also increases in the mammary gland during involution. There is a decrease in the size of the alveoli and lumen during involution, but general alveolar structure is maintained in bovine mammary gland. In rodents, there is extensive tissue degeneration and disintegration of the alveolar structure through apoptosis during mammary gland involution.

A dry period is required between the end of lactation and calving so that milk yield in the next lactation will not be reduced. To induce the dry period, the amount of grain and water supplied to the cow is first limited to reduce milk production, and then milking is stopped 45–50 days before the expected date of parturition. A shorter dry period significantly reduces subsequent milk production. The early dry period is the time of the highest incidence of new intramammary infections. This may be due to the presence of a large volume of milk in the gland that is conducive to bacterial growth, leakage from the teats or the lack of teat-end disinfection. The concentration of total leucocytes increases in mammary secretions during this period.

About 3–4 weeks prepartum, the mammary gland begins the transition from the non-lactating to the lactating state. Transport of IgG into the colostrum begins 2 weeks prior to parturition and the concentrations of the major milk components increase dramatically 3–5 days prepartum.

Model systems for studying mammary gland development and function

In vitro *cell culture systems*

Bovine mammary epithelium cells can be isolated and cultured in a collagen matrix to form a three-dimensional structure. Mammary tissue can be obtained from pregnant cows, minced and dispersed in a mixture of collagenase and hyaluronidase. The epithelial cells are purified by density gradient centrifugation. Alternately, a cloned mammary cell line can be used, such as the MAC-T3 (mammary alveolar cell transfected with large T antigen).

Whole animal studies

An *in vivo* model for studying endocrine effects on the mammary gland is the method of close arterial infusion into the mammary gland. This can be accomplished without general anaesthesia or deep surgical dissection, which minimizes the stress to the animal. The arterial supply to the two contralateral sides of the mammary gland can be catherized separately, so that one side of the gland can be used as the treatment and the other side for the control.

The catheter is placed into the saphenous artery and inserted up to the external iliac artery, which supplies the ipsilateral mammary glands via the external pudendal artery, and the hind limb via the femoral artery. The position of the catheter is determined by real-time ultrasound imaging. Evans blue dye injected through the catheter appears exclusively in the ipsilateral vein (Maas *et al.*, 1995).

Test compounds can also be infused into one half of the udder via the streak canal. The contralateral half of the udder is the control and receives the vehicle alone. This method tests the activity of compounds that are infused into the internal milk secretory system of the udder rather than delivered to the mammary gland via the blood.

Mammary gland function can be assessed by biopsy of mammary tissue from the live animal or by imaging the mammary tissue using magnetic resonance imaging (MRI).

Hormones and mammary gland development (mammogenesis)

A number of autocrine and paracrine factors are important in mammary growth and development (Table 4.1). Mammary epithelial cells grow into the fat pad, and the developing structures will only be formed when the fat pad is present. This may involve specific fatty acids from the fat pad.

Oestrogen stimulates mammary duct growth, and oestrogen and progesterone act synergistically to stimulate lobule–alveolar development. Progesterone is elevated throughout gestation and oestrogen is particularly elevated during the latter half of pregnancy. Thus, mainly ductal and lobular growth occur during the first half of pregnancy, with lobule–alveolar growth occurring during the second half of pregnancy. Prolactin and growth hormone (ST) are needed for the steroid hormones to be effective. Prolactin binds to a receptor on the mammary epithelial cell, which induces dimerization of the receptor and activates Janus kinase 2 (JAK2). JAK2 phosphorylates and activates members of the STAT family of transcription factors (see Section 1.3), which then activate certain genes involved in differentiation of terminal end buds of the mammary ducts. Levels of prolactin, ST and insulin decrease during gestation.

Placental lactogen is a peptide hormone produced by the placenta and is structurally related to prolactin and/or ST, depending on the species. It binds to the prolactin receptor in rodents to stimulate mammogenesis during pregnancy. Levels of placental lactogen in the cow are very low in maternal blood, so it has little effect in cattle.

Mammary-derived growth inhibitor (MDGI) is a 14.5 kDa protein that is produced by mammary epithelial cells. It acts in an autocrine manner to inhibit cell growth and especially to induce differentiation in mammary epithelium. MDGI levels increase in mammary gland 2 weeks prior to parturition and are high in lactating dairy cows. MDGI

Table 4.1. Hormone effects on mammary development and milk production.

Stage of development	Stimulated by	Inhibited by
Mammary development	Oestrogen/progesterone, prolactin, placental lactogen, ST/IGF-I, EGF, FGF, TGFα, MCSF, HGF, Crypto-1	MDGI, TGFβ
Lactogenesis	Prolactin, oestrogen, glucocorticoids, insulin	Progesterone
Galactopoiesis	Thyroid hormones, prolactin, milk removal, ST	Glucocorticoids, oestrogen

Crypto-1 (or Cr-1) is also known as teratocarcinoma-derived growth factor (TDCF-1).

has been shown to be the same as the fatty acid binding protein isolated from heart (H-FABP). FABP is important in intracellular transport and metabolism of fatty acids, cell differentiation and signal transduction.

Epidermal growth factor (EGF) stimulates mammogenesis, but not in cattle. Fibroblast growth factors (FGFs) may be involved in the growth of mammary stromal cells such as fibroblasts. Transforming growth factor-α (TGFα), hepatocyte growth factor (HGF) and macrophage colony stimulating factor (MCSF) have also been implicated in stimulating mammogenesis. TGFβ inhibits mammogenesis in mice and may also be effective in cattle. See Section 4.3 for more information of the EGF family of growth factors.

ST stimulates the growth of ducts during mammary development near puberty and lobulo-alveolar growth during pregnancy. Injection of ST between 8 and 16 months of age increases growth of the parenchyma and total mammary cell numbers in cattle. This occurs via the action of insulin-like growth factor-I (IGF-I) produced either by the liver or locally in the mammary stroma. The negative effects of high feeding level on mammary growth near puberty may be due to increased local production of IGFBP-3, which binds IGF-I to inhibit it.

Oestrogen stimulates proliferation of stromal cells in the mammary gland and stromal cells produce IGF-I. Insulin has little effect on mammogenesis *in vivo*, but administration of very high levels of insulin can mimic the effects of IGF-I. Cortisol targets the endoplasmic reticulum and Golgi apparatus in the differentiation of the lobule–aveolar system in cattle. This is necessary so that prolactin can later induce the synthesis of milk proteins. For a review of hormonal effects on mammary growth and lactation, see Tucker (2000).

Hormones and initiation of lactogenesis

There is considerable species variability in the effects of hormones on **lactogenesis** (the start of milk production) and **galactopoiesis** (the maintenance of milk production).

Prolactin, oestrogen and glucocorticoids initiate lactation, provided there is a well-developed lobule–alveolar system. Oestrogen stimulates the release of prolactin from the anterior pituitary and increases the number of prolactin receptors in mammary cells. There is a surge of prolactin several hours before parturition. If this surge is blocked with bromocriptine, milk yield is reduced. Prolactin increases the translation of milk protein mRNAs, swelling of Golgi membranes and milk protein secretion, along with synthesis of lactose and milk fat.

Injections of glucocorticoids into non-lactating cows with well-developed lobule–alveolar systems induce the onset of lactation. Milk production is increased if prolactin is also present, and there is a synergy between glucocorticoids and prolactin in initiating lactation. Glucocorticoids bind to receptors in mammary tissue to increase the development of the rough endoplasmic reticulum and other ultrastructural changes to increase the secretion of α-lactalbumin and β-casein. Binding to the corticosteroid binding globulin (CBG) reduces the activity of glucocorticoids in serum. During the peripar-turient period, levels of the CBG decrease and free glucocorticoid levels increase, which might explain the lactogenic effects of glucocorticoids. Glucocorticoids also suppress the immune system, which may contribute to the increased incidence of mastitis during early lactation.

Insulin is important in stimulating glucose uptake and the expression of milk protein genes required for lactogenesis. IGF-I also stimulates cell division leading up to lactogenesis.

The importance of the lactogenic complex of the hormones insulin, prolactin and glucocorticoids has been demonstrated using *in vitro* culture of bovine mammary tissue. The specific hormones that are involved in regulating lactogenesis vary among different species.

Progesterone inhibits the initiation of lactogenesis, but has no effect on milk yield once lactation is established. Progesterone has been suggested to work by increasing the mammary threshold to prolactin, by altering the secretion of prolactin from the pituitary or acting as a glucocorticoid receptor antagonist. The level of progesterone is high during gestation and serves to inhibit lactogenesis until just before parturition. The level decreases about 2 days before parturition to remove the inhibition of milk synthesis.

ST does not appear to be involved in the onset of lactation in cattle. ST was not lactogenic when added to slices of mammary tissue cultured *in vitro* or when given to late pregnant cattle in the dry period.

Lactation can be induced hormonally in non-pregnant cattle that have been dry for at least 30 days. They are injected for 7 days with high levels of oestrogen plus progesterone, followed by increasing prolactin and finally treatment with glucocorticoids. After 3 weeks the cows are milked, since milk removal is important in stimulating mammary growth and inducing lactation. This method is not effective in all cows and the level of milk production is lower than if the cow gave birth and started lactation naturally.

Milking different quarters of the udder in dairy cattle at different times prepartum changes the composition of the milk from each quarter at parturition. This suggests that local factors within each quarter of the gland also affect the final activation of milk secretion. The matrix in which the cells are grown also affects mammary cellular functions. The basement membrane is a component of the extracellular matrix and contains proteins such as collagen, laminin and proteoglycans. Epithelial cells in all tissues are in contact with the basement membranes, which allows the cells to develop polarity and secrete milk components at the apical surface of the cell. Mammary cells grown in tissue culture on a collagen matrix can also be induced to form three-dimensional structures resembling alveoli.

Maintenance of lactation (galactopoiesis)

Hormonal effects

Thyroid hormones are required for maximal milk production. During lactation there is decreased conversion of thyroxine (T_4) to the active hormone triiodothyronine (T_3) in liver and kidney, but increased conversion to T_3 in the mammary gland. This would enhance the priority of the mammary gland for metabolites compared to other body tissues. Surgical thyroidectomy or treatment with radioactive iodine, which is sequestered in the thyroid gland to destroy thyroid function, decreases milk yield. Administration of thyroid hormones or iodinated casein (thyroprotein) causes a temporary increase in milk production. However, milk yield after treatment with iodinated casein is below normal, so there is no net benefit.

High levels of exogenous glucocorticoids and oestrogen decrease milk production in an established lactation. Physiological levels of glucocorticoids stimulate milk production in rats, but the results are ambiguous in ruminants. There is no change in CBG during lactation in cattle, and corticoids are not limiting to milk yield.

Prolactin is required for the maintenance of milk production in rats and rabbits, with decreases in milk yield of 50% or more after bromocriptine administration. Decreasing prolactin by bromocriptine administration is much less effective in ruminants, except for the lactating ewe. Suckling induces a threefold release of prolactin over 30 min after milking, which is much less dramatic than the peripartum prolactin surge associated with lactogenesis. Milking is thought to increase prolactin secretion by decreasing prolactin inhibiting factor (PIF) release from the hypothalamus. There is a correlation between milk yield and prolactin in blood 5 min after milking, but milk yield and prolactin levels before or 1 h after milking were not correlated.

Prolactin levels increase with increasing temperature and photoperiod. Prolactin levels are also affected by stress, so blood samples must be taken carefully. Increasing the photoperiod from 8 to 16 h day^{-1} increased milk yield by 6–10% in cattle. Increasing photoperiod also stimulated release of prolactin and IGF-I, but the galactopoietic effect of these hormones has not been proven.

Cows can be pregnant and lactating at the same time. Administration of bovine placental lactogen increases milk yield, but little endogenous placental lactogen is normally released into the maternal circulation, so it will usually have a nominal effect.

Milk removal

Milk removal from the mammary gland is necessary for the maintenance of lactation. Milk secretion rate is less affected by long milking intervals in high-producing dairy cows than in low-producing cows. Increasing milking frequency from 2 to 3 or 4 times per day can increase milk yield to a similar extent as treatment with bST (see below). The rate of milk secretion increases only in the part of the gland that is milked more frequently. Since all sides of the udder are exposed to the same systemic hormones, this suggests that local autocrine factors within the individual parts of the mammary gland affect milk production.

The increased milk production is not simply due to manipulation of the gland, but requires actual removal of the milk. A 7.6 kDa milk whey protein isolated from goats' milk has been proposed as a feedback inhibitor of lactation. This protein acts in an autocrine manner to reduce the rate of milk secretion, stimulate the degradation of newly synthesized casein and reduce prolactin receptor numbers on mammary epithelial cells.

Acute accumulation of milk in the mammary gland also increases intra-mammary pressure, which activates sympathetic nerves to decrease mammary blood flow and limit the supply of hormones and nutrients to the gland.

Effect of bST

The effects of ST on growth and carcass composition are discussed in Section 3.4. The development of recombinant DNA technology made possible large-scale production of bST for use in improving the efficiency of milk production in dairy cattle. ST is not galactopoietic in rats but it is the major galactopoietic hormone in cattle. Blood levels of ST decrease with advancing lactation. Recombinant bST was approved for use in dairy cattle prior to 1990 in the former Soviet Union, Brazil, Mexico, South Africa, Bulgaria and the former Czechoslovakia, and in the USA in 1994. bST has also been shown to increase growth rate in lambs and to increase milk production in ewes and buffalo.

bST increases milk yield by 10% when administered in early to mid-lactation, and by 40% in late lactation. The percentage increase depends on the dose, formulation, nutrition programme, herd health and, most importantly, management and environmental factors. Similar increases in milk yield are found in all breeds. Various effects of parity have been reported. It may be more cost effective to administer bST after peak lactation, when the effect of treatment is greatest. Treated cows become more persistent in milk production, so milk yield decreases at a slower rate than untreated cows after peak production (Fig. 4.3). This also extends the calving interval and results in fewer calves being born per year. Increasing ST by treatment with growth hormone releasing hormone from the anterior pituitary is as galactopoietic in dairy cattle as treatment with ST directly.

During the early transition phase of bST treatment, there is a delay of a few weeks in increasing the voluntary feed intake, although milk yield increases immediately. This puts the cow in negative energy balance, so body fat stores are mobilized and body condition scores are reduced. There may be a slight increase in the milk fat content during this time and a slight decrease in milk protein due to a limited supply of amino acids. Adequate body condition score should be achieved prior to calving.

Cows treated in the long term with bST increase their nutrient intake to support the increased milk production. The absolute nutrient requirements for maintenance and

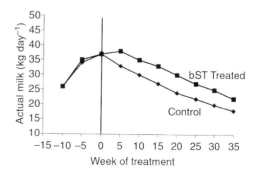

Fig. 4.3. Effect of bST on milk production.

the nutrient requirements per unit of milk produced by the dairy cow are not changed by bST treatment. However, the increased milk production results in an overall increase in feed efficiency, since the requirements for maintenance are a smaller percentage of the overall requirements. Cows treated with bST are like genetically superior cows at the same level of milk production and should be fed according to the level of milk production.

Inadequate nutrition will decrease the response to bST. Using more grain can increase the energy density of the ration, but dietary buffers are then needed to maintain pH balance in the rumen. Feeding ruminally inert fat such as the calcium salts of long-chain fatty acids (Megalac®) also increases the dietary energy. Sufficient levels of high-quality rumen undegradable protein are also needed. The diets needed by bST-treated cows are more expensive, but the income from milk production is increased over feed costs. For more information on the effects of bST on nutrient requirements of dairy cattle, see NRC (1994). For a more general review on the effects of bST, see Burton *et al.* (1994).

Mechanism of action

ST does not bind to receptors in the bovine mammary gland, but acts by partitioning additional nutrients to the mammary gland during lactation for the increased synthesis of lactose, protein and fat. Insulin normally increases the utilization of acetate by adipose tissue for lipid synthesis and thus is involved in partitioning nutrients away from milk synthesis and towards body tissues. Levels of insulin in blood are negatively correlated with milk yield. Especially during early lactation, adipocytes become less sensitive to insulin, and ST acts on the adipocytes to make more energy from fat available to the mammary gland. Activities of acetyl-CoA carboxylase and fatty acid synthase are decreased in adipose tissue by ST treatment. This increases the level of non-esterified fatty acids (NEFA) in plasma, which can be oxidized by peripheral tissues or used for the synthesis of milk fat. Insulin also encourages glucose sparing from peripheral tissues and increased the glucose available for milk production by the mammary gland. The lipolytic effects of bST are most prominent during the first two-thirds of long-term treatment during lactation.

ST also stimulates secretion of IGF-I from the liver, which increases proliferation and survival of mammary cells. However, IGF-I infusion into mammary glands is not galactopoietic, so these mitogenic responses of IGF-I may not be important in the galactopoietic effects of ST. However, IGF-I may compete with insulin for binding to the insulin receptor in peripheral tissues, thus contributing to the insulin resistance caused by ST. There are at least six binding proteins for IGF-I, which either inactivate or enhance the activity of IGF-I or have biological activities of their own. These proteins are present in the blood as well as in the interstitial spaces of the mammary gland.

Delivery

The Monsanto Corporation has developed and markets bST, using the trade name Posilac®. Recombinant produced bST has up to eight amino acids added to the amino terminus, depending on the manufacturing method. This has resulted in four different recombinant bST products, Somagrebove® (American Home Products), Somavubove® (Pharmacia and Upjohn), Sometribove® (Monsanto) and Somidibove® (Elanco).

bST undergoes deamidation in solution, with a tenfold increase in deamidation from 5°C to 37°C. Covalent crosslinks also form between bST molecules and result in hydrophobic aggregates that precipitate at neutral pH and high protein concentrations. This can cause extremely slow and incomplete release of bST from a sustained release formulation. Aggregation can be reduced somewhat using detergents such as Tween 20 or hydroxypropyl-β-cyclodextrin and sorbitol, but the high protein concentrations needed for prolonged release of more than 1 month are still a problem.

bST can be injected daily, or more commonly every 2 weeks using a prolonged release formulation. Suspensions of bST in oil have been made for sustained release and to stabilize the molecule by excluding water.

Aluminium monostearate has been used as a dehydrating or gelling agent with bST dispersed in oil at 10–50% by weight. Microparticles of bST have also been made using waxes, stearates, fatty acid anhydrides, sucrose, sodium sulphate, polysorbate 80 or sodium benzoate and dispersed in a carrier vehicle. Once injected, the microparticles are dissolved and the bST released over time (see Section 2.4). Implants have been made of compressed pellets of bST powder, with exipients such as polyanhydride, polycaprolactone, polyesters, cholesterol and ethylcellulose to slow the release of bST. Implants can suffer from problems of incomplete release of bST from their interior, the necessity of sterilizing them and problems with placing them. For more information, see Foster (1999).

Safety concerns of bST use

bST only slightly alters milk composition by altering the nutrient requirements of the cow. It increases milk production and also the problems normally associated with high milk production. These include increased somatic cell counts and mastitis. These are thought to be a result of the increased milk yield rather than a direct effect of bST. There is also some evidence of increased lameness in bST-treated cows. This might be due to negative effects of bST on connective tissue and bone development (see Section 3.4).

Cows treated with bST can take longer to come into oestrus after parturition than untreated cows, particularly if their body condition is not adequate. Adequate nutrition should be provided to replenish body condition during late lactation or during the dry period. There is no effect on the number of services required per cow once the animal starts cycling, and levels of GnRH, FSH and LH are unchanged by bST treatment. Delaying bST treatment until after peak lactation or until conception has occurred can counteract the negative effects on reproduction in dairy cows.

bST is thought to have no oral activity in adult humans and is destroyed by digestion; however, young infants can absorb intact proteins. Bovine IGF-I has the same sequence as human IGF-I, but levels in milk are only increased by abnormally high doses of bST and are not above those found in human breast milk. Both bST and IGF-I will be at least partially denatured by heat during pasteurization. It is estimated that by 1992, over 41,000 cows were treated with bST in scientific studies. The Food and Drug Administration of the USA reviewed more than 120 studies prior to approval of bST.

Factors affecting milk composition

The genetics and nutrition of dairy cows affect the composition of milk. There are large differences in milk fat content among different breeds of dairy cows, ranging from 5.1% for Jersey to 3.7% for Holstein-Friesian cows. Likewise, Jerseys produce milk with 3.8% protein compared to 3.1% protein for milk from Holstein-Friesian cows. Milk fat and protein content increase in the later stages of lactation, and frequency of feeding and method of feeding also affect milk composition. Undoubtedly, hormones are involved in the underlying mechanisms for these differences in milk composition. All components of milk can be altered to some degree. The amount of fat and the fatty acid composition of milk fat are most easily altered, followed by the protein content, with the lactose content of milk being the most difficult to change. Lactose is the most important osmotic component of milk, so changes in lactose synthesis are accompanied by changes in water volume and thus milk yield. For a review of the biological potential to alter milk composition, see Kennelly and Glimm (1998).

Milk protein

The primary purpose of milk proteins is to provide a source of readily digestible proteins with a balanced amino-acid content for the young. However, the digestion of milk proteins produces a wide array of bioactive peptides that modulate digestive and metabolic processes. This includes peptides with effects on nutrient uptake (phosphopeptides, casomorphins), immune function (immunopeptides, casokinins, casomorphins) and neuroendocrine function (casokinins).

Peptides with antimicrobial properties that are active against bacteria, yeast and fungi are produced by proteolysis of lactoferrin and casein. Antihypertensive peptides that inhibit the angiotensin-converting enzyme to reduce blood pressure are derived from casein. Antithrombotic peptides that inhibit platelet function are derived from casein and lactotransferrin. Casein phosphopeptides form a complex with calcium ions and provide a passive means for increased calcium absorption in the small intestine, as well as recalcification of dental enamel. Immunomodulatory peptides that affect the immune system and the cell proliferation response are derived from casein. Casomorphins are opioid peptides derived from casein. They are absorbed from the intestine to produce analgesic effects and induce sleep. They also inhibit intestinal peristalsis and motility, modulate amino-acid transport, and stimulate the secretion of insulin and somatostatin. Thus, there is a lot of potential for producing functional ingredients and high value-added products from milk. Many of these proteins could be salvaged from by-products of the manufacture of dairy products. For more information, see the review by Clare and Swaisgood (2000).

Increasing dietary protein or decreasing the degradation of dietary protein in the rumen increases the protein content in milk. Most additions of fat to the diet decrease milk protein content. Infusion of glucose along with increasing insulin in blood increases the protein concentration in milk. This may provide a method for manipulating the composition of milk.

Milk fat

The largest proportion of milk lipids is in the form of triglycerides, along with lesser amounts of diglycerides and monoglycerides. The fatty acid composition of can be altered to produce a more desirable product. For example, increasing the content of C18:0, and particularly C18:1, and decreasing the content of C14:0 and C16:0 may help to reduce plasma cholesterol and would result in a softer butter. Conjugated linoleic acid occurs predominantly in meat and dairy products. It has been shown to reduce the growth of many types of tumours and to reduce the risk of atherosclerosis (see Section 3.7). Altering the fatty acid profile of milk has also been reported to lower plasma cholesterol in people consuming the product. Increasing the fat content has traditionally been a focus in dairy breeding programmes. However, altering milk composition through breeding strategies is effective only in the long term.

Volatile fatty acids produced in the rumen, especially acetate, are the major precursors for synthesis of milk fat. Acetate production is increased with high-fibre diets, while low-fibre diets increase the production of propionate, which is a precursor for glucose and lactose synthesis. Milk fat content is decreased as the proportion of concentrate is increased in dairy rations, particularly in late lactation. Milk fatty acids are either derived from dietary long-chain fatty acids (50%), microbial synthesis or from body stores of fat (5–10%), or are synthesized directly from acetate and β-hydroxybutyrate by the mammary epithelial cells (40–45%). Microbial action in the rumen produces branched-chain and odd-carbon-number fatty acids, as well as products of biohydrogenation of polyunsaturated dietary fatty acids. Hormones affect the delivery of nutrients to the mammary gland and the activities of the acetyl-CoA carboxylase and fatty acid synthase enzymes involved in fat synthesis in the mammary gland.

The mammary gland cannot produce fatty acids longer than 16 carbons. The balance between fatty acids synthesized in the mammary gland and those available from the diet or microbial synthesis can be altered by dietary means. Feeding sources of unsaturated fatty acids, such as sunflower and safflower supplements, increases the content of unsaturated fatty acids, particularly C18:1, and lowers the content of C16:0 in milk. Levels of eicosapentaenoic acid (EPA) and docosohexaenoic acid (DHA) in cows' milk can be increased to the levels found in human breast milk by including a fish-meal-containing supplement in the diet of dairy cattle. For information on bioactive lipids in milk, see Molkentin (1999).

Metabolic diseases related to lactation

Ketosis

Ketosis results from a high demand for glucose for milk synthesis at a time when the cow is in negative energy balance. This usually occurs in early lactation when the cow may not obtain sufficient feed to produce enough propionate for glucose synthesis. This leads to the catabolism of fat and protein and the overproduction of ketone bodies and the accumulation of fat in the liver. From 4–12% of dairy cows develop clinical symptoms of ketosis, including depressed appetite, milk yield and body weight, increased milk fat, acetone breath, ketonaemia, ketonuria and fatty liver.

An imbalance in the insulin/glucagon ratio, abnormal liver function, adrenal corticoids, thyroxine and mineral and vitamin deficiencies have been implicated in causing ketosis. Moderate levels of body fat prepartum and increasing feeding rapidly postpartum are desirable, but appetite and intake are normally depressed postpartum. Treatments are aimed at increasing body glucose. This includes intravenous glucose infusion, preferably in combination with polypropylene glycol, which is converted to propionate. Administration of cortisone or ACTH can raise blood glucose but may be detrimental in the long term.

Milk fever

Milk fever (parturient paresis) is an economically important metabolic condition that occurs at the onset of lactation and can reduce the productive life of a dairy cow. Clinical symptoms include reduced appetite and rumen motility, inhibition of urination and defecation, lateral recumbency and eventual coma and death if left untreated. Cows recovering from milk fever have a greater incidence of ketosis, mastitis, dystocia, retained placenta, displaced placenta and uterine prolapse than cows that have never had milk fever.

Milk fever is due to a severe drop in blood calcium, which disrupts neuromuscular function. The concentration of calcium is 12 times higher in milk than in blood. A cow producing 10 l of colostrum requires about 23 g of calcium, which is about nine times the total plasma pool of calcium, while a cow in early lactation milking 40 l day^{-1} requires 56 g of calcium. Calcium in plasma is derived from the diet and from bone storage reserves. At parturition, the cow must bring more than 30 g of calcium per day into the plasma pool. However, it takes about 14 days to condition the effective release of calcium from bone, and the early postpartum cow may not receive enough dietary calcium due to a depressed appetite. This can result in low blood calcium (hypocalcaemia) of less than 6.5 mg 100 ml^{-1}, compared to 10 mg ml^{-1} for normal blood calcium, and can cause milk fever.

HORMONES INVOLVED. Blood calcium levels are increased by the coordinated actions of the calcitrophic hormones parathyroid hormone (PTH) and 1,25-dihydroxyvitamin D_3 [1,25(OH)$_2$D$_3$] and decreased by calcitonin (Fig. 4.4). PTH is secreted from the parathyroid gland in response to low blood calcium; first causing increased calcium reabsorption from the glomerular filtrate in the kidney and then calcium resorption from bone. The total soluble calcium in bone fluids of the typical cow is estimated at 6–10 g and this is increased by an additional 6–8 g during metabolic acidosis. PTH stimulates the rapid transport of the calcium from the bone fluids to the extracellular fluid.

PTH also induces the 1α-hydroxylase enzyme in the kidney that activates vitamin D to 1,25(OH)$_2$D$_3$. The 1,25(OH)$_2$D$_3$ stimulates the synthesis of calcium transport proteins that move calcium across the intestinal epithelial cells, and the magnesium-dependent Ca-ATPase that pumps the calcium out of the cell into the plasma. The 1,25(OH)$_2$D$_3$ also acts synergistically with PTH to increase the release of bone calcium and increase the renal reabsorption of calcium.

Calcitonin release from the thyroid increases calcium deposition in bone by inhibiting the effects of PTH. Calcitonin also stimulates the excretion of calcium and phosphorus in the kidney.

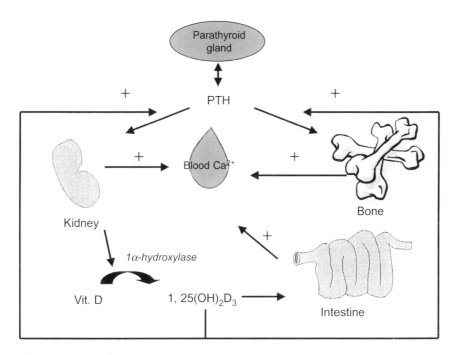

Fig. 4.4. Effect of parathyroid hormone and vitamin D on calcium metabolism.

PREDISPOSING FACTORS. Cold stress in subzero temperatures can induce hypocalcaemia. Milk fever is more prevalent in over-fat cows (fatty liver) than in thin cows, as mobilization of fatty acids can affect blood calcium levels.

Some breeds of cows, especially Channel Island, Swedish Red and White, and Jersey breeds, are more susceptible to milk fever. A lower number of intestinal receptors for $1,25(OH)_2D_3$ are present in Jerseys than in age-matched Holsteins. The decreased receptor number would limit the effectiveness of $1,25(OH)_2D_3$ in stimulating the reabsorption of calcium.

The incidence of milk fever increases dramatically in third and greater lactations. Older cows have increased milk production and greater demands for calcium. However, the ability to resorb calcium from bone and transport calcium across the intestinal epithelial cells, as well as the synthesis of $1,25(OH)_2D_3$, decrease with age. This is due to a decrease in the intestinal receptors for $1,25(OH)_2D_3$ and an increase in the activity of the C24-hydroxylase enzyme that inactivates $1,25(OH)_2D_3$.

TREATMENT AND PREVENTION. Treatment of milk fever consists of supplementing blood calcium levels until the bone and intestinal transport systems adapt to provide the necessary calcium. The intravenous infusion of 23% calcium borogluconate is most widely used, but this can result in high plasma calcium if administered too quickly, causing cardiac arrest. About 25% of cows treated this way require additional treatment. Calcium and phosphorus supplements can also be given around parturition in the form of calcium carbonate, calcium chloride gels, or dicalcium phosphate. A paste of calcium propionate has been used to give a more sustained calcium release and to provide propionate as a gluconeogenic precursor.

Stimulating calcium turnover from bone before calving can reduce the incidence of milk fever. This is done by feeding limiting amounts of dietary calcium (100 g day^{-1}, or 70 g day^{-1} for small breeds) for 2 weeks before calving. Insufficient dietary magnesium reduces calcium mobilization from bone, while excess dietary phosphorus can interfere with vitamin D metabolism and cause milk fever. Dry cow rations need to be balanced for calcium (0.39%), phosphorus (0.24%) and magnesium (0.23%).

Injection of vitamin D_3 or its synthetic analogues within 8 days prepartum may also be effective in preventing milk fever. However, cows treated with vitamin D compounds, including 1α-dihydroxyvitamin D_3, develop hypercalcaemia and are unable to produce endogenous $1,25(OH)_2D_3$ due to inhibition of the 1α-hydroxylase enzyme. More active and longer-lasting vitamin D analogues, such $24\text{-F-}1,25(OH)_2D_3$, have been developed as an experimental implant, but are not used commercially.

PTH infusions or infections can prevent milk fever, with about 20 times more PTH needed for intramuscular than intravenous treatment. A sustained release of PTH via an implant has been suggested.

Anionic salts such as ammonium chloride and magnesium sulphate also stimulate the release of calcium from bone. These salts are rather unpalatable, and must be mixed with feed in two daily feedings and not fed for more than 3–4 weeks. Hydrochloric acid that has been diluted and mixed with molasses can also be used as a source of dietary anions, but HCl is dangerous to handle and corrosive to machinery. Adding anions reduces the dietary cation–anion difference (CAD) to reduce the pH of the blood and urine. The CAD can be calculated as:

$$CAD = (0.38Ca^{2+} + 0.3Mg^{2+} + Na^+ + K^+) - (Cl^- + 0.6SO_4^{2-})$$

This takes into account the average absorption of the different ions in cattle.

A CAD of –50 to –100 mEq kg^{-1} of diet, or urinary pH of between 5.5 and 6.2, has been suggested as optimal. Mild acidosis increases tissue responsiveness to PTH, resulting in increased resorption of calcium from bone and increased synthesis of $1,25(OH)_2D_3$. This may be due to improved function of PTH receptors at lower pH.

An alternative approach to increasing dietary anions is to lower the levels of dietary cations. This is particularly important if the CAD of the diet is greater than 250 mEq kg^{-1}, as it will be difficult to add enough anions to lower the CAD to –100 mEq kg^{-1} without seriously affecting the palatability of the feed. Potassium is the cation present in highest amounts in ruminant forages, and growing conditions that produce consistently low levels of potassium in forages would be beneficial in reducing the cation content of the feed.

The strategies for preventing milk fever can be summarized as follows. If the dietary CAD is below 250 mEq kg^{-1} of feed, anions can be added to the diet to produce mild acidosis and increase the sensitivity of the PTH receptors. If the CAD is above 250 mEq kg^{-1} feed, calcium gels can be administered to increase the passive absorption of calcium. PTH, vitamin D analogues or low calcium diets can also be given to increase the activity of the calcitrophic hormones and increase the available calcium. For more information, see Horst *et al.* (1997).

4.2 Egg Production

This section covers endocrine factors that regulate sexual differentiation and development, the rate of follicular development and egg production, and the formation of the egg shell in poultry.

Sexual development

Female chickens have a unique reproductive structure since they have only one functional ovary. During embryonic development, birds start out with two undifferentiated gonads. As the male matures, it develops two functional testes, while in the female the left ovary matures and the right ovary regresses. The reason why only one ovary develops is not known, but it may be related to weight constraints for flight and the restraints imposed by the additional nutrient requirements of the organ. However, these factors are not important in laying hens. Interestingly, some species of wild birds, particularly hawks, develop two functional ovaries and oviducts. A line of Rhode Island Red chickens with two oviducts has also been described (Wentworth and Bitgood, 1988). This does suggest that it may be possible for two functional ovaries and oviducts to develop in chickens. This has the potential to increase egg production, although the physiological and metabolic constraints of producing twice as many eggs may make this impossible. More likely, egg production may increase slightly and might be more prolonged, with ovulation alternating between the two ovaries.

The control of sex is also important, since the broiler industry prefers males for their improved growth rate and improved lean yield, while the layer industry uses only females. Millions of chicks are therefore culled because they are not the desired sex. Various hormonal and physical manipulations can somewhat alter the sexual phenotype and produce pseudo-sex-reversed birds, but no reliable method has yet been found to permanently alter the phenotypic sex of commercial poultry.

Hormonal effects

During embryonic development, the primordial germ cells (PGCs) settle in the urogenital ridge region of the embryo around 3 days of incubation. Up to 5–7 days of incubation, the gonads are bipotential, meaning that they can form either ovaries or testes. The undifferentiated gonad is composed of an inner medulla, which produces androgens, and an outer cortex, which produces oestrogens. Sex differentiation and regression of the right ovary begins 6.5 days after hatching. In developing genetic females, the gonads produce mostly oestrogens, while androgens predominate in the genetic male. The major hormone present then determines the pattern of gonadal development, with oestrogen playing a key role. Steroid hormones also play a role in the development of secondary sex characteristics as the birds mature. Androgens stimulate comb growth, spur growth, male feathering and copulatory behaviour, while oestrogens cause female feathering patterns and inhibit male phenotypic and behavioural expression.

Gonad development in fish and reptiles can be altered by treatment with exogenous sex steroids; androgens induce the formation of functional testes, while oestrogens induce the formation of functional ovaries. The sensitivity of fish and reptiles to exogenous hormones is the basis of several assays for endocrine disruptor chemicals (EDCs, see Section 6.4). Mammalian gonads are insensitive to exogenous hormone treatment, while the gonads of birds are intermediate in response. Treatment with an aromatase inhibitor prevents the synthesis of oestrogen and causes the development of a functional male phenotype, although the female growth pattern is not altered. Removal of the left ovary or natural regression of the left gonad after infections or tumours induces the right gonad to develop as a semi-functioning testis and for a male phenotype to develop. Castrated males continue to grow like males, while genetically male embryos treated with oestrogen at 4 days of age develop a transient female phenotype.

The aromatase enzyme is expressed in the medulla of the left and right gonads of female embryos, but not in male embryos. The oestrogen receptor is expressed in the left gonad and not the right gonad of both sexes, but the expression in males is restricted to a very early stage of development. The lack of an oestrogen receptor in the right gonad could explain why the right gonad does not develop as an ovary. The oestrogen receptor is transiently present in the left male gonad, which explains why the left gonad can be sex reversed in the male by administration of oestrogen. The oestrogen receptor does not normally play a role in the male, since the aromatase gene is not expressed and oestrogen levels would normally be low.

Genetic effects

In birds, the female is the heterogametic sex (ZW), while the male is homogametic (ZZ). The mammalian Y and avian W chromosomes are both small and contain very few genes. It is possible that gonadal differentiation is controlled by the ratio of sex chromosomes to autosomes, also known as chromosomal dosage, as occurs in *Drosophila*. Alternatively, major sex determining genes may be present on the W chromosome that triggers the development of the ovary. This gene, termed *ASW* for avian sex-specific W linked, has recently been cloned and mapped to the W chromosome. It is related to the HIT (histidine triad) family of proteins and the putative protein kinase C inhibitor (PKCI).

The mode of action of *ASW* has not been determined, but it may act to induce the formation of the ovary in an analogous manner to the *SRY* gene on the mammalian Y chromosome, which leads to the development of testes (see Section 5.1). In mammals, female is the 'default sex', and the activity of the *SRY* gene is required to cause the undifferentiated gonad to form a testis. Conversely, in birds the default is the formation of the male testis. The *SRY* gene has not been found in birds, but the *SOX9* gene, which is related to *SRY*, is expressed at day 6 of incubation in the testis, while expression in the ovary is weak. *SOX9* may have a role in testis function, but in contrast to mammals it is expressed after the *AMH* gene (see below), so it is not likely to play a role in sex determination. For further information, see Capel (2000).

Both Müllerian and Wolffian ducts are present during early development before sexual differentiation has occurred. When an ovary is present and the synthesis of oestrogen occurs, the Müllerian ducts develop into oviducts, uterus and cloaca, and the Wolffian ducts regress. In the presence of a testis, the Wolffian ducts develop into seminal vesicles, ductus deferens and epididymis, and Müllerian ducts regress. The development of the Müllerian ducts is inhibited by anti-Müllerian hormone (AMH, also known as Müllerian inhibiting substance, MIS) that is produced by the developing testis. AMH in chicken has been cloned and is a 644-amino-acid glycoprotein that is a member of the TGFβ family. It may also inhibit aromatase activity. The action of AMH is inhibited by oestrogen, and the lack of oestrogen receptors in the right ovary results in regression of the right Müllerian duct. AMH may be controlled by *SOX9* as well as by other genes encoding transcriptional factors involved in sexual determination, including *DAX1*, *SF1*, *WT1* and possibly *DMRT1* (see Section 5.1). *DAX1* is expressed at similar levels in both male and female gonads during differentiation and may be more important in regulation of steroidogenesis. *SF1* is expressed by day 3.5 of incubation and, in contrast to mammals, is higher in female than in male gonads, with expression up-regulated in developing ovaries after the onset of differentiation. *WT1* is expressed at similar levels in male and female gonads. *DMRT1* is expressed at 3–4 days of incubation in the genital ridge and Müllerian ducts. It is located on the Z chromosome and is expressed in higher levels in ZZ embryos than in ZW embryos, making it a good candidate for sex determination in birds. For more information on sexual differentiation in birds, see Shimada (2002), and for information on the control of spermatogenesis, see the review by Thurston and Korn (2000).

A more complete understanding of the pattern of gene expression during sexual differentiation may lead to methods for reliably altering the phenotypic sex of poultry.

Regulation of follicular development and egg production

Commercial chickens (pullets) begin to lay eggs as they approach sexual maturity at 18–20 weeks of age. However, not all hens begin to lay eggs at the same time. Egg production increases to a maximum of about 90% of hens producing an egg every day over about 2 months, as all the hens come into lay, and then gradually decreases as the hens get older. An ideal production curve would be when maximum egg production is reached immediately and then maintained indefinitely (Fig. 4.5).

Follicular development is under the control of the pituitary gonadotrophins, but the precise mechanisms are not well understood. Multiple follicles begin to mature at the same time in the avian ovary. A high percentage of small follicles do not grow to a fully differentiated stage, but undergo atresia and are resorbed. Those that grow to 10 mm in diameter are recruited to the stage of accumulating yellow yolk. Between 4 and 8 yellow growing follicles are present at any time, and these are arranged in a hierarchy of size, based on their stage of maturation.

Follicular growth and yolk deposition have been studied by feeding birds gelatin capsules containing fat-soluble dyes, such as Sudan black or Sudan red. As the yolk is deposited in the growing follicle, the dye forms a coloured ring. This can be seen by

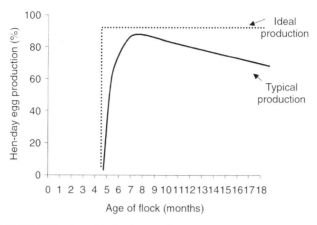

Fig. 4.5. Changes in egg production of laying hens with age.

slicing the yolk after the egg has been hard-boiled.

Before recruitment as yellow follicles, the numerous small follicles produce dehydroepiandrosterone (DHEA), androstenedione and oestrogens (Fig. 4.6). Following recruitment, the yolk-filled follicles produce progesterone as well as decreased quantities of oestrogens. As the follicles continue to mature, they lose the ability to produce androgens and oestrogens and the largest follicle produces large quantities of progesterone. This surge of progesterone stimulates the release of LH, which acts by positive feedback to increase progesterone production by the largest follicle, which ultimately results in ovulation. Plasma levels of LH and progesterone peak 4–6 h prior to ovulation. LH stimulates steroidogenesis in all follicles, with androgens and oestrogens produced by the small follicles.

Ovulation is caused by a surge in LH concentrations and the timing of ovulation is controlled by the circadian rhythm, which is normally set by the light–dark cycle. In the absence of a light–dark cycle, eggs are laid at

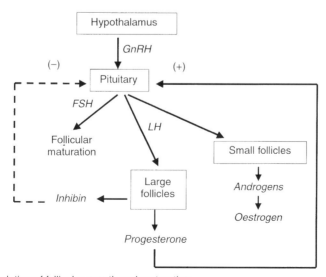

Fig. 4.6. Regulation of follicular growth and maturation.

all times throughout the day, although some hens may establish a circadian rhythm from other environmental cues. As little as 1.25 h of darkness is sufficient to establish the circadian rhythm from the photoperiod. The transition from light to dark is important and eggs are laid 12–18 h after darkness begins.

The length of the ovulatory cycle and follicular maturation can range from 23 to 28 h in different birds. However, hens in normal production are kept on a light–dark cycle of 14L–10D totalling 24 h, which fixes the period of time that LH is released and ovulation can occur. Birds that take longer than 24 h to mature a follicle will therefore not always have a follicle ready for ovulation when levels of LH have peaked. This results in a non-egg-laying day or a pause in the sequence or 'clutch' of eggs that are laid. If hens are kept on a 28-h day, virtually all hens will be able to mature a follicle in this period and one egg will be laid for each light–dark cycle. A bird that has a rate of follicular development that is less than 24 h will not ovulate the mature follicle until the LH surge. Selection for birds with rapid rates of lay, using short 22-h photoschedules, has been used to increase the rate of egg production, but lower egg weights were obtained. For more information on factors affecting the ovulatory cycle of hens, see the review by Etches (1990).

There are two forms of GnRH in the chicken (Table 4.2), with GnRH-I being the most prevalent. Immunization against GnRH-I but not GnRH-II inhibits ovulation. Mammalian GnRH is also active in chickens.

As the hen ages, there is decreased gonadotrophin production from the anterior pituitary gland in response to the GnRH. This results in a reduced number of follicles reaching the final stages of maturity, so that a mature follicle is not always available for ovulation and the rate of lay decreases. However, since fewer follicles are developing, they receive a proportionally greater amount of yolk, causing an increase in egg size.

FSH is involved in maintaining the hierarchy of size in the developing follicles and the rate of follicle atresia. Levels of FSH are decreased by inhibin, which is produced by the four largest preovulatory follicles (F1–F4), in particular the largest F1 follicle. Inhibin is a gonadal glycoprotein that is a member of the transforming growth factor-β (TGFβ) family of peptides (see Section 5.1). It is comprised of two subunits, α and β, while activin is a related protein that is comprised of two β-subunits and has opposite physiological effects on FSH. Inhibin reduces the secretion of FSH from the anterior pituitary gland without affecting LH production. It has also been shown to have paracrine effects in the ovary, acting as a competitive FSH receptor antagonist in mammals. Follistatin is a soluble protein that binds and inhibits activin; it also binds inhibin with less affinity but its physiological role with inhibin is not completely understood. The IGFs also have important effects on the control of reproductive function in the ovary. They appear to act as paracrine/autocrine regulators of follicular growth and differentiation.

Application

The expression of inhibin in the F1–F4 follicles and plasma levels of inhibin are higher in hens that lay at a low rate compared to those that lay at a high rate. The α-subunit of inhibin has been cloned and sequenced, and this information has been used to develop inhibin

Table 4.2. Structure of mammalian GnRH, chicken GnRH-I and chicken GnRH-II (from Etches, 1990).

	Amino-acid number									
	1	2	3	4	5	6	7	8	9	10
mGnRH	Pyro-Glu	His	Trp	Ser	Tyr	Gly	Leu	Arg	Pro	Gly-NH$_2$
cGnRH-I	Pyro-Glu	His	Trp	Ser	Tyr	Gly	Leu	Gln	Pro	Gly-NH$_2$
cGnRH-II	Pyro-Glu	His	Trp	Ser	His	Gly	Trp	Tyr	Pro	Gly-NH$_2$

conjugates for immunization of hens. Immunization against inhibin increased the number of follicles that were recruited into the preovulatory hierarchy in chickens and advanced the onset of lay and increased the rate of egg production in Japanese quail (Moreau *et al.*, 1998).

Eggshell formation

Poor eggshell quality, resulting in cracked shells, is a major source of losses to the layer industry. Poor shell quality occurs due to environmental factors, such as temperature and stress, as well as nutritional factors, such as the availability of dietary calcium and levels of minerals such as phosphorus and chloride. Poor shell quality along with decreased egg production is a major cause of culling in older hens. The egg shell is also important for successful development of the chick embryo, as it provides protection from damage, infection and desiccation, and provides a source of calcium for skeletal development. The development of a thicker shell is undesirable, since this would reduce the exchange of gas and water and make it more difficult for the embryo to hatch. Rather, it is the mineralization process, which involves both shell matrix proteins and calcium salt crystals, that affects shell strength.

After ovulation, the ovum is fertilized in the infundibulum and then albumin is added in the magnum. The shell membranes are added further along in the isthmus (Fig. 4.7). Then about 6 g of calcium carbonate (calcite) and a thin layer of calcium phosphate (apatite) are laid down in the 18 hours while the egg is in the shell gland or uterus. The egg shell consists of an organic matrix (3.5%) and mineral layers (95%). The molecular constituents involved in the mineralization process have been studied, with the goal of improving eggshell quality, since the structural organization within the mineral layer is the most important factor affecting shell strength.

The egg shell is divided into six layers (Fig. 4.8). The innermost layers are the membranes that enclose the yolk and albumin. The outer membrane is anchored to the first calcified (cone) layer. This is followed by the palisade layer, which consists of an array of crystals arranged perpendicular to the shell surface and then a vertical layer of crystals. An organic cuticle covers the outer surface of the egg shell and contains the majority of shell pigments.

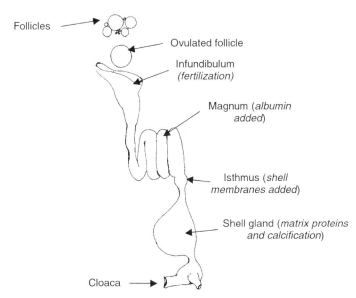

Fig. 4.7. Diagram of the reproductive tract of poultry.

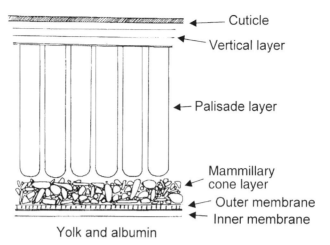

Fig. 4.8. Diagram of the various layers in the shell.

Shell matrix

The matrix is composed of a mixture of proteins and polysaccharides that directs the nucleation and controls the growth of crystals in the mineral layer, thus affecting the mechanical strength of the shell. The shells of eggs at the end of a hen's laying cycle can still contain the same amount of calcium, but eggshell strength can be decreased due to poor structural organization in the mineral layer.

Some of the proteins involved in regulating mineralization are acidic and are located at the interface between the crystal and the matrix, while others are found within the crystals. Crystal growth is initiated by deposition of calcium carbonate on organic aggregates that are present on the outer surface of the eggshell membranes. Crystal structure is affected by protein–mineral interactions and competition for space from crystals growing from adjacent centres of nucleation. The shell membranes are necessary for shell deposition and affect the pattern of mineral growth. This has been shown in experiments in which replacement of part of the shell of a snail by eggshell membranes resulted in an avian pattern of mineral formation in the snail shell.

Several proteins are secreted into the uterine fluid and are present in the egg shell, including ovocleidin-17 (OC17, a specific uterine protein), osteopontin (OPN, a bone-matrix phosphorylated glycoprotein), serum albumin, ovalbumin, lysozyme and ovotransferrin. OPN is expressed in the shell gland and secreted into the lumen only during the period of shell calcification. OPN synthesis is induced by the mechanical strain of the egg in the gland and it is not expressed if the egg is removed prematurely from the shell gland. The eggshell matrix also contains proteoglycans, primarily keratan sulphate and dermatan sulphate. These macromolecules influence the organization of crystal growth by controlling the size, shape and orientation of calcite crystals. They are present in uterine fluid at different concentrations, depending on the stage of shell calcification, while levels of calcium and bicarbonate remain high. Changes in matrix shell proteins have also been shown during the growth of shells from molluscs and echinoderms.

Functional studies of shell-matrix proteins have involved studying the effects of these proteins on delaying calcium carbonate precipitation and affecting the size, shape and orientation of crystals formed *in vitro* from a supersaturated salt solution. Soluble extracts of egg shells, or uterine fluid collected towards the completion of the shell, delay precipitation and cause aggregation of the crystals. Uterine fluid in which the large molecules are removed by ultrafiltration has no effect on crystal formation. For further information on the eggshell matrix, see the review by Nys *et al.* (1999).

Calcium metabolism

The average egg shell contains 2.3 g of calcium and, at a rate of production of 250 eggs year^{-1}, the hen turns over 580 g of calcium each year. Few species of animals turn over this much calcium as a percentage of body weight. The calcium used for shell formation comes from the blood. Plasma calcium level rises from 100 μg ml^{-1} before the onset of lay to 200–270 μg ml^{-1} throughout egg production. Most of the calcium in plasma is associated with organic complexes, very-low-density lipoprotein (VLDL) and vitellogenin, with only 20% of the calcium present in an ionized form.

An adequate supply of calcium is essential for proper shell formation, with normal layer rations containing 3.5% calcium. Calcium is needed during the night when shell formation occurs. Feed intake increases in the 2 hours before the onset of darkness, and hens will readily pick out large particles of oyster shell or limestone pellets, and store these in the gizzard, where soluble components will be leached by the HCl present. The passage of calcium through the digestive tract can be regulated to optimize the absorption of calcium. A coarse particle size of calcium improves shell quality by increasing the period when calcium is available from the diet. High levels of saturated fatty acids in the diet lead to the formation of calcium soaps and decrease calcium bioavailability. High levels of dietary phosphorus decrease calcium availability, with 0.3% available phosphorus being adequate for normal performance and bone integrity. Feeding a calcium-deficient diet results in decreased plasma LH, decreased egg production and ovary regression by 6–9 days.

The additional calcium that is required for shell formation is mobilized from storage in bone (Fig. 4.9). The calcium required for the formation of a single egg shell is equivalent to about 10% of the total calcium stored in the skeletal system and, in the absence of adequate dietary calcium, the bones can be demineralized to cause 'cage layer fatigue' and bone breakage. Calcium storage in bone develops as oestrogen levels increase at sexual maturity and is in a labile form of calcium phosphate, called medullary bone, that is present in the long bones. Vitamin D stimulates the mobilization of calcium from bone by osteoclasts, as well as calcium uptake from the intestine, by increasing the synthesis of the calcium-binding protein calbindin. Vitamin D$_3$ is converted into 25-hydroxycholecalciferol in the liver and then further hydroxylated to the active form 1,25-hydroxycholecalciferol in the kidney. If blood calcium levels decrease, parathyroid hormone (PTH) is released from the parathyroid

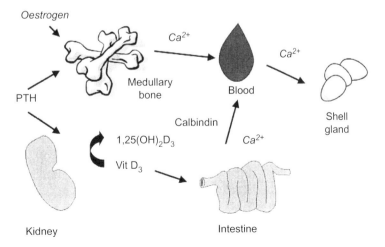

Fig. 4.9. Calcium mobilization in birds.

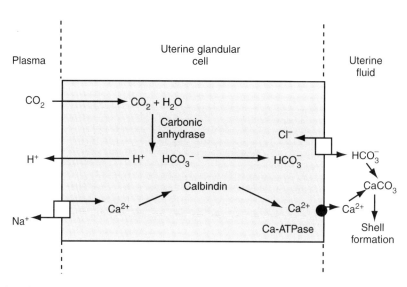

Fig. 4.10. Ionic fluxes through uterine glandular cells (after Nys *et al.*, 1999).

glands. PTH stimulates osteoclast activity directly and also increases the 1-hydroxylation of 25-hydroxycholecalciferol in the kidney, which increases the removal of calcium from bone. Calcium deposition in bone occurs by osteoblasts, which have increased activity when the shell is not being deposited.

During shell formation, the uterine cells secrete a fluid into the shell gland that is supersaturated with calcium and bicarbonate. The changes that occur in the uterine fluid during calcification are summarized in Table 4.3. Sodium and chloride levels are high in the uterine fluid at the beginning of calcification and then decrease by 18 h as they are reabsorbed by active transport into the uterine cells. The level of potassium increases and the pH decreases during calcification. Calcium absorption from the blood and secretion in the uterine fluid occurs by active transport via a calcium-ATPase pump and a Ca^{2+}/Na^+ exchanger (Fig. 4.10). A calcium-binding protein, calbindin, is present in the uterine cells and this protein is identical to a vitamin D-dependent protein found in the intestine. Calbindin may be involved in Ca^{2+} transport or in protecting the cells from high levels of intracellular calcium. Levels of calbindin increase in uterine cells at sexual maturity in response to oestrogen stimulation of oviduct development and are not dependent on vitamin D. Calcium secretion and levels of calbindin mRNA vary during the ovulatory cycle and increase during shell formation. Shell deposition only occurs when the presence of a yolk in the uterus is synchronized with ovulation. This suggests that hormonal factors linked to ovulation and follicular maturation regulate calbindin synthesis.

Bicarbonate is produced by the action of carbonic anhydrase in the uterine cells (Fig. 4.10). The level of carbonic anhydrase increases with oviduct development and also at the onset of egg production. A vitamin D

Table 4.3. Summary of ionic changes in uterine fluid shell formation.

	Start of calcification	End of calcification
Na	144	80
Cl	71	45
K	12	60
Ca	6–10	
Bicarbonate	60–110	
P_{CO_2}	100	
pH	7.6	7.1

Values are reported in mmol l^{-1}, except for P_{CO_2}, which is in mmHg.

response element is present in the carbonic anhydrase II gene; however, the activity of carbonic anhydrase is associated with the development of the oviduct induced by sex steroids, so the exact hormonal regulation of carbonic anhydrase remains to be established.

Chloride levels greater than 0.2% affect shell quality negatively, particularly for older hens. High environmental temperature decreases shell quality because of reduced calcium consumption as well as increased respiration rate, which causes respiratory alkalosis and reduces the CO_2 and CO_3^{2-} available for shell formation. Supplying carbonated water, sodium bicarbonate and vitamin C may also improve shell quality, particularly in hot environments. Increasing the concentration of CO_2 in the air also improves shell quality, but this is not practical in commercial egg production.

Applications

The levels of plasma hormones, such as parathyroid hormone (PTH) as well as PTH-related peptide, oscillate during the egg cycle and may be involved in regulation of matrix protein synthesis. However, the hormonal control of matrix protein synthesis has not been established. Further work in this area could lead to biochemical markers for eggshell mineralization and shell quality.

Hens with poor eggshell quality tend to have lower plasma levels of 1,25-hydroxycholecalciferol, and this could possibly be used in genetic selection programmes for improved shell quality.

4.3 Wool Production and Endocrine Defleecing

Introduction

This section describes various dietary and endocrine manipulations that can be used for defleecing sheep. The endocrine factors (cytokines) that are involved in the function of wool and hair follicles, in particular the epidermal growth factors, are discussed.

The traditional method of mechanical shearing of sheep is labour intensive, requires considerable skill, is time consuming and is a major annual cost to a wool-growing enterprise. Simple, low-cost and effective methods for defleecing sheep would thus provide considerable benefits. Some ancestral breeds of sheep, such as the Wiltshire Horn, naturally shed their wool seasonally. Shedding occurs during the spring and early summer when daylength is increasing. Wool growth in modern domestic breeds varies seasonally, possibly due to similar mechanisms. Changes in the cellular activity in the mature wool follicle affect the rate of fibre production and fibre diameter. Follicles undergo a growth cycle, which consists of periods of fibre growth (anagen) followed by follicle regression (catagen), quiescence (telogen) and re-initiation of fibre growth (proanagen). This cyclic growth is regulated by growth factors.

The matrix cells of the dermal papilla at the base of the follicle produce the keratin proteins that are the main components of the wool fibre (Fig. 4.11). At least five areas are potential targets for the action of defleecing agents. These are:

1. The follicle bulb in which active cell division is occurring.
2. The keratogenous zone where cells produced in the bulb grow and are modified to form the wool fibre and root sheath. Oxidation of thiols in proteins occurs here and RNA and DNA are degraded and resorbed at the top of this zone.
3. The zone of final hardening of the wool fibre, where oxidation of the thiol groups is completed.
4. The inner root sheath, which forms the outer cuticle of the fibre.
5. The intercellular cement in the fibre.

Defleecing methods

Copper deficiency results in wool with low intrinsic strength, and zinc deficiency results in partial or complete shedding of the wool. Chelating agents such as EDTA can be given to reduce the available copper and zinc and to cause defleecing, but the doses required are near lethal. Single oral doses of various thallium salts (10–14 mg kg^{-1} BW) or the anti-cancer drug cyclophosphamide (25–30 mg

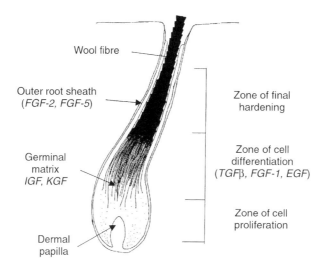

Fig. 4.11. Diagram of epidermis and follicle anatomy.

kg^{-1} BW), which arrests cell division, can cause defleecing, but these compounds are toxic. Hair loss in mice is induced by the selenium-containing amino acids, selenocystathione and selenocystine, but these compounds have not been tested in sheep, due to their cost and toxicity.

Producing imbalances in available amino acids can also weaken the wool fibres. This can be done by infusion of an amino-acid mixture without lysine or methionine into the abomasum. Infusion of methionine into wheat-fed sheep leads to distortion and partial degradation of the wool fibres. Abomasal infusion of a complete amino-acid mixture or infusion of methionine into roughage-fed or grazing sheep increases wool growth.

Intravenous infusion (80 mg kg^{-1} BW for 2 days) or oral doses (400–600 mg kg^{-1} BW) of the amino acid mimosine (Fig. 4.12) to sheep decreases wool growth and causes defleecing. Mimosine, which is present in high levels in the tropical legume *Leucaena leucocephala*, arrests cell division in the follicle bulb and wool fibre growth for 10–12 days within 2 days after dosing. The rate of regrowth and fibre diameter is increased after defleecing. Lower doses cause a partial break in the fibres without shedding the fleece. The analogue isomimosine is also effective in defleecing. Mimosine is effective under controlled laboratory conditions, but is not completely reliable in practice. A large dose is needed for defleecing (16 g for a 40 kg sheep), which is difficult to administer, close to the lethal dose and the effectiveness is influenced by the nutrition of the sheep. For more information on chemical defleecing, see Reis and Panaretto (1979).

Hormones from the thyroid, adrenal, pineal, pituitary and gonads can affect hair and wool loss in mammals. Administration of ACTH, cortisol and cortisone has been shown to inhibit wool growth, but long-term exposure to high levels of hormone is needed. Shedding is induced in Merino sheep with synthetic corticosteroids such as dexamethasone given at 8.5 mg kg^{-1} BW$^{0.75}$ (metabolic body weight) and flumethasone given at 1.3 mg kg^{-1} BW$^{0.75}$ as a constant intravenous infusion over 8 days. Plasma levels of dexamethasone and flumethasone were 40–50 and 8 µg ml^{-1}, respectively, from this

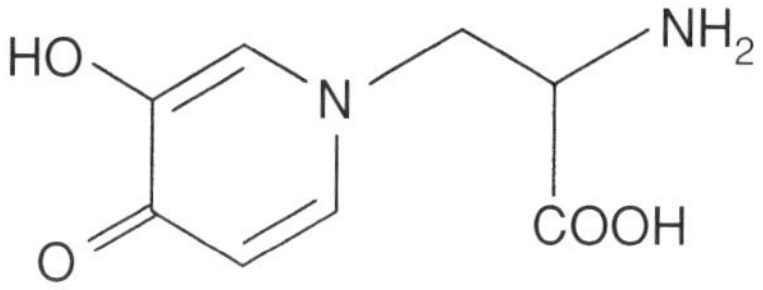

Fig. 4.12. Structure of mimosine, an amino acid that causes defleecing.

treatment. This dose is expensive and the extent of shedding varies among animals and differs in various anatomical areas on the animals.

Endocrine defleecing can be accomplished in modern breeds of sheep using epidermal growth factor (EGF). This process, known as Bioclip®, was developed by CSIRO Animal Production and The Woolmark Company in Australia, and has been marketed by Bioclip Pty Ltd in Australia since 1998. In the first stage, known as donning, sheep are prepared by removing undesirable fleece, such as head and shank wool or stained wool, to improve the quality of the final wool that is harvested. The sheep are then fitted with a fleece retention net and injected with EGF in the inguinal (inside back leg) region. Doffing occurs at day 14–18 after EGF treatment, when the net is removed along with the wool. The sheep are then treated with a protective coat until day 28, when there is sufficient wool regrowth to protect the sheep from climatic extremes. Reported costs for using Bioclip® are Aus$4.50 per animal, which includes $0.90 in preparative work.

Endocrine defleecing has the advantages of improving wool quality by eliminating second cuts on the wool, thus reducing the variability in fibre length, the amount of short fibres and mechanical damage, and improving carding yield. The ends of the wool fibres are more rounded and thus the wool is softer and more comfortable than wool harvested by mechanical shearing. The defleecing process does not affect non-wool fibres, such as kemp and hair fibres, which remain on the sheep and do not contaminate the final fleece. This improves the processing and dyeing of the wool yarn. Avoiding shearing cuts to the skin also decreases the contamination of the wool with skin pieces and increases the value of the sheep leather.

Model systems used to study function of follicles

A simplified *in vitro* model system to study follicular function is the microdissection of follicles from skin samples followed by growth in culture (for example, see Bond *et al.*, 1998). *In vivo* model systems in which the follicle activities are exaggerated, such as synchronized cycles of shedding sheep, can also be used. Whole animals can be treated with growth factors and depilation (removal of wool) force measurements then used to determine whether the strength of the wool fibres has been affected. Skin samples can be removed from the animal after treatment and various cell parameters can be measured. Growth factor expression can be knocked out or growth factors can be overexpressed in transgenic animals. Naturally occurring mutations that affect fibre growth, such as *angora, waved* and *rough*, can also be studied to determine the role of growth factors in these phenotypes.

Growth factor effects on hair and wool follicles

A large number of growth factors have been implicated in the regulation of wool follicle function (Table 4.4). These factors act locally within the follicle to regulate cell function in an autocrine or paracrine manner. Growth factors bind to cells within the wool follicle bulb to affect cell proliferation and keratinocyte differentiation, which affect the rate of growth and characteristics of the wool fibre. Growth factors also bind to the extracellular matrix (ECM), which consists of fibrous proteins, glycoproteins and proteoglycans. Growth factors influence ECM synthesis and degradation or are activated by ECM components and affect the wool follicles embedded in the ECM.

Growth factors affect the morphogenesis of wool follicles in the fetus and also the proliferation and differentiation of cells in the mature follicles. The development of follicles in the fetus determines the density, size and form of follicles, which affects the extent of wool production and wool fibre characteristics in the adult.

Insulin-like growth factors

The insulin-like growth factors are single-chain polypeptides of about 70 amino acids. They act as mitogens, morphogens and differentiation and cell-cycle progression factors. ST stimulates IGF-I production by the liver,

Table 4.4. Growth factors affecting follicular function (after Nixon and Moore, 1998).

Growth factor	Receptor(s)	Function
Insulin like growth factor-I (IGF-I)	IGFR-I, insulin receptor	Required for follicle growth in culture; expression varies with follicle cycle
Insulin-like growth factor-II (IGF-II)	IGFR-I and II	IGF-II mRNA and receptor present in follicles
Acidic fibroblast growth factor (FGF-1)	FGFR-1	Associated with differentiating keratinocytes
Basic fibroblast growth factor (FGF-2)	FGFR-2	Associated with basement membrane of follicle; inhibits hair growth in mice
Keratinocyte growth factor (FGF-7 or KGF)	FGFR-2	Dermal papilla to germinal matrix communication
Fibroblast growth factor-5 (FGF-5)	FGFR-1	Cycle-dependent expression in peripheral follicle cells
Transforming growth factor-β (TGFβ)	TGFβR-I and II	Follicle regression and inhibition of cell proliferation
Epidermal growth factor (EGF)	EGFR	Inhibits cell proliferation in wool follicle and causes catagen in follicle and fibre weakness
Transforming growth factor-α (TGFα)	EGFR	Mutants have enlarged follicles and wavy fur
Hepatocyte growth factor (HGF)		Keratinocyte growth and motility
Parathyroid hormone related protein (PTHrP)	PTHR	Antagonists stimulate hair growth in mouse
Platelet derived growth factor (PDGF-A and -B)	PDGF-Rα and β	Mitogenic in connective tissue sheath of hair follicles
Vascular endothelial growth factor (VEGF)		Autocrine action in human dermal papilla cells

as discussed in Chapter 3. IGF-II is present in higher amounts than IGF-I in sheep plasma and skin, but its role is less well understood.

Insulin-like growth factors stimulate keratinocyte proliferation and prolong the growth phase of hair follicles. IGF-binding proteins have been found in the dermal papilla, suggesting that they may modulate the effects of IGF on fibre growth. Mice with null mutations for IGF-I and IGF-II have hypoplasia of the epidermis and reduced follicle density. Conversely, mice overexpressing IGF-II had skin hypertrophy, while transgenic sheep expressing IGF-I had improved wool production. However, whole-body or skin-patch infusions of IGF-I in sheep had no effect on wool growth.

Fibroblast growth factors

The key members of the FGF family involved in the maintenance and control of fibre growth are acidic FGF (FGF-1), basic FGF (FGF-2), keratinocyte growth factor (KGF or FGF-7) and FGF-5. Basic FGF is found in the outer root sheath of the wool follicle and is associated with the ECM between the outer root sheath and the dermis. It may provide a mitogenic stimulus for the follicle bulb cells. Acidic FGF is concentrated in the upper bulb cells and is associated more with cell differentiation than proliferation. It appears to regulate fibre differentiation by affecting the expression of particular keratin genes and the pattern of differentiation of cells in the follicle. KGF is synthesized within the dermal papillae, while its receptor is present in epithelial follicle cells in the bulb. Treatment with KGF stimulates hair growth in mice, while KGF knockout mice have matted and greasy coats, similar to the naturally occurring *rough* mutation. KGF is thus important for normal fibre development. FGF-5 is localized to the outer root sheath and is increased during anagen and decreased after catagen. Gene knockout resulted in an exceptionally long coat, analogous to the *angora* mutation. FGF-5 thus appears to be involved in follicle regression.

Transforming growth factor-β

The TGFβ family includes three isoforms of TGFβ along with bone morphogenic proteins (BMPs) and the inhibins and activins. They are synthesized as a large precursor that is cleaved to form a 112-amino-acid growth factor and a binding protein. TGFβ receptors are serine/threonine kinases (see Section 1.3). Binding of the growth factor to the Type II receptor results in phosphorylation and activation of the associated Type I receptor. The receptors are located in the follicle matrix and outer root sheath. TGFβs regulate cell growth and differentiation, and are some of the earliest signals found during development of skin and hair follicles. They generally inhibit growth by suppressing mitosis of epithelial cells, inducing differentiation or inducing apoptosis. The roles of the individual TGFβs are not clear, but they are involved throughout follicle morphogenesis. Transgenic overexpression of BMPs in the outer root sheath and TGFβ1 in the epidermis result in decreased epithelial cell proliferation and follicle formation. TGFβs also regulates cell proliferation in follicles during the hair growth cycle, with highest levels at catagen.

EGF family of growth factors

Skin and hair follicles synthesize a number of the EGF family of peptide growth factors (Table 4.5), including EGF, transforming growth factor-α (TGFα), amphiregulin and heparin binding (HB)-EGF, which all bind the EGF receptor. Members of the EGF growth factor family have a 36–40-amino-acid conserved motif and three disulphide bridges, which form three peptide loops. EGF stimulates cell proliferation, while TGFα may be the fetal ligand for the EGF receptor and results in a transformed phenotype by regulating cell migration and differentiation. Amphiregulin contains a putative nuclear translocation signal and stimulates the growth of normal keratinocytes and fibroblasts while inhibiting growth of several breast cancers. Heparin-binding EGF has an N-terminal extension and can bind the extracellular matrix. Cripto is another member of the EGF family of growth factors. It induces branching morphogenesis in mammary epithelial cells and inhibits the expression of various milk proteins.

EGF stimulates the proliferation of cell types of ectodermal or endodermal origin. EGF consists of 53 amino acids but it is also active as a 150-amino-acid prohormone when

Table 4.5. EGF superfamily of growth factors.

EGF type	Receptor type	Function
EGF	ErbB1 (EGFR)	Promotes mitosis in mesodermal and ectodermal tissue
TGFα	ErbB1	Transforms normal cells; proposed fetal ligand for EGFR
Amphiregulin	ErbB1	Stimulates growth of normal fibroblasts and keratinocytes; inhibits growth of carcinoma
Neu differentiation factors (NDF) neuregulins and hereregulins	ErbB3 and ErbB4	
Betacellulin	ErbB1 and ErbB4	
Epiregulin	ErbB1 and ErbB4	
Heparin-binding EGF	ErbB1 and ErbB4	
Urogastrone		Inhibits secretion of gastric acid
Cripto		Mammary gland development

it is anchored to the cell membrane with the EGF sequence located in the extracellular space. The EGF peptide is identical to urogastrone, which is a peptide isolated from the urine of pregnant women that blocks the secretion of gastric juices. A major site of EGF production is the salivary gland, and production is stimulated by androgens. EGF is released in the saliva and is thought to play a role in wound healing in animals that lick their wounds. The studies using EGF to inhibit wool growth used EGF prepared from submaxillary glands of adult male mice. The activity of EGF is determined with a bioassay that measures its effectiveness in causing precocious incisor eruption and eyelid opening in newborn mice. EGF was first named based on the thickening and keratinization of the epidermis on the back and eyelid caused by injection of salivary gland extracts.

EGF RECEPTOR. The EGF receptor (EGFR) is a 180 kDa polypeptide that is part of a larger family of receptors known as ErbB, which are the receptors for the v-*erb-B* oncogene (see Table 4.5). The EGFR was identified in 1980 as the first member of the tyrosine kinase receptor type (see Section 1.3). Shortly afterward, the receptors for insulin, IGF-I and PDGF (platelet-derived growth factor) were also shown to be tyrosine kinase receptors. The free receptor is a single polypeptide chain. Binding of EGF to the receptor causes the receptor to form a dimer, which is in many cases a heterodimer of different receptor subtypes. Whether a homodimer or heterodimer is formed and which receptors are involved, dramatically affects the signalling properties of the receptor.

When the dimer is formed, the catalytic site of one chain phosphorylates a number of tyrosines in the adjacent chain. These phosphotyrosines serve as recognition sites for binding of signalling proteins that have a SH2 (src homology region 2) region. An example of these signalling proteins is phospholipase C (PLCγ1), which is activated by binding to the receptor, thereby increasing calcium mobilization and activating protein kinase C. PLCγ1 also has a SH3 domain, which allows it to bind to the cytoskeleton and provide access to the phosphatidylinositol substrate in the cell membrane. EGFR signalling is involved in regulating cell development and motility, protein secretion, wound healing and tumorigenesis. For more information, see the review by Wells (1999).

EFFECTS OF EGF ON FOLLICLES. EGF reduces hair growth in mice by reducing the number of proliferating cells in the follicle bulb and reducing the rate of follicle development. Mice treated with EGF produce hair that is shorter, finer and wavy. Inactivation of the TGFα gene in knockout mice also produces a wavy coat, while overexpression of this gene results in a thickened epidermis and stunted hair growth.

EGF treatment of sheep produces different effects in the follicle than in the surrounding epidermal cells. EGF treatment causes weakening in the unhardened wool fibres and a partial disruption of fibre growth in the inner root sheath cells in the keratogenous zone of the follicle bulb. A gradual inhibition of mitotic activity then occurs in the bulb cells over a period of 2–3 days, with a catagenic regression of the follicle. This is followed by asynchronous regeneration of the follicles 4–8 days after infusion of EGF. However, EGF has the opposite effect on cells in the epidermis and sebaceous gland, where cell proliferation is increased by 2 days after infusion of EGF. The mechanism behind the contrasting effects of EGF in the follicles and epidermis is not known. It may be the result of a massive disruption in the normal homeostatic mechanisms that regulate the activities of the different cell types in the skin.

Subcutaneous injections or intravenous infusions of 3–5 mg of EGF caused a transient regression of the wool follicles from the actively growing stage, and resulted in shedding of the entire fleece. Treatment with 1–3 mg of EGF resulted in a zone of weakness in the wool fibres 3-4 weeks later but the fleece was not shed. The weakness of the wool fibres is assessed by measuring the depilation force, which is the force required to pluck wool staples from the midback region of the sheep.

Following treatment with EGF, there were changes in the regrowth wool. The content of high-sulphur proteins increased from 19% to 30%, while the content of high-tyrosine proteins decreased from 12% to 5%. There were also alterations in the proportions of other components. The maximum change in composition was observed around 4 weeks, with the composition returning to normal by 10 weeks. The effects were dose dependent and were similar to those seen when chemical defleecing agents were used. This suggests that the changes in wool composition are characteristic of wool growing from newly regenerated or regenerating follicles.

Commercial use of EGF for endocrine defleecing uses Met-EGF®, which is a genetically engineered form of EGF produced in a recombinant *E. coli* expression system prepared by Pitman-Moore Ltd, New South Wales, Australia. It results in easy and consistent fleece removal.

OTHER EFFECTS OF EGF. Other effects of EGF are the requirement of sufficient time for regrowth of the wool, food rejection and anorexia, decreased gastrin secretion and gut motility, and an increase in abomasal pH. There can be erythema around the eyes and muzzle, and some rupturing of dermal blood vessels. In pregnant ewes, there can be a delayed onset of oestrus, abortions, increased GH and placental lactogen, decreased thyroxine, but no effect on pregnancy rate or lambing. In rams, there can be decreased sperm motility for about 12 weeks and increased plasma cortisol. Because of the negative effects of EGF on reproduction, it is recommended that producers wait for 5 weeks after using EGF before breeding.

Summary of growth factors affecting fibre growth

FGF-5 and TGFβ have been identified as markers of catagen, and they inhibit follicle growth. EGF and FGF can influence fibre characteristics. Fibre curvature may be controlled by factors that result in an asymmetric distribution of keratinocytes in the follicle bulb. Infusion of EGF stimulates epidermal thickening and also induces catagen in wool follicles prior to fleece shedding. For practical applications, it is easiest to treat all sheep about 12 weeks prior to breeding. EGF treatment does produce a variety of behavioural and physiological side-effects.

Continued research on the ultrastructure of wool follicles and the detailed mechanisms of the effects of different growth factors is needed. Information on the cellular mechanisms affected by EGF infusion is incomplete. The side-effects of Met-EGF and the reasons why Met-EGF is less variable need to be studied further. For further information, see Steensel *et al.* (2000) and Stenn and Paus (2001).

Questions for Study and Discussion

Section 4.1 Mammary Gland Development and Milk Production

1. Describe the structure of the mammary gland. Outline the structural changes that occur during sexual development, pregnancy and the drying off period. What model systems are used to study mammary gland function?
2. Describe the endocrine factors that regulate mammogenesis.
3. Describe the endocrine factors that regulate the initiation of lactogenesis.
4. Describe the factors that regulate galactopoiesis. Describe the role of bST in increasing milk production.
5. Describe factors that can affect the content of milk protein and milk fat.
6. Describe factors affecting ketosis and milk fever. What methods are used to control these metabolic diseases of lactation?

Section 4.2 Egg Production

1. Describe factors regulating sexual differentiation and gonad development in birds.
2. Describe the hormonal regulation of follicular development in poultry.
3. Describe the structure of the egg shell and the role of the shell matrix.
4. Discuss the regulation of calcium metabolism in shell formation.

Section 4.3 Wool Production and Endocrine Defleecing

1. Describe various nutritional and endocrine methods for defleecing sheep. What are the advantages and disadvantages of each method?
2. Describe the model systems that can be used to study the function of wool follicles.
3. Outline the roles of growth factors, particularly IGFs, FGFs and TGFβ, in wool follicles.
4. Describe the roles of the EGF family of growth factors and their receptors in the function of wool follicles.

Further Reading

Mammary gland development and milk production

Burton, J.L., McBride, B.W., Block, E., Glimm, D.R. and Kennelly, J.J. (1994) A review of bovine growth hormone. *Canadian Journal of Animal Science* 74, 167–201.

Clare, D.A. and Swaisgood, H.E. (2000) Bioactive milk peptides: a prospectus. *Journal of Dairy Science* 83, 1187–1195.

Foster, T.P. (1999) Somatotropin delivery to farmed animals. *Advanced Drug Delivery Reviews* 38, 151–165.

Horst, R.L., Goff, J.P., Reinhardt, T.A. and Buxton, D.R. (1997) Strategies for preventing milk fever in dairy cattle. *Journal of Dairy Science* 80, 1269–1280.

Kennelly, J.J. and Glimm, D.R. (1998) The biological potential to alter the composition of milk. *Canadian Journal of Animal Science* 78(Suppl.), 23–56.

Knight, C.H. and Wilde, C.J. (1993) Mammary cell changes during pregnancy and lactation. *Livestock Production Science* 35, 3–19.

Maas, J.A., Trout, D.R., Cant, J.P., McBride, B.W. and Poppi, D.P. (1995) Method for close arterial infusion of the lactating mammary gland. *Canadian Journal of Animal Science* 75, 345–349.

Molkentin, J. (1999) Bioactive lipids naturally occurring in bovine milk. *Nahrung* 43, 185–189.

National Research Council (1994) Effect of somatotropin on nutrient requirements of dairy cattle. In: *Metabolic Modifiers: Effects on the Nutrient Requirements of Food-Producing Animals*. NRC, pp. 23–29. Available at: www.nap.edu/openbook/0309049970/html

Serjrsen, K., Purup, S., Vestergaard, M. and Foldager, J. (2000) High body weight gain and reduced bovine mammary growth: physiological basis and implications for milk yield potential. *Domestic Animal Endocrinology* 19, 93–104.

Tucker, H.A. (2000) Hormones, mammary growth, and lactation: a 41-year perspective. *Journal of Dairy Science* 83, 874–884.

Egg production

Capel, B. (2000) The battle of the sexes. *Mechanisms of Development* 92, 89–103.

Etches, R.J. (1990) The ovulatory cycle of the hen. *Critical Reviews in Poultry Biology* 2, 293–318.

Moreau, J.D., Satterlee, D.G., Rejman, J.J., Cadd, G.G., Kousoulas, K.G. and Fioretti, W.C. (1998) Active immunization of Japanese Quail hens with a recombinant chicken inhibin fusion protein enhances production performance. *Poultry Science* 77, 894–901.

Nys, Y., Hincke, M.T., Arias, J.L., Garcia-Ruiz, J.M. and Solomon, S.E. (1999) Avian eggshell mineralization. *Poultry and Avian Biology Reviews* 10, 143–166.

Shimada, K. (2002) Sex determination and sex differentiation, *Poultry and Avian Biology Reviews* 13, 1–14.

Thurston, R.J. and Korn, N. (2000) Spermiogenesis in commercial poultry species: anatomy and control. *Poultry Science* 79, 1650–1668.

Wentworth, B.C. and Bitgood, J.J. (1988) Function of bilateral oviducts in double oviduct hens following surgery. *Poultry Science* 67, 1465–1468.

Wool production and endocrine defleecing

Bond, J.J., Wynn, P.C. and Moore, G.P.M. (1998) The effects of fibroblast growth factors 1 and 2 on fibre growth of wool follicles in culture. *Acta Dermato-Venereologica (Stockholm)* 78, 337–342.

Nixon, A.J. and Moore, G.P.M. (1998) Growth factors and their role in wool growth: a review. *Proceedings of the New Zealand Society for Animal Production* 58, 303–311.

Reis, P.J. and Panaretto, B.A. (1979) Chemical defleecing as a method of harvesting wool from sheep. *World Animal Review* 30, 36–42.

Steensel, M.A.M. van, Happle, R. and Steijlen, P.M. (2000) Molecular genetics of the hair follicles: the state of the art. *Proceedings of the Society of Experimental Biology and Medicine* 223, 1–7.

Stenn, K.S. and Paus, R. (2001) Controls of hair follicle cycling. *Physiological Reviews* 81, 449–484.

Wells, A. (1999) Molecules in focus – EGF receptor. *International Journal of Biochemistry and Cell Biology* 31, 637–645.

5
Endocrine Manipulation of Reproduction*

5.1 Manipulation of Reproduction in Mammals

This section focuses on the potential for the endocrine manipulation of reproduction, with a particular emphasis on females, since with the advent of artificial insemination this is usually the limiting factor in animal production systems. Unifying concepts are first presented, followed by discussion of species-specific information, as appropriate. This information is then used as background for understanding currently available methods of manipulating the system and for devising potentially useful methods for future application.

The process of sexual differentiation and maturation is first described, including the sexual differentiation of different tissues, the genes that are involved in these processes and the regulation of meiosis in germ cells. A description of the regulation of the oestrous cycle, pregnancy and parturition and the induction of puberty follows this. Opportunities and possibilities for the manipulation of the oestrous cycle are then discussed. This includes a description of the different hormonal preparations that are available, methods for detection of oestrus, the induction and synchronization of oestrus and the techniques of superovulation and embryo transfer. This is followed by a discussion of methods for maintaining pregnancy or inducing parturition. Finally, problems that occur during the postpartum interval, inducing puberty and advancing cycling in seasonal breeders are discussed.

Sexual differentiation and maturation

Sex determination is controlled by the genetic sex of the animal, which is fixed at the time of conception. In the process of sex differentiation, the genetic sex directs the development of either ovaries or testes, which then determines the phenotypic sex as the animal matures. Usually, a male genotype will produce a male phenotype and a female genotype will produce a female phenotype. However, in some fish and reptiles the phenotypic sex can differ from the genotypic sex due to environmental factors such as temperature. The administration of sex steroids at a critical point during development can also permanently alter the phenotypic sex in some species.

There is usually one pair of sex chromosomes, either XX/XY or ZW/ZZ for female/male determination. In the XX/XY system, the female is the homogametic sex, while in the ZW/ZZ system, the male is the homogametic sex and the female is the heterogametic sex. The ZW/ZZ system is found in birds, reptiles and some fish and amphibians. Other systems of polygenic sex determination involving multiple chromosomes with

---

*This chapter reviews hormonal manipulation of reproduction in mammals and in farmed fish. Chapter 4 has already covered certain aspects of reproduction in poultry.

sex determining genes are also known in some lower vertebrate species. Polygenic sex determination produces male:female ratios that are different from the 1:1 ratio seen with pure sex chromosome systems.

Differentiation of the gonads and ducts

Sex differentiation relates to the development of the phenotypic sex from the genetic sex. This includes the migration of primordial germ cells, the development of gonadal ridges and the differentiation of the gonads into ovaries or testes, which is controlled by sex-determining genes. In the absence of sex-determining genes on the Y chromosome to initiate testis formation, the undifferentiated gonads in the early embryo develop into ovaries and lead to a female phenotype. In mammals, organization of spermatic cords during testicular development begins earlier in gestation than organization of the primordial follicles during ovarian development.

In the undifferentiated stage of gonadal development, both Wolffian and Müllerian ducts are present and these later develop into the male or female accessory sex organs. The Müllerian ducts differentiate to form the Fallopian tubes, uterus and vagina, while the Wolffian ducts form the epididymis, ductus deferens and seminal vesicles. The Sertoli cells of the fetal testes produce anti-Müllerian hormone (AMH, also known as Müllerian inhibitory substance, MIS), which is a glycoprotein related to the TGFβ family of peptides. AMH acts locally to induce regression of the Müllerian ducts and to inhibit the expression of aromatase, but is not required for testis determination (see also Section 4.2 for information on sexual development in birds). The development of male external genitalia is dependent on androgens produced by the testis. In the absence of androgens, the female phenotype develops. Testosterone is converted to the ultimate androgen, dihydrotestosterone (DHT), by the 5α-reductase enzyme to stimulate the differentiation of male genitalia. In addition to the androgenic effects of testosterone and DHT, these steroids also produce anabolic effects on various tissues (see Section 3.2).

Differentiation of the brain

The undifferentiated mammalian brain is inherently female, but androgens act on the brain to programme male behaviour and activity of the hypothalamus in controlling pituitary function. Sexual differentiation is a permanent change due to gonadal steroids, which alters the sensitivity of a behaviour to activation by sex steroids later in life. Sexual differentiation of behaviour includes both masculinization, which is the development of male behaviours such as mounting and copulation, and defeminization, which is the loss of sexual receptive behaviour, lordosis and the surge of luteinizing hormone (LH) in response to oestradiol. The release of pituitary trophic hormones, such as somatotrophin and the gonadotrophins, is pulsatile in females and more uniform or tonic in males (see Section 3.4).

The neonatal testis produces androgens that act at a critical time in development to programme or 'imprint' a male pattern of function in the hypothalamus. This critical period, which is species specific, may be related to the stage of neuronal development in the hypothalamus and can be affected by thyroid hormones. Generally speaking, in animals with longer gestation periods that are more developed at birth, the critical period occurs prenatally, while for those with shorter gestation periods that are less mature at birth the critical period extends into the early postnatal period. The sexual dimorphic development of the brain results in permanent differences in 'hard wiring' of the brain. This includes differences in neuronal growth, cell death, synthesis of neurotransmitters, the synaptic connections of nerve cells, and in the size of particular regions, such as the sexually dimorphic nucleus of the preoptic area (POA). The POA is involved in regulating the release of gonadotrophins due to feedback from oestrogen and progesterone.

The effect of androgens on the differentiation of the hypothalamus actually occurs via oestrogens, which are synthesized from testosterone by the aromatase enzyme in the brain (see Section 1.2). The administration of dihydrotestosterone, which cannot be aromatized, does not masculinize the brain; however, exogenous oestrogens will have this effect. The fetal female brain is protected from the minor levels of circulating oestrogens by the fetoneonatal oestrogen binding protein (FEBP) that is present in high levels in the circulation. FEBP has a high specific binding affinity for oestrogen and prevents it from reaching the brain. Synthetic oestrogens that have a low affinity for FEBP will also masculinize the brain. In birds, oestrogen results in a female pattern of brain differentiation, while androgens are important in the control of song in male birds. While oestrogens are required for ovary development in non-mammalian vertebrates, they are not required for the initial differentiation of the reproductive tract in mammals.

The development of a number of other tissues is also sexually dimorphic. For example, there are marked sex differences in steroid metabolism in the liver, and this can affect the metabolism and clearance of drugs, hormones and xenobiotics. In rats, the pattern of steroid metabolism in the liver is regulated by the pattern of growth hormone secretion from the pituitary, and removal of the pituitary results in a male pattern of steroid metabolism. The pattern of growth hormone secretion is due to imprinting of the hypothalamus by exposure to androgens during neonatal life. A pulsatile pattern of growth hormone release occurs in male rats and can be mimicked by intermittent injections, while a more constant release of growth hormone occurs in females and can be mimicked by a constant infusion technique. Growth hormone regulates the expression of various steroid metabolizing cytochrome P450 (CYP) genes in the liver, by a mechanism that does not appear to involve IGF-I. For more information, see the article by Gustafsson (1994) and other articles in the book by Short and Balaban (1994).

Sex differentiation in cattle, sheep and pigs

Sexual differentiation occurs prenatally in sheep and cattle, but is completed postnatally in swine. The pattern of secretion of testicular steroids differs among cattle, sheep and swine. Boars have three periods of elevated steroids during development, while rams and bulls have only two.

Differentiation of the gonads in cattle occurs from day 45 to day 70 of gestation. Female fetuses that are treated with exogenous testosterone show masculinization of the gonads. Sexual differentiation of the brain in cattle occurs after day 60 of gestation, as treatment with exogenous testosterone before this sensitive period does not affect brain function later in development. Steroid synthesis in female ovaries is low from day 70 to day 85, while levels of androstenedione are high in male fetuses until day 100, thus providing steroid precursors for male programming of the hypothalamus.

Testosterone production by the fetal testis in sheep increases from day 35 of gestation to a peak at day 70. Masculinization of the gonads occurs from day 35 to day 45, while male differentiation of the brain occurs from day 50 to day 80. The reduced steroid production by the ovary from day 45 to day 80 protects the female brain from masculinization. Male lambs need additional exposure to sex steroids during the critical period from 4 to 8 weeks after birth in order for male sexual differentiation to be completed.

Boars have three periods of enhanced testicular development. Levels of testosterone are increased from day 35 to day 40 of gestation and differentiation of the gonads is complete by day 45. The second period occurs shortly after birth, when the number of Leydig cells increases. During this early postnatal period, the testis is actively producing steroids due to a lack of negative feedback control of steroids on LH secretion in males at this time. The third period of testicular development occurs at puberty when spermatogenesis occurs. Boars castrated within the first 2 months after birth show female behaviours after treatment with oestrogen as adults, including receptivity to mature boars. However, males castrated after 6 months of age show little female behaviour after oestrogen treatment. This suggests that male programming of the brain in boars occurs during the prepubertal period at 3–5 months of age.

Over 90% of female twins from mixed-sex pregnancies in cattle are freemartins and the syndrome is also known in goats, sheep and pigs. A freemartin is a sterile XX/XY chimera produced from a female twin that has a placenta that is fused with a male twin. The blood systems of the embryos are joined through vascular anastomoses in the fused placenta of the twins. This allows the transfer of AMH, androgens and possibly blood cells from the male to the female fetus, interfering with development of the ovaries and paramesonephric ducts. The gonads of the freemartin range from modified ovaries to testes-like structures that can be retained in the abdomen. The hypothalamus is also masculinized by exposure to androgens and the animal can display male patterns of behaviour.

The risk of freemartinism is greatly increased in sheep with litters of four or more lambs. Placental fusion occurs rarely in mice, and a band of connective tissue separates the two placentas and prevents the formation of a freemartin. This may be a protective mechanism that has evolved in litter-bearing species and may explain the very low incidence of freemartins in pigs. For more information on sex differentiation in cattle, sheep and pigs, see the review by Ford and D'Occhio (1989).

Sex-determining genes

The key sex-determining genes SRY (sex-determining gene on Y) and SOX9 (SRY box) have been identified in mammals. These genes are involved in testis formation and in the development of the male phenotype. In the absence of the SRY gene, ovaries develop. SRY is located on the short arm of the Y chromosome and codes for a protein called the testis-determining factor (TDF), which has an 80-amino-acid HMG (high-mobility group) domain that can bind to and bend DNA. This region also binds calmodulin (see Section 1.3) and contains nuclear localization signals. Mutations in the HMG domain are found in some XY sex-reversed individuals. In mice, the TDF also has a large glutamine-rich repeat domain that is involved in transcriptional activation. SRY is implicated in the differentiation of Sertoli cells, the migration of primordial germ cells from the mesonephros to the genital ridges and the proliferation of cells within the genital ridges. The exact role of SRY in testis differentiation is not known, but SRY is thought to trigger each of these events, possibly by activating a secondary gene, such as SOX9.

SOX9 is up regulated in the genital ridges of male embryos and down regulated in females. The SOX9 protein has an HMG domain and two transcription activation domains and binds to the AMH promoter to increase the expression of the AMH gene. The expression of SOX9 may be reduced by the anti-testis gene DAX1 (DSS-adrenal hypoplasia critical region of the X chromosome), which is expressed more in females than in males. Steroidogenic factor 1 (SF-1) also plays a role in regulating SOX9 expression and in the activation of AMH gene expression by SOX9. In mammals, levels of SF-1 are maintained in males but not in females after gonadal differentiation has begun. Since the AMH gene is not required for testis differentiation, SOX9 and SF-1 proteins must act on other genes besides AMH during testis differentiation. The DMRT1 (DM domain related transcription factor 1) gene is related to sex-determining genes in other species and is expressed in the developing testis in XY males. It is also present on the chicken Z chromosome and is a good candidate for a sex-determining gene.

The WNT genes code for growth and differentiation factors that appear to play a role in the development of the ovaries and oocytes. WNT-4 may act as a suppressor of Leydig cell function in females and is necessary for the formation of Müllerian ducts. WNT-4-deficient females show a partial sex reversal. WNT-7α is involved in the further development of the Müllerian ducts to form the oviduct and uterus. For more information on genes that affect sex differentiation, see the reviews by Haqq and Donahoe (1998), Heikkila *et al.* (2001) and Koopman *et al.* (2001).

Regulation of meiosis in germ cells

Male gametes, or spermatozoa, are produced from a continuously replenishing population of spermatogonia. Meiosis begins in male germ cells at puberty and continues throughout the reproductive life of the animal. This ensures an abundant supply of male gametes for fertilization. In contrast, the process of folliculogenesis in females produces only a restricted number of oocytes, with meiosis occurring during specific stages of development. The fetal ovary contains all the available oocytes at birth, as a pool of primordial follicles that are arrested in the late prophase or diplotene stage of meiosis. These follicles begin to grow again as the female nears puberty and ovulation begins. An understanding of the factors that regulate meiosis in germ cells could lead to applications for controlling fertility.

During development of the gonads, the fate of the germ cells depends on the surrounding cells in the tissue. Oocytes in the cortex of the developing ovary are the first to reach the diplotene stage and be enclosed in follicles. In the testis, the prespermatogonia are enclosed in testicular cords when the testis differentiates, and meiosis does not begin until puberty. Meiosis in both sexes is thought to be triggered by meiosis activating substance (MAS), and to be prevented within the testicular cords by meiosis preventing substance (MPS). MAS is found in preovulatory follicular fluid and in adult testis, and can activate meiosis in both male and female germ cells and across various species (humans, cattle, mice). MAS has been identified as two steroids in the cholesterol biosynthesis pathway (Fig. 5.1), namely 4,4-dimethyl-5α-cholest-8,24-diene-3β-ol and 4,4-dimethyl-5α-cholest-8,14,24-triene-3β-ol. When oocytes surrounded by cumulus cells are treated with FSH, MAS is produced by the cumulus cells and acts in a paracrine manner to stimulate meiosis in the oocyte.

Growth differentiation factor 9 (GDF-9), a member of the TGFβ family, is an oocyte-specific factor made throughout folliculogenesis. GDF-9 is required for the development of follicles but not for oocyte growth and differentiation in mice.

It may be possible to control fertility with MAS analogues. These compounds should not affect the normal synthesis or actions of steroid hormones. MAS agonists would be expected to increase the production of spermatogonia and the maturation of oocytes, while MAS antagonists would act as contraceptive agents. Inhibition of the 14α-reductase or stimulation of 14α-demethylase (CYP51) would increase levels of MAS in tissues. For more information, see the reviews by Byskov et al. (1998) and Albertini and Carabatsos (1998).

Regulation of the oestrous cycle

In this section, the natural mechanisms regulating the oestrous cycle are presented. This information is expanded in a later section that describes methods for manipulating the cycle.

Reproductive performance in dairy cattle has decreased with increasing milk production in dairy herds. Information on factors affecting the development of follicles, oestrous behaviour, ovulation and corpus luteum development and regression may be used to devise methods for improved reproductive performance. In dairy cattle, the goal is to manipulate the reproductive cycle to maximize milk production. In beef cattle, the goal is to maximize the number of calves produced by rebreeding the female within a set interval. The production of one calf per year for dairy and beef cows is considered optimal. Similar optimal goals for sows and ewes would be 30 piglets or two lambs per year, respectively.

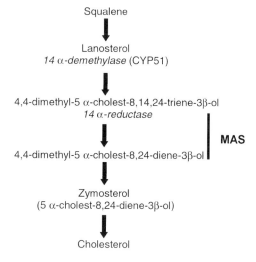

Fig. 5.1. Cholesterol biosynthesis pathway and formation of meiosis activating substance (MAS) (Byskov et al., 1998).

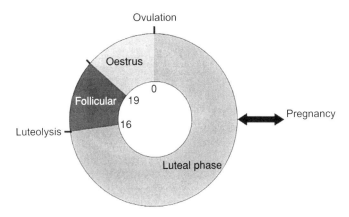

Fig. 5.2. Outline of the oestrous cycle.

Overview of the oestrous cycle

In the absence of pregnancy, the mature female undergoes a continuous series of reproductive cycles (Fig. 5.2) in which a group of ovarian follicles mature (follicular phase), the female becomes receptive to mating (oestrus or heat), and the dominant follicle ovulates and a corpus luteum (CL) is formed (luteal phase). The number of follicles ovulated and the CL formed is a characteristic of a species or sometimes a strain within a species. If fertilization of the egg and implantation of the embryo does not occur, the CL regresses (luteolysis) and the cycle repeats itself. The cycle lasts for approximately 21 days for cattle, pigs and goats, from 19–25 days for horses and 16–17 days for sheep.

Follicular development

The maturation of ovarian follicles is driven by the gonadotrophic hormones luteinizing hormone (LH) and follicle stimulating hormone (FSH) that are released from the anterior pituitary (Fig. 5.3). The release of the gonadotrophic hormones is driven by gonadotrophin releasing hormone (GnRH) produced by the hypothalamus. GnRH is delivered to the anterior pituitary via the hypothalamic–hypophyseal portal blood vessels (see Section 1.4). Both GnRH and the gonadotrophins are released in a pulsatile manner, with the pulse frequency and amplitude dramatically affecting biological activity. Variations in the frequency of pulsatile GnRH release have differential effects on LH and FSH production. Many external factors affect the activity of the pulse generator in the central nervous system and thus affect the activity of the reproductive system. The factors include nutrition, stress, suckling, presence of males, season and visual and olfactory cues.

LH and FSH are glycoprotein hormones that are each comprised of two peptide chains, a common α-chain of 90 amino acids and a β-chain that is hormone specific. Pig LH-β has 119 amino acids and pig FSH-β has 107 amino acids. The gonadotrophins have multiple roles, including controlling the development of ovarian follicles, ovulation, and formation and function of the corpus luteum, and regulating the production of gonadal hormones. In the male, LH stimulates steroidogenesis in Leydig cells and FSH binds to Sertoli cells. Inhibin, activin and follistatin (activin binding protein) are produced by the gonads and regulate the release of FSH by the pituitary. Inhibin and activin are members of the TGFβ family of polypeptides. Inhibin reduces the production of FSH, while activin increases FSH production independently of GnRH. Follistatins are a family of monomeric glycoproteins that bind to activin and prevent the stimulation of FSH production. For more information on inhibin, activin and follistatin, see Findlay (1993) and for activin and its receptor, see Peng and Mukai (2000).

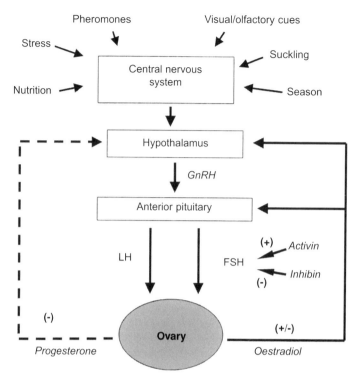

Fig. 5.3. Hormonal regulation of the oestrous cycle.

At birth, the mammalian ovary contains a large number of primary follicles (about 500,000 in cows) that gradually grow and mature as the animal matures. These primary follicles consist of an oocyte surrounded by a single layer of granulosa cells and interstitial tissue, which keep the ovum suspended in the first stage of meiotic division. As the follicle grows, the oocyte increases in diameter and is surrounded by a zona pellucida. The number of layers of follicular cells also increases and a second mass of theca cells develops. The vast majority of follicles will undergo atresia after maturation, since only a limited number of ovulations occur (over a 15-year period a cow can ovulate at most only 300 follicles). Growth of the follicles up to 3 mm in diameter is independent of FSH, while FSH is required for growth of the follicles from 3 to 10 mm in diameter. A group of follicles produce activin and are 'recruited' to begin maturation due to a transient increase in FSH. The size of this group can vary from 50 follicles in pigs to 5–10 in cattle and 1–4 in horses. The granulosa cells of the larger developing follicles produce oestradiol, using androgens that are produced by the theca cells. The granulosa cells of the large developing follicles also produce inhibin, while the other developing follicles produce follistatin, which together inhibit the release of FSH from the pituitary.

IGF-I activity also increases as the follicles develop, due to decreased levels of IGF-binding proteins. IGF-I may play a more important role in the growth of small follicles to recruitment, rather than in the growth of large follicles. Treatment with exogenous somatotrophin increases IGF-I and the number of small follicles, and also increases the number of corpora lutea in cattle and large follicles in pigs. Somatotrophin acts both by increasing cell proliferation and decreasing apoptosis and atresia of follicles.

As FSH levels decline, one follicle (or more in litter-bearing species such as pigs) acquires LH receptors on its granulosa cells and becomes the dominant follicle. Growth of this dominant follicle to a preovulatory follicle is dependent on pulses of LH. The growth of the other subordinate follicles is depressed due to the lack of FSH and they undergo atresia. If there are insufficient pulses of LH, the dominant follicle also undergoes atresia, while a surge in LH results in its ovulation. A new wave of follicular growth (Fig. 5.4) occurs about every 10 days in cattle, so there will be 2–3 follicular waves during each oestrous cycle. Sheep have 3–4 waves, goats have 4–5 waves and horses have 1–2 waves in each cycle. Waves of follicular growth start before puberty and continue even throughout most of pregnancy, although during these times the follicles do not ovulate and become atretic. There is apparently little evidence for waves of follicular growth in pigs.

Prolificacy in sheep is due to large genetic differences in ovulation rate among different breeds of sheep. Increased ovulation in the Booroola breed is due to a single gene, while in the Finn and Romanov breeds, increased ovulations are due to multiple genes. Waves of follicular development occur in all genotypes. The increased ovulation rate in Finn ewes is due to an extended period of follicular recruitment, which allows follicles to be maintained from one wave to the next.

Oestrus and ovulation

The dominant follicle produces oestradiol, which acts on the brain to induce oestrous behaviour or 'heat'. Oestradiol also stimulates the production of oestradiol receptors in the hypothalamus and pituitary. Oestradiol then acts by positive feedback on the pituitary to increase LH production, while reducing FSH levels by negative feedback. This leads to a surge in LH release that culminates in rupture of the follicle and release of the ovum (**ovulation**) and formation of the corpus luteum (CL or yellow body) on the ovary.

Luteal phase

The CL develops from the follicle after ovulation and is present during a large part of the oestrous cycle known as the luteal phase. After ovulation, the theca cells degenerate and the granulosa cells hypertrophy and luteinize to form lutein cells. The lutein cells produce progesterone, which inhibits GnRH secretion by the hypothalamus and thereby decreases pulsatile LH release. The CL grows rapidly and progesterone production increases at the beginning of the cycle (days 3–12 in the cow and days 2–8 in the ewe and sow) and then remains constant until day 15–16, when regression (**luteolysis**) begins unless fertilization occurs. The presence of a functional CL during the luteal phase of the cycle prevents ovulation, and any follicles that mature during the luteal phase will undergo atresia (except in the mare). Elevated levels of cortisol can also block or delay the preovulatory LH surge.

Oestrogen acts on the uterus to increase

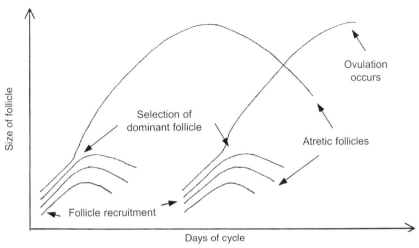

Fig. 5.4. Waves of follicular growth and development.

the number of receptors for oestrogen and oxytocin. If fertilization and implantation do not occur, high levels of progesterone and oxytocin from the ovary stimulate the uterus to secrete prostaglandin $F_{2\alpha}$ ($PGF_{2\alpha}$). $PGF_{2\alpha}$ causes regression of the CL by interfering with LH action on the CL and increases the production of oxytocin by the ovary. A new follicular phase then begins as the follicles mature to ovulation. For more information on follicular development and reproductive cycles in farm animals, see the reviews by Wiltbank (1998) and Driancourt (2001).

Pregnancy

If fertilization occurs, the embryo implants in the uterine endometrium and produces proteins that prevent regression of the CL. A functional CL is necessary for the maintenance of pregnancy in all farm animals. For pigs, about four embryos are required to sustain a pregnancy. In sheep, cattle and goats, a group of antiluteolytic proteins are produced by the conceptus. These were named trophoblast protein-1 (TP-1) and later renamed interferon-τ (IFN-τ) due to the sequence similarity with type-I interferon. IFN-τ decreases the production of $PGF_{2\alpha}$ by the uterus. The decrease in $PGF_{2\alpha}$ suppresses luteolysis, so the CL is maintained and continues to produce progesterone.

Progesterone is required for the maintenance of pregnancy and is produced by the ovary, placenta and adrenal gland. The relative importance of these sources depends on the species and the stage of gestation. Progestogens alter the ionic permeability of the muscles of the myometrium to decrease the excitability of the cells and they also serve as precursors for the synthesis of other steroids. The placenta secretes increasing amounts of oestrogens during the second and third trimesters. High levels of progesterone and oestrogen reduce GnRH release from the hypothalamus and LH release by the pituitary. Supplementing with exogenous progesterone may be useful in maintaining pregnancy.

Parturition

The increased production of corticosteroids by the fetus signals the beginning of parturition, and reducing the production of cortisol using corticotrophin releasing hormone (CRH) receptor antagonists (see Section 6.3) delays parturition. Increased levels of corticosteroids stimulate the production of oestrogens by the placenta. Oestrogens stimulate the production of $PGF_{2\alpha}$ by the uterus and increase the number of oxytocin receptors in the uterus. Oxytocin is released by the posterior pituitary in response to the fetus entering the birth canal. Oxytocin increases the production of $PGF_{2\alpha}$ and both hormones increase contractions in the uterus.

Puberty and seasonality

An animal reaches the puberty stage of sexual maturation when it is able to display complete sexual behaviour and produce and release gametes. During the onset of puberty, both the amplitude and frequency of LH pulses from the pituitary increase because the hypothalamic–pituitary axis becomes less responsive to negative feedback from oestradiol. Regular oestrous cycles then begin in females with the maturation of the hypothalamic–pituitary axis.

The onset of puberty is influenced primarily by reaching a target age and weight, although the age at puberty varies with the breed, nutritional status and environmental effects such as photoperiod and temperature. Puberty occurs at around 6–7 months of age in sheep, goats and pigs and at around 12 months of age in cattle.

There are seasonal effects on reproductive efficiency in cattle, but cattle are not considered to be true seasonal breeders. Season plays a role in the induction of puberty in cattle, with heifers born in the autumn reaching puberty before those born in the spring. Animals born in the autumn are from 6–12 months of age when spring arrives, and the increased photoperiod during the spring stimulates the hypothalamic–pituitary–ovarian axis. Seasonal differences in photoperiod can affect the period of postpartum anoestrus by 10–35 days. Cows calving in the period of lengthening photoperiod from late spring to early autumn have a shorter period of anoestrus than animals that calve in the winter, when the photoperiod is decreasing.

The domestic pig is capable of producing piglets throughout the year, although there is reduced fertility in the late summer and early autumn. During this time, the seasonal breeding European wild boar is in seasonal anoestrus. Seasonal infertility results in a reduced farrowing rate, delayed puberty in gilts and a prolonged weaning to oestrus interval.

True seasonal breeders include sheep, goats and horses, as well as buffaloes, which are important in Asia, and caribou and reindeer, which are raised in northern regions. Sheep, goats and deer are short-day breeders, with increased reproductive activity during the shorter days in the autumn, while horses are long-day breeders, with increased reproductive activity during the longer days in the summer. Seasonal differences in anoestrus in wild species allow the birth of young when food is abundant and survival of the young is more likely. During the anoestrous period, low levels of oestradiol inhibit the secretion of LH from the pituitary. The development of follicles can continue in some species during the period of anoestrus, but the levels of LH are insufficient for the final stages of follicular development. In other seasonal breeders, follicular development ceases during the anoestrous period. Testis size and sperm production is also reduced in rams during the non-breeding season. Season affects the time to reach puberty in sheep, with ewe lambs born in the autumn not showing oestrus until after the summer anoestrus.

During the breeding season, the negative feedback of oestradiol on LH secretion is reduced. The change in photoperiod increases LH pulse frequency, and FSH levels increase to stimulate the production and maturation of large, active follicles. Eventually, a dominant follicle develops and ovulation occurs. The first oestrus that occurs in mares during the spring transition period after the winter anoestrus is typically erratic and long in duration and it is difficult to determine the time of ovulation.

Seasonal differences in photoperiod are detected by the pineal gland, which receives light impulses from the retina of the eye by way of the optic nerve. The pineal gland releases melatonin (N-acetyl-5-methoxytryptamine, Fig. 5.5) during periods of darkness, but not during light periods, and the pattern of melatonin release regulates the pulsatile release of GnRH from the hypothalamus. In this way, the pattern of melatonin release regulates seasonal rhythms in both short- and long-day breeders, with increased melatonin inhibiting GnRH release in long-day breeders and stimulating GnRH release in short-day breeders. Melatonin does not act directly on the GnRH neurones, but acts indirectly through other neurones that finally synapse on the GnRH neurones. Melatonin affects the negative feedback of steroids on GnRH secretion. For more information on melatonin and its effects, see the reviews by Malpaux *et al.* (1999) and Zawilska and Nowak (1999).

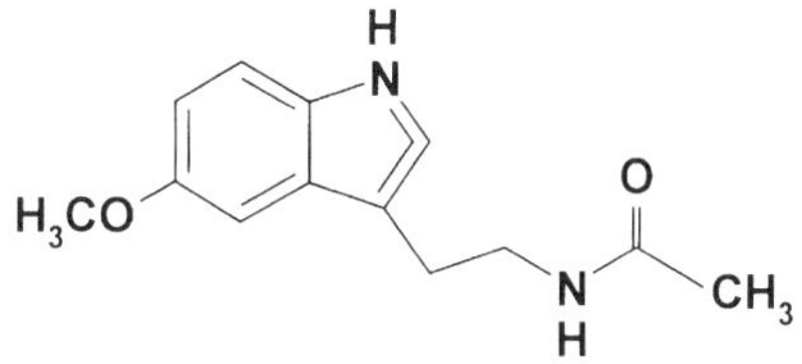

Fig. 5.5. Structure of melatonin.

Regulation of LH production

Dopamine is involved in the negative feedback of oestrogen on neurones that produce GnRH (Fig. 5.6), and injection of dopamine antagonists (pimozide, domperidone, perphenazine, sulpiride) increases LH secretion in seasonally anoestrous ewes and mares. Opioids, such as β-endorphin, are also involved in the reduced gonadotrophin secretion in seasonally anoestrous mares, and administration of the opioid antagonist naloxone advances the first ovulation of the year. Opioids are also thought to reduce the production of GnRH in cycling animals following ovulation, while glutamate, nitric oxide (NO), carbon monoxide and neuropeptide Y stimulate neurones that produce GnRH.

NO is a gaseous chemical messenger that acts locally in tissues and is involved in vas-

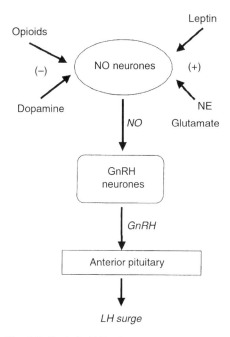

Fig. 5.6. Control of LH release.

cular relaxation, activation of the immune system, platelet function and as a neurotransmitter. NO is synthesized from arginine by the calcium-sensitive enzyme, nitric oxide synthase (NOS). NOS exists in a number of different tissue-specific isoforms, including a neuronal nNOS, an inducible iNOS in immune cells and endothelial eNOS. NO is involved in the LH surge mechanism and also acts in the ovary to affect steroidogenesis, ovulation and luteolysis. NO is produced by nNOS in specific neurones and diffuses to adjacent GnRH neurones. In the GnRH neurones, NO activates soluble guanylate cyclase to increase levels of cGMP, and cyclooxygenase to increase synthesis of prostaglandins ($PGE_{2\alpha}$), which leads to increased release of GnRH. Leptin acts via increased NO in the hypothalamus and pituitary to increase release of LH. Glutamate and noradrenaline (norepinephrine) neurones also stimulate NO production. In the ovary, the activity of NOS increases during the preovulatory LH surge, stimulating prostaglandin production and the inflammatory process at ovulation. For more information, see the review by Dhandapani and Brann (2000).

Regulation of steroidogenesis

Primarily, LH and FSH regulate the production of steroid hormones by ovarian cells, although somatotrophin also plays an important role. Somatotrophin may act by potentiating the effects of gonadotrophins through upregulation of gonadotrophin receptors. Somatotrophin increases the synthesis of some steroidogenic enzymes during the follicular and luteal phases to increase the secretion of oestradiol and progesterone. Somatotrophin increases the synthesis of steroid acute regulatory protein (StAR), which is responsible for translocating cholesterol into the mitochondria. It also increases the synthesis of cytochrome P450scc (CYP11A), which catalyses the side-chain cleavage step of cholesterol to form pregnenolone; CYP17, which converts pregnenolone and progesterone to androgens; and aromatase (CYP19, CYParom), which converts testosterone to oestrogen. These effects are mediated in part by increased production of IGF-I and in part by direct action of somatotrophin. Somatotrophin can be produced by ovarian tissue, to act locally, as well as by the pituitary.

Reducing the enterohepatic circulation can decrease levels of steroid hormones. Steroid hormones are metabolized in the liver and eliminated in the bile as conjugates with sulphate or glucuronic acid (see Section 1.2). These conjugates can be removed by bacterial metabolism in the intestine and the free steroids reabsorbed into the circulation. Increasing the liver metabolism of steroids, by treating with inducers of cytochrome P450, such as phenobarbital, or decreasing the absorption of steroids from the intestine, using binding agents or mineral oil, will increase the elimination of steroids and lower plasma steroid levels.

Manipulation of the oestrous cycle

During the normal reproductive cycle, a group of ovarian follicles matures during the follicular phase, the female becomes receptive to mating during oestrus, the dominant follicle ovulates and a corpus luteum (CL) is formed during the luteal phase. The repro-

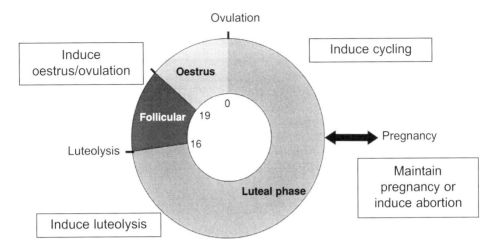

Fig. 5.7. Opportunities for manipulation of the oestrous cycle.

ductive cycle can be manipulated in a number of ways to improve reproductive efficiency (Fig. 5.7). In the cycling female, luteolysis can be induced, oestrus and ovulation can be induced or regulated, and the number of follicles that ovulate can be increased. The pregnant female can be treated to either maintain pregnancy or to induce parturition. The noncycling female can be treated to induce cycling, as in inducing puberty and ending seasonal anoestrus. For more information, see Wright and Malmo (1992).

Hormone preparations for manipulating reproduction

A number of hormone preparations are commercially available for the manipulation of fertility. This includes GnRH analogues, gonadotrophin preparations, synthetic progestogens, oestrogens, corticosteroid, and prostaglandins. In the United States, only GnRH and its analogues and $PGF_{2\alpha}$ and its analogues are approved for use on lactating dairy cows, although other hormonal treatments can be used in heifers and beef cattle.

The commercially available GnRH and gonadotrophin preparations are summarized in Table 5.1. Natural GnRH is a decapeptide (5-oxo-Pro-His-Trp-Ser-Tyr-Gly-Leu-Arg-Pro-Gly), also known as gonadorelin. Potent analogues with higher receptor-binding affinities and longer half-lives include deslorelin ([D-Trp6]-GnRH) and buserelin ([D-Ser(tBu)6, Pro9 NEt]-GnRH). Treatment with 2.5 mg of GnRH IV produces pulses of LH, while increasing the dose to 250 µg produces a surge in LH release that can cause ovulation of mature follicles.

Pregnant mare serum gonadotrophin (PMSG), also known as equine chorionic gonadotrophin (eCG), has primarily long-acting FSH activity, but also some LH activity, and is used for oestrus induction and superovulation in cows and sheep. A number of FSH preparations of pituitary origin are available. Human chorionic gonadotrophin, isolated from the urine of pregnant women, has LH activity and is used to induce ovulation of mature follicles. LH from pituitary origin is also available. PG600 is a combination of PMSG and hCG that is used for puberty induction in gilts.

The commercially available progestogens, oestrogens and corticoids are summarized in Table 5.2 and the structures of orally active progestogens are given in Fig. 5.8. The main progestogen is progesterone, which primes the brain for oestrous behaviour after oestrogen exposure, suppresses the secretion of GnRH and is required for the maintenance of pregnancy. Progestogens are used in treatment programmes to induce or synchronize ovulation. Pretreatment with progestogens is

Table 5.1. GnRH and gonadotrophin preparations.

Category	Hormone	Biological name or activity	Commercial name and company
GnRH	Gonadorelin	Natural GnRH	Cystorelin® (Elf Sanofi), Factrel®, (Ayerst) and Fertagyl® (Intervet)
	Deslorelin	([D-Trp6]-GnRH)	Ovuplant® (Ayerst)
	Buserelin	([D-Ser(tBu)6, Pro9 NEt]-GnRH)	
	Nafarelin		
GnRH antagonists	Nal-Glu, Cetrorelix		
Gonadotrophins	PMSG/eCG	Long-acting FSH activity, some LH activity	Equinex® (Ayerst) and Folligon® (Intervet)
	Pituitary FSH		FSH-P® (Schering) and Folltropin-v® (Vetrapharm), Ovagen® (ICP), NIH-FSH-S8®
	hCG	LH activity	Chorulon® (Intervet)
	Lutropin-V	Pituitary LH	Lutropin-V® (Vetrepharm)
	PMSG and hCG combination		PG600® (Intervet)

necessary for the development of LH receptors in preovulatory follicles and for normal luteal function after ovulation is induced with GnRH or LH.

Synthetic progestogens that can be given orally include melengestrol acetate (MGA), 6-methyl-17-acetoxy-progesterone (MAP) and 6-chloro-8-dehydro-17-acetoxy-progesterone (CAP). Subcutaneous implants of progestogens, progesterone-releasing intravaginal devices (PRIDs) and controlled internal drug releasing devices (CIDRs) are popular methods for producing a sustained release of progesterone over a 7–10 day period. Intravaginal sponges containing MAP are used in ewes. Altrenogest is used in mares and fluorogestone acetate (FGA) is used to induce oestrus in ewes.

Table 5.2. Progestogens, oestrogens and corticoids.

Category	Hormone	Biological name or activity	Commercial name and company
Progestogens		Orally active progestogens	MGA, MAP (Provera), CAP, Altrenogest® (Regu-Mate® from Hoechst), fluorogestone acetate, Levonorgestrel®, Desogestrol®
	Norgestomet	Progestogen implant	Synchro-mate-B® (Ceva laboratories)
Oestrogen		Long-acting oestrogen	ECP (Upjohn)
Corticosteroids	Flumethasone, dexamethasone	Short-acting analogues	
	Dexamethasone trimethylacetate, triamcinolone acetonide	Long-acting analogues	

MGA, melengestrol acetate; MAP, 6-methyl-17-acetoxy-progesterone; CAP, 6-chloro-8-dehydro-17-acetoxy-progesterone; ECP, oestradiol cypionate.

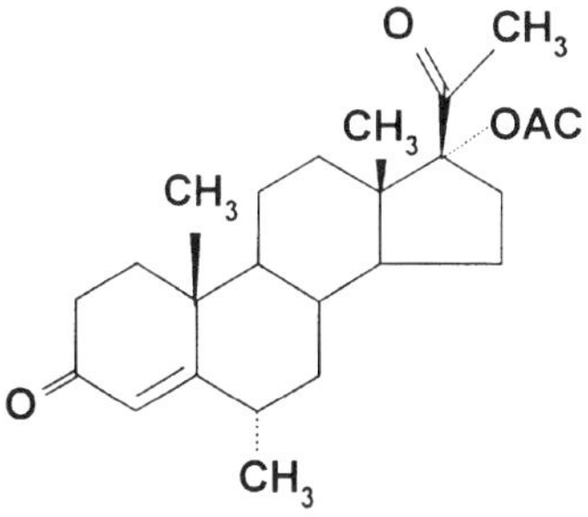

17-Acetoxy-6-methylprogesterone (MAP)

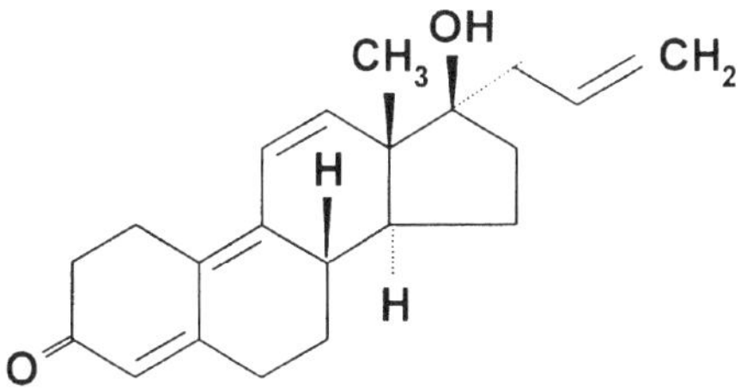

Altrenogest

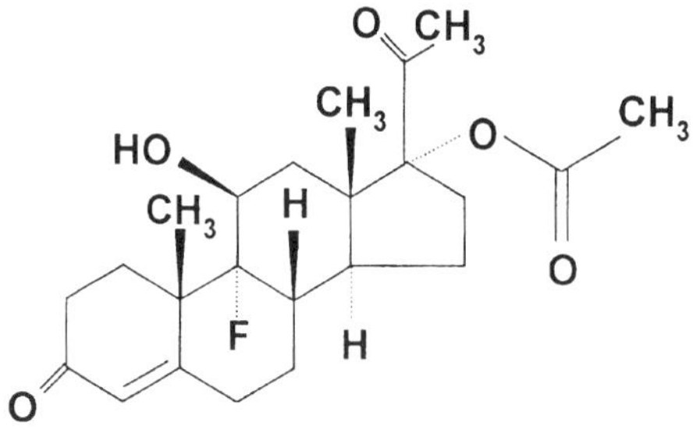

Fluorogestone acetate

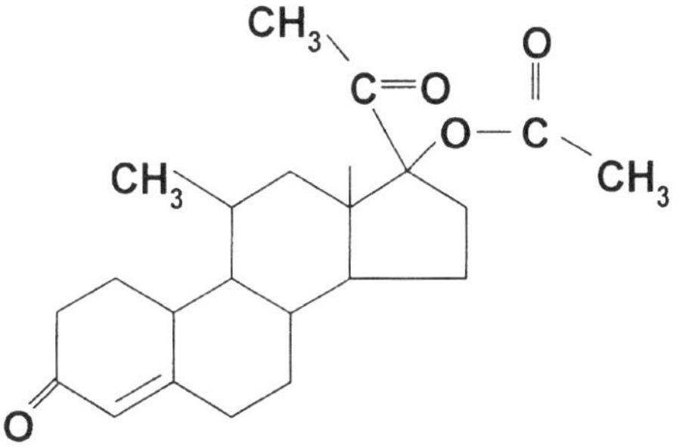

Norgestomet

Fig. 5.8. Structures of some orally active progestogens.

Oestrogens are luteolytic and are given before progestogen treatment in programmes to synchronize oestrus. Synthetic oestrogens are available that are metabolized more slowly than oestradiol. Corticosteroids can be used for induction of parturition. Short-acting analogues include flumethasone and dexamethasone, and long-acting forms include dexamethasone trimethylacetate, triamcinolone acetonide and suspensions of flumethasone or betamethasone (Fig. 5.9).

Several steroid hormone antagonists have been developed (Table 5.3). Steroid hormone antagonists bind to the normal ligand-

Table 5.3. Hormone antagonists.

Category	Hormone	Biological name or activity	Commercial name and company
Antiprogestogen		Registered for veterinary use	Aglepristone® (Virbac)
	Onapristone (ZK98299) Aglepristone (RU46534)		
Oestrogen antagonists	Tamoxifen, clomifene	Type I oestrogen antagonists with mixed antagonist–agonist effects	
		Type II anti-oestrogens with no agonist effects	ICI 164,384, ICI 182,780
Antiandrogens		Cyproterone acetate, flutamide	

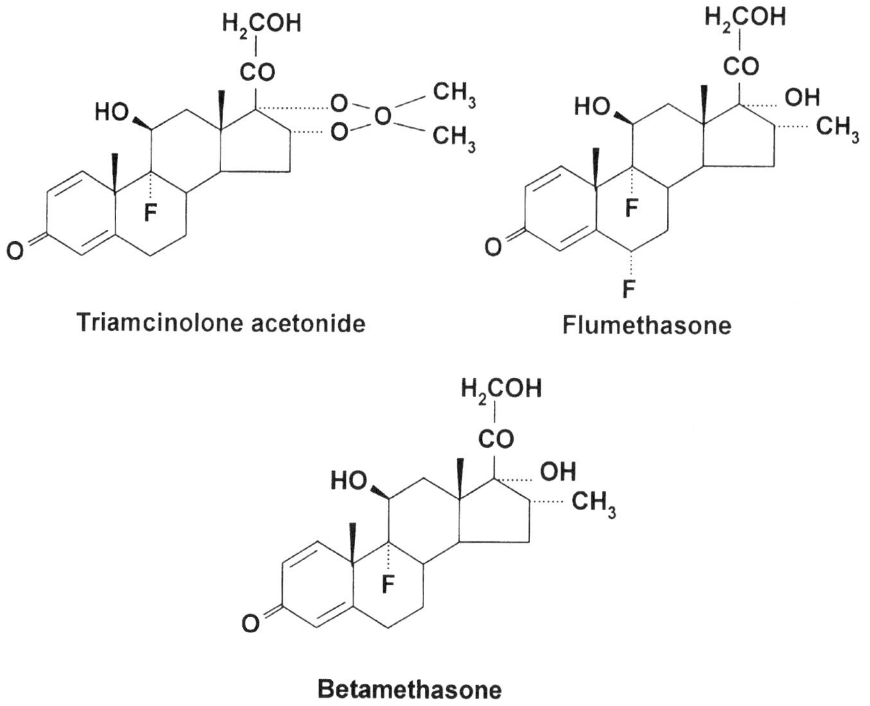

Fig. 5.9. Structures of some synthetic corticosteroids.

binding domain of a receptor, but they induce a different conformation of the receptor which results in decreased hormone-dependent gene expression. Antiprogestogens (Fig. 5.10) include mifepristone, aglepristone and onapristone; the binding affinity varies with species and tissue type. The hydrophobic side-chain at C17 is thought to be responsible for high-affinity binding to the receptor, while the additional side-group at C11 causes a change in conformation of the receptor that inhibits the activity. Antiprogestogens can be used to prevent implantation of the fertilized ovum, to cause resorption or abortion of established pregnancies or to hasten parturition at the end of pregnancy.

Anti-oestrogens and anti-androgens (Fig. 5.11) are used to treat hormonal-responsive cancers in humans, such as breast and prostate cancer. Tamoxifen and clomiphene have mixed antagonist–agonist effects, depending on the species and tissue type. For example, in humans tamoxifen acts as an anti-oestrogen in the mammary gland and an oestrogen agonist in the uterus.

Type II anti-oestrogens include ICI 164,384 and ICI 182,780, which are 7α-alkyl-amide analogues and do not have oestrogen agonist activity.

Anti-androgens, such as cyproterone acetate and flutamide, inhibit the development of secondary sex glands in laboratory animals and male sexual activity in some species. Daily treatment with cyproterone acetate has been shown to prevent taint in boars (see Section 3.3). For more information on steroid hormone antagonists, see Hoffmann and Schuler (2000).

Prostaglandin $F_{2\alpha}$ and its analogues (Table 5.4 and Fig. 5.12) are used for their luteolytic action (regression of the CL) or stimulation of the myometrium to induce parturition. This includes $PGF_{2\alpha}$ itself, available as dinoprost, as well as a number of synthetic prostaglandins. Synthetic prostaglandins have fewer side-effects, such as sweating and abdominal cramps, than $PGF_{2\alpha}$. Similarly, sows injected with $PGF_{2\alpha}$ to induce parturition have increased nest-building behaviour, while if cloprostenol is

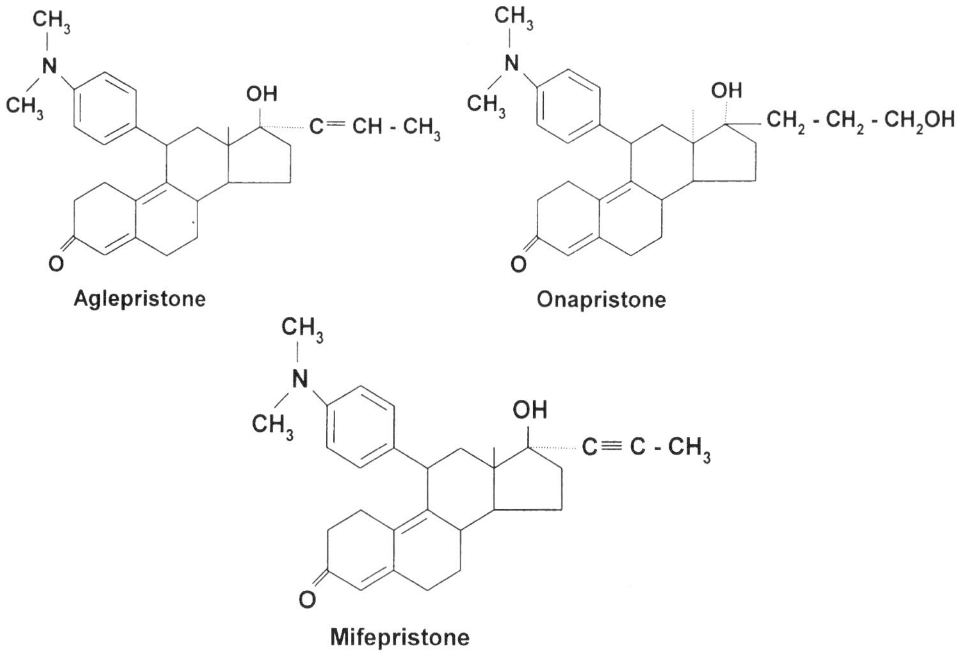

Fig. 5.10. Structures of some synthetic antiprogestogens.

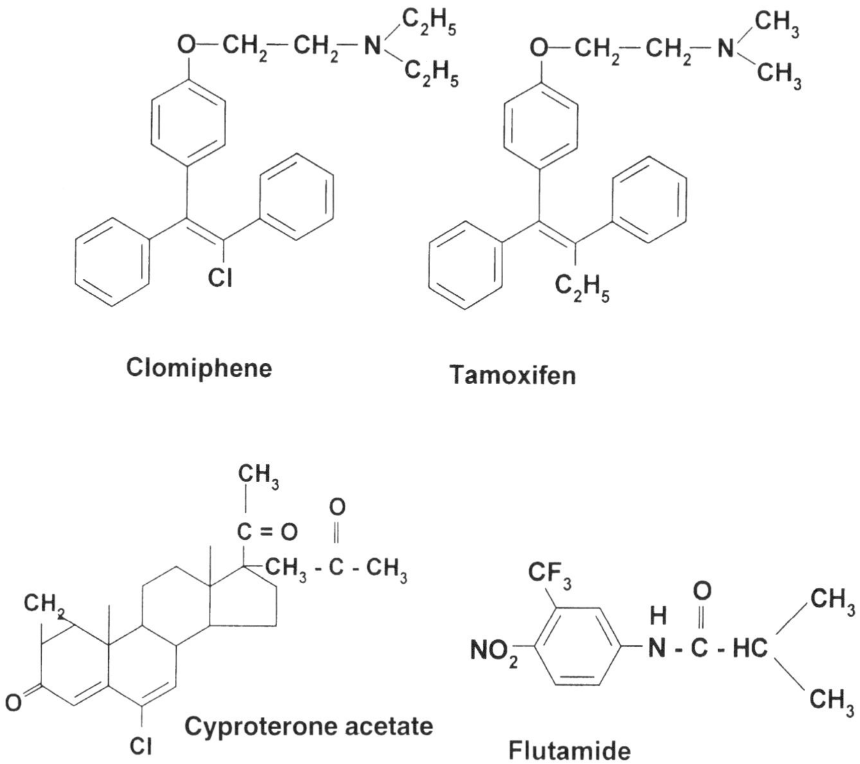

Fig. 5.11. Structures of some oestrogen and androgen antagonists.

Table 5.4. Prostaglandins.

Category	Hormone	Commercial name and company
PGF$_{2\alpha}$	Dinoprost	Lutalyse® (Upjohn)
Synthetic prostaglandins	Cloprostenol	Estrumate® (Mallinckrodt)
	Fenoprostalene	Bovilene® (Syntex)
	Alphaprostol	Alfavet® (Hoffman–La Roche)
	Fluprostenol	Equimate® (Miles)
	Prostalene	Synchrocept® (Syntex)

used to induce parturition, nest-building behaviour does not increase.

Use of hormone agonists to control fertility

GnRH analogues, progestins and androgens can act synergistically by negative feedback to reduce the release of gonadotrophins from the pituitary and thereby reduce fertility. The orally active progestogen melengestrol acetate (MGA) is used for suppressing oestrus and improving the rate of gain and feed efficiency in heifers. GnRH agonists (e.g. deslorelin, nafarelin, Table 5.1) have a higher affinity for GnRH receptors and a longer half-life in the circulation than GnRH itself. Chronic treatment with GnRH agonists initially causes a dramatic increase in the release of LH and FSH, followed by the down regulation of GnRH receptors on pituitary gonadotroph cells and inhibition of pulsatile LH release. The lack of LH blocks ovulation, but the ovary is still responsive to exogenous LH. Implants of GnRH agonists have been developed and can be used to prevent pregnancy in heifers under extensive management systems for up to 10 months. The procedure can also be used to control the time of ovulation by injection of LH in animals treated with GnRH agonist, which allows artificial insemination (AI) at a fixed time. This approach can be used to optimize ovulation rate and recovery of embryos in multiple ovulation and embryo-transfer protocols.

Bulls implanted with GnRH agonists

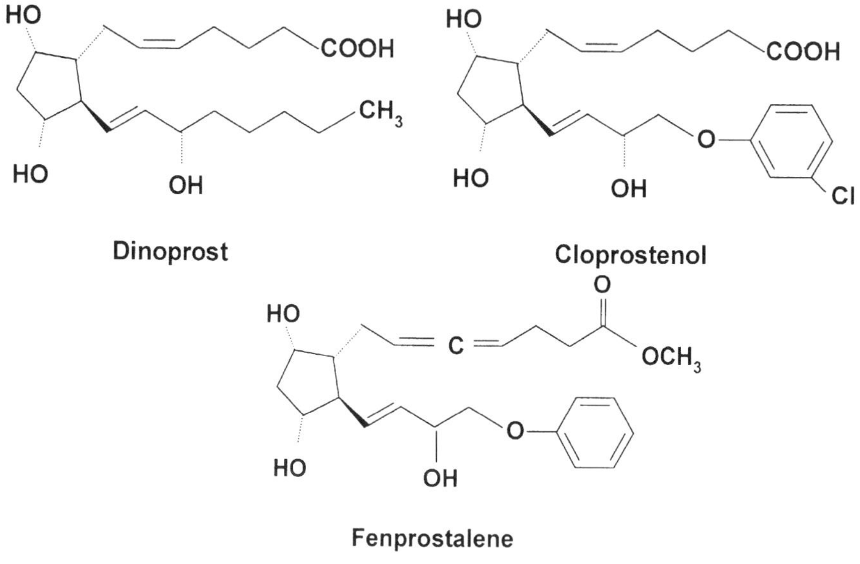

Fig. 5.12. Structures of some prostaglandin F$_{2\alpha}$ analogues.

have higher levels of testosterone for the duration of treatment. However, males of other species, such as the pig, have decreased levels of testosterone after treatment with potent GnRH agonists, which overstimulate the pituitary gland and shut down gonadotrophin release. GnRH antagonists (e.g. Nal-Glu and Cetrorelix, Table 5.1) have been used to reduce sperm counts in humans, but these peptides must be injected. Orally active non-peptide GnRH antagonists have also been developed and could be potentially used as contraceptive agents in livestock or captive animals.

Treatment of males with high levels of androgens reduces the secretion of LH and FSH by negative feedback on the pituitary, and thus androgens can act as contraceptive agents. For example, treatment of boars with testosterone propionate is reported to decrease the intensity of boar taint, and testosterone analogues are being investigated as male contraceptives in monkeys and humans. Testosterone itself is rapidly degraded by the liver and is thus not practical for use as a contraceptive agent. Testosterone esters, such as testosterone enanthate and testosterone undeconate, are effective in reducing sperm counts in humans. The 7α-methyl derivative of 19-nortestosterone, which cannot be reduced by the 5α-reductase enzyme, prevents dihydrotestosterone-like effects such as prostate hyperplasia. Combinations of testosterone esters and progestins (medroxyprogesterone acetate, Levonorgestrel® or Desogestrol®, Table 5.2) are more effective contraceptive agents in humans. Implants of the progestins MGA or Levonorgestrel® are used to control fertility in zoo mammals. The use of a 'male pill' containing the anti-androgen cyproterone acetate with testosterone undeconate has also been studied in humans. For more information, see Amory and Bremner (2000).

Transdermal patches can be used as an alternative to weekly injections or daily oral dosing for delivering combinations of androgens and progestins for contraceptive purposes, but can result in skin irritation. Other complications from some steroid contraceptives include weight gain and a decrease in serum HDL-cholesterol.

Methods for detection of oestrus

During oestrus or 'heat', the female will accept a male, and during 'standing heat' will stand to be mounted. In pigs, this lordosis can be more easily detected by pushing forward on the rear of the sow. The efficiency of detection of oestrus in cattle can be less than 50%, so either methods to improve heat detection, or strategies to control the time of ovulation can be used to improve conception rates.

OESTROUS BEHAVIOUR. Cows need adequate space to display mounting behaviour, with softer footing such as grass, dirt or straw. Mounting behaviour is decreased on concrete floors or floors that are too slippery or too coarse. Cows with sore feet or legs have less mounting activity. Cows that are themselves in heat or have recently been in heat are most likely to mount a cow in heat. As the number of cows in heat increases up to 3–4, the number of mounts per cow increases dramatically.

The extent of mounting can be used as an indicator of oestrus. The tail-head area can be scarred and dirty from mounting by herd mates. A more accurate measure is to use brightly coloured enamel-based paint on the tail-head and to cover the paint strip with a contrasting colour of chalk. The extent of cover of tail paint and chalk is scored using a scale of 5, for no signs of oestrus and full presence of paint and chalk, to 0 for standing oestrus and absence of paint and chalk. Pressure-activated heat mount detectors that change colour from the weight of the mounting animal can also be fixed to the tail-head. There are more sophisticated pressure sensors available that either display the frequency of mounting behaviour directly or send signals to a remote device that records the mounting behaviour. Vasectomized teaser bulls, which can be fitted with a chin ball marking harness, are particularly useful in detecting heat when there are only a few cows in heat. Cows or heifers treated with testosterone or oestradiol can also be used as an alternative to vasectomized bulls.

Standing heat can often be brief in duration and is not always observed since it occurs mostly in the early morning and late evening, but various secondary signs can be used to detect oestrus. During oestrus, vaginal mucus is discharged that is clear and stringy, and the vulva is reddened, but this can be difficult to detect in cows. Blood stains on the tail or vulva are indicative of a recent heat. Cows in heat are restless, bellowing and trailing other cows. Pedometers fixed to the leg of the cow can be used to measure an increase in walking activity as a measure of oestrus. Feed intake and milk yield can also decrease during oestrus. Changes in perineal odours occur near oestrus and these can be detected using trained sniffer dogs or potentially using an 'electronic nose', which has electronic sensors that change electrical characteristics when exposed to volatile compounds.

Follicular development can be estimated by palpating the ovaries via the rectum or using real-time ultrasound equipment with a transducer placed in the rectum. Pregnancy can be confirmed by this type of examination at about 40 days after breeding. For more information on the detection of oestrus in cattle, see the review by Diskin and Sreenan (2000).

MILK PROGESTERONE. The level of progesterone in milk can be used to evaluate oestrus and pregnancy in dairy cattle. The level of progesterone rises slowly for the first 4–6 days after ovulation and reaches a maximum at days 10–17. It falls sharply at day 18–19 in non-pregnant cows due to luteolysis, but remains elevated in pregnant cows since the CL continues to function.

The accuracy of pregnancy diagnosis using milk progesterone is only about 80%, due to factors such as errors in oestrus detection, differences in cycle length, uterine disease, ovarian cysts and early embryonic mortality. It can more reliably be used to determine if a cow is *not* pregnant. This involves comparing progesterone levels in milk samples taken at the time of insemination to progesterone levels in milk at 21–24 days later.

A high level of progesterone in milk can confirm the lack of oestrus and that a cow should *not* be inseminated. This can be particularly useful in high-producing cows or cows under heat stress, when there are poor outward signs of oestrus. A low progesterone level suggests that the cow might be near oestrus, but does not confirm that it is at the optimum stage for insemination. Low progesterone levels can also be due to inactive ovaries or the presence of follicular cysts. Luteal cysts can be distinguished from follicular cysts by a high level of progesterone. The presence of a functional CL can be confirmed by accompanying high progesterone levels in cows that will be used as embryo transfer recipients.

Milk progesterone can be evaluated using commercially available kits for on-farm use and could potentially be measured with an on-line system during milking in the future.

Induction and synchronization of oestrus

Synchronization of oestrus involves manipulating the ovarian cycle in order to mate the female at a predetermined time. This allows for greater management control of reproduction, improved reproductive efficiency and a predetermined parturition time. Synchronizing oestrus reduces the number of checks required for determining when it is present and increases the intensity of oestrus when a group of cows are in oestrus at the same time. Timed artificial insemination (AI) can also be used on all cows in a dairy herd at the same time when protocols designed to bring all cows into oestrus within a narrow window of time are used. Oestrus synchronization is also necessary for embryo-transfer programmes.

Effective methods for the synchronization of oestrus are also necessary in order for AI to be used in the beef cattle industry, although AI is used to a much smaller extent in beef cattle than in dairy cattle. This is because of the extensive nature of beef cattle operations, while dairying is an intensive operation. Increased used of AI would allow the use of superior genetics in the beef industry, as has been done in the dairy industry for some time. Fertility is the most important trait for beef cattle, since beef breeders get most of their income from calves born into the herd.

The fertility of the first oestrus after parturition ('foal-heat' in mares) is low, and the timing of ovulation can be erratic, so animals can be treated with luteolytic agents at this time to induce regular oestrus. Oestrogen (as oestradiol cypionate, Table 5.2) is also used to induce oestrus in 'jump' mares used as teasers for semen collection from stallions.

Methods for oestrus synchronization must produce a high proportion of females in heat at a predetermined time, maintain high rates of fertility and have no undesirable side-effects. In addition, they should be easy to perform and be of low cost.

STRATEGIES FOR SYNCHRONIZING OESTRUS. The presence of a functional CL during the luteal phase of the cycle prevents ovulation. Progesterone produced by the CL delays the maturation of LH-dependent follicles and primes the brain for oestrous behaviour. In the absence of pregnancy, prostaglandin $F_{2\alpha}$ causes luteolysis and decreases progesterone levels. This restores the LH pulse frequency and amplitude, allowing the subsequent development of the dominant follicle, leading to oestrus, ovulation and normal luteal function. Treatment of animals having a functional CL with prostaglandin $F_{2\alpha}$ will thus result in oestrus and ovulation a few days thereafter.

Schedules for oestrus synchronization have been devised based on $PGF_{2\alpha}$, a combination of $PGF_{2\alpha}$ and GnRH and a combination of progestins and $PGF_{2\alpha}$.

Prostaglandin $F_{2\alpha}$-based systems. Injection of $PGF_{2\alpha}$ will induce regression of the CL and not adversely affect subsequent oestrous cycles. However, $PGF_{2\alpha}$ is not effective on newly established CL, and cows injected on days 1–5 of the oestrous cycle are non-responsive. There are a number of waves of follicular development in each oestrous cycle, and cows injected on days 7 or 15 of the oestrous cycle have a highly developed follicle ready to ovulate after CL regression. $PGF_{2\alpha}$ treatment on days 7 or 15 will therefore induce oestrus 3 days after injection. A practical approach (Fig. 5.13) is to inject $PGF_{2\alpha}$ twice, 14 days apart, so that the cows will be responsive to $PGF_{2\alpha}$ at least at the second injection and come into oestrus shortly thereafter. Alternatively, cows can be injected at the beginning of 1 week and those that come into oestrus later in the week are inseminated. The remaining cows are injected with $PGF_{2\alpha}$ at 14 days after the first injection and then inseminated when oestrus is detected or at the latest 4 days after the last injection of $PGF_{2\alpha}$. However, not all cows respond to $PGF_{2\alpha}$ treatment, particularly before day 12 of the oestrous cycle. Improved synchronization of oestrus is obtained with an additional injection of $PGF_{2\alpha}$ 2 weeks before the initial injection. Cows are rebred 21 days after the first insemination if they are seen to be in oestrus. This decreases the number of days open and the net cost per cow. However, this process requires detection of oestrus, keeping accurate records and the identification of individual animals.

GnRH and the Ovsynch® protocol. Treatment with $PGF_{2\alpha}$ results in regression of the CL, but does not synchronize the growth of follicles or affect the preovulatory surge of LH. Injection of GnRH at 7 days before treatment

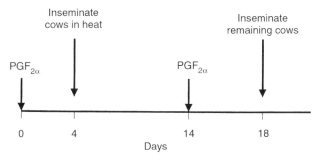

Fig. 5.13. Prostaglandin-based system of oestrus synchronization.

Fig. 5.14. Ovsynch® protocol for oestrus synchronization.

with PGF$_{2\alpha}$ synchronizes follicular growth and oestradiol secretion with luteolysis and improves the timing of oestrous behaviour. The Ovsynch® protocol (Fig. 5.14) utilizes GnRH and PGF$_{2\alpha}$ in a procedure that allows for AI without oestrus detection in lactating dairy cows. The protocol can be used at any point in the oestrous cycle and is particularly useful when the efficiency of oestrus detection is poor. The Ovsynch® protocol without oestrus detection results in pregnancy rates that are similar to multiple treatments with PGF$_{2\alpha}$ every 14 days and insemination at detection of oestrus. In addition, the Ovsynch® protocol reduces the number of days postpartum to first insemination and the number of days open.

The initial treatment with GnRH will ovulate a dominant follicle if it is present and initiate a new follicular wave, or a new follicular wave may be occurring spontaneously. The injection of PGF$_{2\alpha}$ given 7 days later will cause regression of the CL. A second injection of GnRH is given 48 h later and will cause the dominant follicle that has grown from the first injection of GnRH to ovulate. Cows are inseminated 16–20 h later so that capacitated sperm are present at the time of ovulation. In some cases, cows will be in oestrus shortly after treatment with PGF$_{2\alpha}$ and these cows should be inseminated at that time.

The Ovsynch® protocol does not improve conception rates in heifers compared to insemination at detected oestrus. However, the Ovsynch® protocol does remove the difficulty of detecting oestrus and therefore reduces the number of days open in heat-stressed cows. It is also useful in treating cows with either ovarian or follicular cysts without identifying the type of cyst, since the GnRH treatment will remove the follicular cyst and PGF$_{2\alpha}$ treatment will remove the luteal cyst (see below). Cows with inactive ovaries in which follicular development has ceased do not respond well.

Progestin-based systems. Treatment with a progestin for 7 days before treatment with PGF$_{2\alpha}$ improves the synchronization and detection of oestrus and conception rates. Short-term exposure to progestins will also induce the onset of oestrous cycles in a proportion of anoestrous cows and heifers. Early use of progestins for oestrus synchronization utilized treatments in excess of 14 days. These resulted in abnormal follicular growth and decreased fertility, and were linked to deleterious effects on sperm transport and viability in the reproductive tract of ewes. Short-term treatment with progestin of less than 10 days' duration is now used along with a treatment to cause regression of the CL, such as PGF$_{2\alpha}$ or oestrogen (Fig. 5.15).

Progestin is normally administered to dairy cows using implants (CIDR or PRID). For beef cattle, norgestomet is available in the Synchro-Mate-B® implant or melengestrol acetate (MGA) is given in the feed (Table 5.2). The implant is placed at the base of the ear

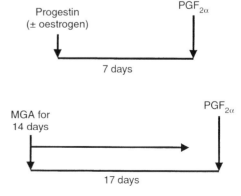

Fig. 5.15. Progestin-based protocols for oestrus synchronization.

and an injection of oestradiol valerate and norgestomet is given when the implant is inserted. The CL regresses spontaneously while the implant is present. The injection of oestrogen and progestin is designed to cause early regression of the CL in cows in the early stage of the cycle. Alternatively, MGA can be fed to beef cattle for 14 days, followed by an injection of $PGF_{2\alpha}$ 17 days later. The treatment with MGA will initiate oestrus in anoestrous females and synchronize oestrus in cyclic females, but will result in a low-fertility oestrus in the 7 days after MGA is withdrawn. The $PGF_{2\alpha}$ treatment occurs in the latter part of the next oestrous cycle, which will cause a high rate of CL regression and synchronization of oestrus. This procedure induces oestrus in about 80% of yearling heifers and is used for this purpose. However, since the protocol takes about 34 days to achieve synchronized oestrus, it cannot be used with postpartum cows that calved late in the season without affecting the start of the new breeding season.

A number of protocols have been developed using progestins to induce puberty and synchronize oestrus in beef heifers. A progestin treatment (norgestomet implant) can be used between the GnRH treatment and the $PGF_{2\alpha}$ treatment (GnRH–P–PGF). A further modification of an additional GnRH treatment 48 h after the $PGF_{2\alpha}$ treatment (GnRH–P–PGF–GnRH) induces a final LH surge and a highly synchronous time of oestrus, which allows the use of timed AI without oestrus detection. Another approach uses an initial injection of oestrogen and a progestin implant (CIDR) for 7–8 days, followed by an injection of $PGF_{2\alpha}$ when the CIDR is removed and a second injection of oestrogen 24–48 h later (E–CIDR–PGF–E). The first oestrogen injection will cause regression of FSH-dependent follicles and the CIDR implant with progestin induces turnover of LH-dependent follicles. Once these follicles become atretic, a new wave of follicular development begins. The injection of $PGF_{2\alpha}$ will cause any CL present to regress, and the final oestrogen injection will induce oestrus and ovulation. The induction of puberty and oestrus is more effective the closer the animals are to a naturally occurring oestrus.

Controlled breeding programmes result in significant financial savings to dairy producers. The additional costs for drugs and labour are more than offset by the savings from decreased labour for oestrus detection, lower costs from fewer days open and fewer replacement heifers needed. Producers can also control the time of calving, to take advantage of seasonal variations in pricing and constraints to production. Beef cattle producers can use controlled breeding programmes to induce puberty and synchronize oestrus in heifers by the start of the breeding season, so they will calve early and improve lifetime production. For more information on the manipulation of the oestrous cycle in dairy and beef cattle, see the reviews by Thatcher *et al.* (1998) and Day (1998).

Synchronization of oestrus in pigs can be accomplished by treatment with progestogen for 14–18 days (Altrenogest® in the feed, Table 5.2) and then breeding at oestrus.

Ewes during the breeding season can be injected with $PGF_{2\alpha}$ twice 9 days apart and then treated with an intravaginal sponge impregnated with MAP (Table 5.2) for 12 days. Ewes will exhibit oestrus 48–72 h after the removal of the sponge. In the non-breeding season, PMSG treatment (Table 5.1) is also necessary to induce ovulation.

Superovulation and embryo transfer

A number of *in vitro* techniques can be used to improve reproductive efficiency and increase the contribution of valuable females to the gene pool. Large numbers of oocytes or embryos can be collected from female donors and transferred to recipient females. Immature oocytes can be matured *in vitro*, fertilized, grown to the blastocyst stage and then transferred to recipient females.

Multiple ovulation and embryo transfer (MOET) involves the recovery of multiple embryos after superovulation and transfer of the embryos to synchronized recipients of lesser genetic merit. To induce superovulation, FSH or PMSG is given in the middle of the cycle (days 8–14 in cattle) to increase the number of follicles that mature into dominant follicles, and to reduce the regression of follicles. A single injection of PMSG (which has a longer half-life than FSH), or twice-daily injections of FSH for 2 days, is required. This is followed by treatment with $PGF_{2\alpha}$ on day 3 to induce luteolysis, and then twice-daily FSH injections are continued to induce superovulation until day 4, when oestrus occurs. Twice-daily inseminations are given, starting 12 h from the start of oestrus. Sperm transport can be reduced by superovulation, particularly when PMSG is used, so inseminations directly into the uterus can be used to improve fertility. Embryos are collected by flushing the uterus non-surgically at 6–8 days after oestrus in cattle and horses, but surgical recovery is used for pigs, sheep and goats. Embryos are assessed for the stage of development and for evidence of physical damage, and good-quality embryos can then be transferred to recipients or frozen. Recipient animals should be checked for the presence of a CL, since a CL is required to maintain the pregnancy after transfer of the embryos.

The ovulation rate and number of transferable embryos can be improved by treatment with anti-PMSG serum at oestrus or after the LH surge has occurred. This suppresses the second wave of follicular development and reduces the number of follicular cysts. A low LH content in the FSH preparation is desirable, since LH interferes with follicular development. The presence of a large dominant follicle reduces the number of follicles that will mature and ovulate in response to exogenous gonadotrophins. There is also the possibility that the animals will produce antibodies against the non-homologous gonadotrophins that are used for superovulation, so that later treatments with gonadotrophins will not be successful.

IN VITRO PRODUCTION OF EMBRYOS. Large numbers of embryos can also potentially be produced from immature oocytes that have not yet committed to atresia. The pig ovary has over 200,000 primordial follicles, while the cow has about half that number. Oocytes can be collected at slaughter, or by aspiration of immature follicles from both mature and immature females. This allows the production of embryos from donors with poor fertility or those that do not respond to superovulation. This is particularly useful for the recovery of genetic material from slaughtered animals. It also has potential for the conservation of rare breeds. Oocytes can also be recovered from young prepubertal females, thus shortening the generation time and increasing the number of potential offspring from valuable individuals. After recovery, the oocytes are matured and fertilized *in vitro* and grown to the blastocyst stage before freezing or transfer to a recipient animal.

The efficiency of blastocyst production from *in vitro* production systems is highly variable and lower than with oocytes recovered after insemination. Blastocyst production depends on the quality of the gametes used and human technical skills. Culture conditions are also important, and the optimum culture conditions vary among different species. Good-quality oocytes that are surrounded by intact cumulus cells and have a homogeneous-looking cytoplasm should be used. *In vitro* maturation (IVM) of oocytes includes both nuclear maturation, seen as the development of a metaphase II spindle, and cytoplasmic maturation, which allows the formation of a male pronucleus after fertilization. IVM has been performed using a number of different systems, including co-culturing the oocytes with oviductal epithelial cells or granulosa cells, or supplementation of media with follicular or oviductal fluid. Defined culture media have also been used, and are preferred since they reduce the potential for contamination with bacteria and viruses from cells in co-culture. Adding gonadotrophins (LH and FSH) along with epidermal growth factor (EGF) or IGF-I to the maturation medium can stimulate oocyte maturation in defined media. One-third of porcine oocytes that are treated with EGF (10 mg ml^{-1}) develop to the blastocyst stage after *in vitro* fertilization (IVF).

A major problem with IVF of porcine oocytes is polyspermy, which occurs when several sperm penetrate each oocyte and the oocyte then fails to develop properly past the blastocyst stage. This is rarely observed with *in vivo* insemination prior to ovulation, since there is a limited number of sperm arriving simultaneously at the oocyte and the zona pellucida effectively blocks the penetration of more than one sperm. The success of fertilization is seen ultimately in cleavage of the oocyte. Both fresh and frozen–thawed sperm have been used for IVF in cattle and swine. Sperm cells can be sorted to some degree into separate pools of X- and Y-bearing sperm and used to produce offspring of the desired sex. Alternatively, a few cells can be removed from an embryo and the sex can be determined using X- and Y-specific DNA probes with PCR.

Embryos are a convenient and cost-effective form for providing new genetics to animal production systems. Since embryos can be produced that are free of disease, they are especially suitable for international trade. Embryo recovery and transfer can also be used to eliminate disease by repopulating herds with uninfected animals. Embryos that are recovered from infected animals can be treated with trypsin and washed to remove bacteria and viruses so they are free of the disease, since the zona pellucida protects the embryo from infection. The embryos are then transferred to clean individuals to produce uninfected animals at birth. For more information on embryo production and transfer in cattle and pigs, see Marquant-Leguienne and Humblot (1998) and Day (2000).

Maintenance of pregnancy

The embryonic loss rate, which is the difference between the fertilization rate and birth rate, is between 30 and 40% in cattle. Embryonic loss is also the major factor affecting litter size in pigs. The highest rate of embryo loss occurs during the time of maternal recognition of pregnancy. A single treatment with $PGF_{2\alpha}$ between 18 and 28 days postpartum can improve uterine involution and may reduce the number of services to conception. The fertility and pregnancy rate from oestrus induced by $PGF_{2\alpha}$ is about 10% higher than naturally occurring oestrus, possibly due to improved quality of the ova with the shorter period of luteolysis.

Treatment with GnRH once between 11 and 13 days after insemination may delay luteal regression or cause ovulation of the dominant follicle in the next wave and form an accessory CL. This will increase progesterone production and improve the rate of implantation of the embryo and maternal recognition of pregnancy. Treatment with hCG on day 15 improves pregnancy in heat-stressed cows but not normal cows. Direct treatment with progestogens after insemination can also increase the secretory function of the uterine endometrium and improve the implantation of the embryo.

Pregnancy rates can be increased in mares by delaying the postpartum oestrus. This is accomplished using progestogen treatment either alone or in combination with oestrogen, starting immediately after foaling. Administration of $PGF_{2\alpha}$ can also be used to induce the next oestrus after foal heat.

Early embryonic mortality is increased with very poor body condition, high temperatures in the first week after mating and severe stress.

Induction of abortion/parturition

Progesterone is required for the maintenance of pregnancy and the CL, placenta and adrenals produce the progesterone at different times during pregnancy. In the mare and ewe, progesterone is produced in the latter half of pregnancy by the placenta, while in the cow, sow and goat the CL persists throughout pregnancy and produces progesterone. Parturition is the process of delivering a fetus and placenta from a pregnant uterus. It is triggered by the fetus, which releases corticosteroids due to activation of the fetal hypothalamic–pituitary–adrenal axis as the fetus matures. The fetal corticosteroids increase oestrogen production and decrease progesterone production by the placenta, which in turn stimulates prostaglandin production by the uterus. Oestrogen also increases the number of oxytocin receptors and sensitizes the uterine myometrium for contractions. In CL-dependent species, the prostaglandin causes regression of the CL and removes the source of progesterone.

Treatment with corticosteroids can thus induce parturition if a functional placenta exists. Prostaglandin treatment is needed to induce parturition when there is a nonfunctional placenta, as in the case of fetal mummification.

Abortion is induced in animals entering feedlots, after inappropriate matings and due to pathologic pregnancy. Up to 5 months of pregnancy, abortion is induced in cattle using $PGF_{2\alpha}$, while combinations of $PGF_{2\alpha}$ and corticosteroids are used late in gestation. Calves born more than 3 weeks premature have poor viability.

Feeding inhibitors of prostaglandin synthase, such as aspirin or indomethacin, can reduce the levels of prostaglandins. This approach may be useful for inhibiting oestrus or maintaining pregnancy. In postpartum ewes, indomethacin treatment increased LH pulse frequency and amplitude and resulted in an early return to oestrus.

Oxytocin is also used to induce parturition, for the expulsion of a retained placenta and the acceleration of uterine involution. It is also used to induce contraction of the myoepithelium cells in the mammary gland and cause milk let down. Bromocryptine is used to inhibit prolactin secretion and prevent spurious lactation.

Postpartum interval

There is normally a period of 30–76 days of anoestrus before the first oestrus occurs postpartum in cattle, with dairy cattle ovulating sooner than beef cattle. Ovarian cycling can be limited by inadequate LH pulse frequency, so that follicles do not mature properly. There can also be a decrease in the normal preovulatory LH surge from positive feedback by oestradiol. The decreased LH production is due to reduced function of the hypothalamus–pituitary axis. The first ovulation postpartum often results in a CL that regresses early and results in a short cycle of 8–10 days. This may be due to the need for progesterone priming of the follicles for normal CL function after ovulation. Progesterone priming of the brain is also necessary for normal oestrous behaviour that is induced by oestradiol.

In order to shorten the period of anoestrous, the frequency of LH pulses has to be increased so that the follicles will reach the final stages of maturity. The presence of continuous low levels of progesterone, that do not affect pulsatile LH release but prevent the LH surge, results in large, persistent follicles containing ageing oocytes. Other factors, such as high blood cortisol levels, may also block the LH surge. Persistent large follicles can form ovarian cysts, which can be treated by injection of GnRH. Luteal cysts that produce high levels of progesterone can also form and can be treated by injection with $PGF_{2\alpha}$. In addition to limitations on ovulation, the uterus of the postpartum cow requires 30–40 days to complete involution, so cattle should not be bred before 60 days after calving.

The duration of postpartum anoestrus is affected by nutrition, suckling, seasonal factors and the presence of males. Some animals, particularly seasonal breeders, are very non-responsive to treatments designed to induce the ovarian cycle and are said to be in 'deep anoestrus'.

Suckling more than 2–3 times per day, or increased frequency of milking, increases levels of endogenous opioid peptides in the brain. This decreases the production of GnRH by the hypothalamus and the release of LH by the pituitary. The decrease in LH prevents the final maturation of the follicle to the preovulatory stage and increases the number of days to first oestrus and ovulation. However, suckling also improves uterine involution and increases the conception rate. Sows need to nurse for about 10 days or they will develop follicular cysts, persistent oestrus and have poor uterine involution. Removing the piglets or exposure to a boar will stimulate follicular growth in sows.

Cystic ovarian disease

About 30% of postpartum cows develop ovarian cysts, which are structures greater than 2.5 cm in diameter that last more than 10 days and are follicular or luteal in origin. Follicular cysts are more common than luteal cysts and are associated with a high level of production, reduced nutrient status, uterine infections, increased age, stress and season, with a higher incidence in the autumn and winter. They are caused by inadequate LH response to oestradiol and failure of ovulation and are associated with anoestrus postpartum. Stress-induced release of ACTH and corticosteroids can also reduce LH release and cause cyst formation.

High levels of progesterone in the blood or milk can indicate the presence of luteal cysts. Follicular cysts can be treated with hCG or GnRH to induce luteinization of follicular cells, and luteal cysts can be treated with $PGF_{2\alpha}$ to cause luteolysis.

Effects of nutrition

The nutrient (energy) balance of the animal is the difference between the availability of nutrients from body reserves and feed intake, and the nutrient requirements of the animal for lactation, metabolism, growth and activity. A reduced energy balance can delay puberty and increase the period of postpartum anoestrus. This may be signalled by a low level of blood glucose, which increases the negative feedback by oestradiol on GnRH release by the hypothalamus and decreases the level of LH. The level of IGF-I is also decreased during negative energy balance, which, together with decreased LH, contributes to decreased follicular growth and maturation. Poor nutrition results in a reduced size of follicles and more rapid turnover of the dominant follicle. Oestrous cycles in mature cycling pigs can be disrupted in severe and prolonged feed restriction, while less severe conditions will delay puberty and increase the period of postpartum anoestrus.

Undernutrition results in a decrease in plasma levels of insulin, IGF-I and leptin, and an increase in growth hormone. IGF-I acts with the gonadotrophins to stimulate follicular development, and a low level of IGF-I in follicular fluid is associated with a low ovulation rate. Insulin stimulates nutrient uptake and prevents apoptosis in granulosa cells, reduces atresia of follicles and increases ovulation rate. Leptin (see Section 3.8) may act to signal the reproductive system that sufficient body fat is present to support a pregnancy. Leptin treatment has been shown to increase sexual development and levels of LH and FSH in male and female rodents. Levels of leptin are correlated with size of the fetus, and appropriate levels of leptin are required for the maintenance of pregnancy.

Dairy cattle are normally in a period of negative energy balance during the postpartum transition period, when maximum milk production is reached before maximum feed intake has occurred. The time to first ovulation postpartum is a function of the number of days the cow is in the lowest energy balance; this averages about 33 days in Holsteins in the USA. There is also a period of negative energy balance in beef cattle, even though milk production is much lower than in dairy cattle. Improved feed intake or body condition score will stimulate puberty in young animals and decrease the period of anoestrus in older animals.

Correcting the negative energy balance and improving body condition score will reduce the period of anoestrus in transition dairy cows. These factors contribute to an increase in conception rates after 60 days postpartum. Follicular growth is decreased in lactating cows compared to dry cows. Treatment with bST stimulates the growth of smaller follicles, while feeding the calcium salts of long-chain fatty acids stimulates growth of the largest follicles. Feeding high-energy diets and a high-quality protein also shortens the time to first ovulation in transitional mares that have a low amount of body fat and low body condition score. For more information on the effects of nutrition on cycling in beef heifers, see Schillo et al. (1992) and for pigs, see Prunier and Quesnel (2000).

Effects of stress

Seasonal periods of heat stress above 32°C increase the length of the oestrous cycle and can reduce plasma levels of oestradiol before oestrus, and thereby decrease the intensity and duration of oestrus. This lowers fertility and oestrus detection rates. Injection of GnRH at oestrus has been suggested as a method to improve the LH surge and ovulation rate. Progesterone production by the CL is also decreased during heat stress.

Elevated body temperature also increases embryonic mortality. Heat stress decreases feed intake, growth rate and milk production. It can also lower libido and sperm quality in males, resulting in lower fertility. More subtle changes in temperature may also play a role in the timing of the circannual reproductive rhythm of seasonal breeders. Bulls also show seasonal effects on fertility, with improved sperm production during long days and decreased sperm production during high temperatures. Other forms of generalized stress (see Section 6.3) can also contribute to the intensity of seasonal infertility.

Inducing puberty

Exposure to a male increases the pulsatile release of LH and decreases the age at which puberty is reached in cattle, sheep, goats and pigs. The presence of a male also shortens the period of anoestrus by acting on the hypothalamus to reduce the negative feedback of oestrogen on LH release. The 'ram effect' is used to induce ovulation in ewes. Male sheep and goats produce pheromones in sebaceous glands that are released on the hair. Male pigs release sex pheromones in the saliva that induce puberty and cause oestrous females to stand for mating. Further details are given in Section 6.2. In addition to male pheromones, visual, auditory and tactile cues are less important components of the male effect on stimulating female reproduction.

Growth hormone plays an important role in sexual maturation and may accelerate puberty by activating the LH pulse generator or potentiating the action of androgens.

Advancing cyclicity in seasonal breeders

In seasonal breeders, such as sheep and horses, regular oestrous and ovulation cycles can be induced at the beginning of the breeding season. The official date of foals for each breeding season is set as 1 January by many breeding associations, so it is advantageous to breed mares early in the year to have an age advantage over foals born later in the season. Increasing the photoperiod using artificial lighting is used to induce oestrus in mares. About 14–16 hours of light exposure for 8–10 weeks is used, starting at the beginning of December to induce oestrus by the middle of February. Supplemental lighting is given at the beginning and end of the natural daylight. A 1–2-hour light period 9.5 h after the beginning of the dark period has also been suggested as a method to induce oestrus, suggesting that this light-sensitive period plays an important role in sensing circannual changes in photoperiod (see also regulation of egg production, Section 4.2).

Melatonin is involved in the regulation of circadian rhythms, including the sleep–wake cycles in humans. In the absence of external signals, these rhythms would be free running, but they are normally synchronized to the endogenous melatonin and photoperiod signals. Administration of exogenous melatonin or inhibition of melatonin synthesis by exposure to bright light can be used to entrain these physiological rhythms to new signals. This can be used to treat sleep–wake disorders in workers on different shifts, from changes in time zone (jet lag), in blind individuals who are unable to perceive light and in elderly people with insufficient melatonin production. In amphibians, fish and some reptiles, melatonin causes aggregation of melanin granules in melanocytes, resulting in blanching of the skin. Seasonal anoestrus can be reduced using melatonin for short-day breeders or increased lighting through artificial lighting programmes for long-day breeders.

A decrease in thyroid hormones occurs during seasonal anoestrus in ewes, but thyroid hormones do not seem to be involved in seasonal anoestrus in mares. Melatonin also inhibits prolactin release in ewes and administration of exogenous prolactin decreased the time to first ovulation in seasonally anoestrous mares.

Exogenous melatonin can be given to ewes by injection or feeding to mimic the effects of short days and stimulate gonadal activity. Long treatment periods are needed and a commercial implant of melatonin (Regulin®, Schering Pty Ltd) is available. Boluses of slow-release melatonin that remain in the rumen for extended periods of time can also be used.

Mares can be treated with progestogens (injection of progesterone in oil or feeding Altrenogest® for 10–15 days) to induce oestrus. Progestogen treatment suppresses LH release and may allow LH stores to increase in anoestrous mares, so that sufficient LH will be released to induce ovulation after progestogen treatment has ended. The effects of progestogen treatment are additive to the effects of artificial lighting in inducing oestrus. Treatment with progestogens is more effective in mares during the latter part of the transition period when larger follicles are present, compared to the early transition period when there are only small follicles. Oestradiol treatment can be used together with the progestogen treatment, to further reduce the variation in follicular development and ovulation. Treatment with dopamine antagonists (domperidone, 1.1 mg kg^{-1} orally or sulpiride 0.5 mg kg^{-1} intramuscularly, see Fig. 5.6) reduces the negative feedback on LH release and results in earlier time of ovulation in mares, especially when combined with light treatment. The period of treatment is longer for mares that are in deep anoestrus compared to mares in transition.

There is an extended period of oestrus in mares, with ovulation occurring from 1 to 10 days after oestrus begins. Oestrus can be induced using PGF$_{2\alpha}$ as described above for cattle, but the time of ovulation is still variable. The most effective method is treatment with progestogen/oestradiol followed by PGF$_{2\alpha}$ to remove any persistent CL. Ovulation of large preovulatory follicles (35 mm in diameter) is then induced with hCG. The GnRH analogue deslorelin (Table 5.1) can also be used as an implant to shorten the time to ovulation in mares. For more information, see Pratt (1998) and Nagy and Daels (2000).

Immunological control of reproduction

Immunization against inhibin increases ovulation rate in sheep and horses by increasing FSH levels. Immunization against oestradiol or androstenedione can also be used in sheep to increase the ovulation rate by removing the negative feedback of gonadal steroids on the hypothalamus and pituitary. A commercial product (Fecundin®, Glaxo Animal Health) uses a dextran adjuvant that can be given as a single injection and produces a moderate antibody response. Synchronization of oestrus is not necessary, but otherwise fertilization rates and embryo survival rates can be decreased. In anoestrus ewes, follicular development is stimulated and multiple ovulations occur when the ewes are exposed to a ram.

Immunization against GnRH produces a form of non-surgical castration by neutralizing the effects of GnRH and inhibiting the production of gonadotrophins and gonadal steroids. This effect is temporary if passive immunization is used. GnRH is available for antibody binding while it is in the hypothalamic–pituitary portal blood vessels on its way from the site of production in the hypothalamus to the anterior pituitary gland. In order to generate antibodies, GnRH, or a GnRH analogue, is coupled to a large molecule, such as albumin, keyhole limpet haemocyanin (KLH) or diphtheria or tetanus toxoid (see Section 2.1). Adjuvants other than Freund's complete adjuvant are used commercially, since Freund's causes severe local inflammation at the injection site.

Immunization against GnRH in males reduces testicular size and function and can be used to control aggression, reduce the incidence of male-associated odours and reduce fertility. A commercially available vaccine (Vaxstrate®), based on a carboxyl-GnRH analogue coupled to ovalbumin that was designed for beef heifers, is also effective in male goats and sheep. Bull calves immunized against GnRH had less carcass fat, increased loin eye area and improved weight gain and feed conversion efficiency compared to steers. This is probably due to the increased exposure to testicular steroids in immunized calves compared to castrates. Immunized lambs produced carcasses that were similar to those of castrates but had similar back fat to intact control lambs. Immunized bucks and boars had decreased odour score but similar carcass weights as controls. A tandem repeat of GnRH linked to KLH produced a more consistent immune response in boars. The effects of immunocastration on carcass composition depend on the vaccination protocol and age of the animal. Immunization protocols have been developed that consistently reduce boar taint and also do not adversely affect the improved feed efficiency of intact boars or the lean content of boar carcasses compared to that of castrates (see Section 3.3). For more information on immunization against GnRH in males, see Thompson (2000).

The fertility of females can be reduced by vaccination with a key target protein that permanently affects the reproductive integrity of the ovary. The zona pellucida (ZP) is a glycoprotein matrix surrounding the oocyte that has receptor sites for sperm binding. If the ZP is masked or altered, fertilization will not occur. Vaccination with purified ZP glycoproteins produces antibodies that inhibit sperm binding and penetration of the ZP and results in immunocontraception. Antibody titres normally fall over a period of time after vaccination and this can lead to an eventual increase in fertility in mature animals that have been immunized against important reproductive proteins such as ZP glycoproteins. Immunosterilization will result if the immune reaction causes irreversible damage, such as the atrophy of follicles or the destruction of oocyte/granulosa cell complexes. For more information, see Fayrer-Hosken *et al.* (2000).

5.2 Endocrine Manipulations in Aquaculture Fish

Fish are an important part of the human diet, and fish meal is an important commodity as a component of animal feeds. Aquaculture is increasingly needed since natural sources are unable to supply the demand for this commodity. Improved reproductive management, growth performance and feed efficiency of aquaculture systems can be achieved by endocrine manipulations.

Control of reproduction

Sex is determined by the presence of sex chromosomes (either XX/XY or ZW/ZZ) in most species of fish of commercial interest. Sex differentiation in fish can occur by direct differentiation of the primordial gonad into either a testis or an ovary. This occurs in 'differentiated' species such as the medaka, coho salmon, common carp and European sea bass. In 'undifferentiated' species, such as the guppy, hagfish and European eel, the gonad first develops as an ovary-like structure and then forms a testis in some of the fish.

The incubation temperature of reptile eggs determines the gonadal sex. At higher temperature, the increased synthesis of oestrogen by the aromatase enzyme down-regulates steroidogenic factor 1 (SF-1) expression and results in the female phenotype. Conversely, when oestrogen production is reduced, SF-1 expression increases and males are produced. This sexual dimorphism in SF-1 expression is reversed in mammals compared to birds and reptiles. Changes in oestrogen synthesis in response to temperature occur first in the brain and later in the gonads.

Sex reversal

Exogenous steroids can be used for sex reversal in many fish, amphibians and reptiles. Exposing the young fry to exogenous hormones can readily alter the phenotypic sex of fish. Sex control is used in aquaculture to produce the sex of fish that grows more rapidly; this is female in the salmonids and the cyprinids, while it is male in the cichlids. Sexual maturation in males is inhibited in production systems in order to reduce problems with poor carcass quality and aggressive and territorial behaviour in sexually mature fish. Both males and females are sterilized to allow larger fish to be reared and to prevent poor feed efficiency, as nutrients are normally diverted into gonadal growth as the fish matures. In sea ranching, rearing sterile fish prevents interbreeding with wild fish stocks by captive fish that escape.

Sex steroids regulate the differentiation of the gonads in fish. Oestrogens are responsible for ovarian differentiation; treatment of genetic female salmon with an aromatase inhibitor will cause them to develop into functional males. Environmental factors can also affect sexual differentiation by inducing or repressing the activity of certain genes, such as that coding for aromatase. Similarly, treatment of genetic female chickens with aromatase inhibitors early in development results in the development of testes rather than ovaries (see Section 4.2). Similarly, treatment of genetic male marsupials with oestrogen early in development causes the development of ovaries.

Fish can be directly feminized by treatment of a group of sexually undifferentiated fish with oestrogen, but half of these phenotypic females will have a male genotype. There may also be some concern from consumers about the direct use of steroid treatment in fish destined for human consumption.

HORMONAL TREATMENTS FOR SEX REVERSAL. Acute administration of hormones to fish can be achieved by injection of the hormone (intramuscularly or into the body cavity), or immersing the fish in water containing the hormone. Feeding hormone-treated food, or implantation of silastic capsules, cholesterol pellets or osmotic minipumps has been used for chronic administration of hormones. Addition of hormone to the water is useful for treatment of larval stages, when the hormone can be absorbed via the gills or integument. Rearing density is important with this method. Administration of the hormone in the feed can be used once external feeding has begun and this avoids disturbing fragile larvae.

Oestrogenic compounds, particularly oestradiol, are used for feminization of fish, while 17α-methyl testosterone is used for masculinizing fish. High doses or long exposure times to the hormone result in sterilization. In Atlantic salmon, doses of >20 mg kg^{-1} of 17α-methyltestosterone for the first 600 degree days of feeding result in sterilization. A number of endocrine disruptor chemicals have also been shown to affect sexual development in fish (see Section 6.4). Other oestrogenic compounds, such as DES, are not used since they are carcinogenic in humans. In the European Union, Council Directive 96/22/EC regulates the use of sex steroids for sex control in finfish aquaculture.

Treatment with oestrogens can sometimes alter male secondary sex characteristics without complete sex reversal. High levels of oestrogen can also reduce survival and growth rate. Salmonids require the lowest exposure (a combination of dose and treatment duration) to steroids for sex reversal, while cyprinids require considerably more exposure. In many species of salmonids, steroids are administered by immersion of the newly hatched fry in water containing the steroid. This likely allows the steroids to accumulate in the yolk sac, where they will continue to affect the developing fish until day 20.

Exogenous steroid hormones are most effective during the 'labile period' of sexual development, when the gonads are still undifferentiated. The timing of this labile period is species specific and somewhat dependent on the hormone used and the dose. For example, the labile period for oestrogen occurs before the labile period for androgens. Treatment times as short as hours have been used. In coho and chinook salmon, one or two 2-h immersions in steroids (400 μg l^{-1}) during the alevin stage at 8–13 days after hatching is sufficient to cause sex reversal. In tilapia, 30–40 p.p.m. of 17α-methyltestosterone is given in the food for 30 days at the swim-up stage to produce all male fish for grow out. Treatment outside the labile period requires higher doses of hormone or longer periods of treatment to achieve sex reversal. The optimum treatment protocols for sex reversal use the minimum dose of natural hormone during the labile period, so that the shortest exposure time, and consequently the least effect on survival or growth rate, can be used.

Steroids are metabolized and cleared through the liver (see Section 1.2) and this occurs rapidly, so levels of exogenous steroids would be undetectable a few days after treatment. Treatment of fish with steroids for sex reversal differs from use of steroids in beef cattle for weight gain (see Section 3.2). Cattle are usually treated with synthetic steroids, the treatment is continued for months and ends very near slaughter. In contrast, fish are usually treated with natural steroids, the treatment is short term and treatment ends long before the fish are marketed, even with the direct method for sex reversal. However, it is likely that indirect methods for sex reversal would be more acceptable to consumers than direct methods.

INDIRECT METHODS. All-female fish can also be produced in two generations using an indi-

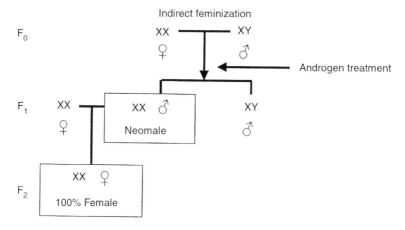

Fig. 5.16. Indirect method for feminization (adapted from Piferrer, 2001).

rect feminization procedure (Fig. 5.16). This indirect procedure does not involve direct treatment of the fish used for consumption. In this procedure, a group of genetic females are first masculinized by treatment with androgens. These 'neomales', which carry only X chromosomes, are identified using sex-specific DNA probes and bred with normal females. The resulting offspring are all female. The stock is maintained by masculinizing a small number of fish to provide more neomales. The untreated fish are grown out and marketed or used as female brooding stock. Thus, the indirect method of producing all-female stock is more time consuming but does not expose the marketable fish to steroids. It is used commercially for the culture of several salmonids, including rainbow trout and chinook salmon, and is applicable to fish species with the XX/XY sex determination system.

All-male stocks can also be produced by indirect treatment with oestrogens (Fig. 5.17). Sexually undifferentiated fish are

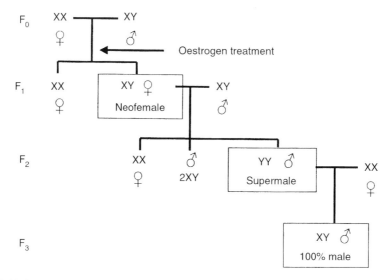

Fig. 5.17. Method to produce all-male stocks (adapted from Piferrer, 2001).

first treated with oestrogens and the 'neo-females' (genetic males but phenotypic females) are bred with normal males. One quarter of these progeny will be YY 'supermales'. These YY males are identified and bred with normal females to produce 100% XY male progeny. YY males have been produced in a number of species, including medaka, goldfish, rainbow trout and Nile tilapia, but are not viable in other species. In ZW/ZZ sex determination systems, as in the black molly, oestrogen treatment of undifferentiated fish produces ZZ neofemales. These are bred to normal ZZ males to produce all male offspring.

Sterile fish can be produced by inducing triploidy. Triploids can be produced indirectly by crossing tetraploids with diploids, or directly by preventing the extrusion of the second polar body after fertilization. Direct triploid production accomplished by heat, pressure or anaesthetic chemical treatment. For example, triploids can be induced in Atlantic salmon eggs by exposure to 30°C for 6–10 min at 20–30 min after fertilization. Pressure treatment is more costly but less variable and can be used to produce triploids over a wider period of time than heat treatment. Exposure of eggs to hyperbaric nitrous oxide gas also produces triploids, but the treatment must begin immediately after fertilization and is intermediate in effectiveness between heat and pressure treatment. For more information, see Piferrer (2001).

Induction of spawning

Environmental and hormonal manipulations can be used to manipulate spawning time (Fig. 5.18). Most fish spawn seasonally in the wild in response to environmental cues that occur with the seasons. In temperate and subtropical species, changes in water temperature and photoperiod dramatically affect gonadal growth and spawning. This allows the progeny to be produced in the spring, when the conditions for their survival are more favourable. However, in captivity, it is necessary to control spawning time in order to provide marketable fish throughout the year. In the red drum, the age to sexual maturity can be reduced by more than half by rearing the fish at high water temperatures. In salmonids, photoperiod is the main environmental cue for gonadal maturation and spawning. The annual cycle of temperature and photoperiod can be condensed into a few months to stimulate gonadal growth and spawning. The change in the direction of the photoperiod (from short days to long days and vice versa) is more important than the actual daylength. Exposure of rainbow trout, which normally spawn in late autumn, to long days early in the cycle followed by short days in late spring, will advance spawning time by 3–4 months.

In addition to environmental cues, spawning can be induced by treatment with a crude preparation of fish gonadotrophin or hCG, but species-specific preparations of

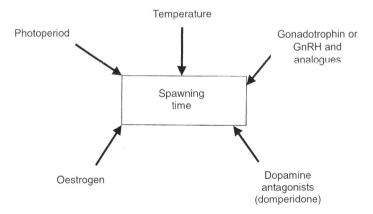

Fig. 5.18. Environmental and hormonal effects on spawning.

gonadotrophins may be needed in some cases. Thyroid hormone also acts in synergy with gonadotrophins to promote the final maturation of oocytes. GnRH can also be used to induce spawning and it has several advantages over gonadotrophins, since it is not species specific and is an easily synthesized small molecule. More active GnRH analogues have also been developed, and a number of slow-release delivery systems are being investigated. These compounds are also orally active, but the dose required is 10 times higher than that required by injection. The most potent GnRH analogue is the salmon GnRH analogue [D-Arg^6, Pro^9 NEt]-sGnRH (sGnRH-A), although analogues of mammalian GnRH (mGnRH) such as [D-Ala^6, Pro^9 NEt]-mGnRH (GnRH-A) are also very active. The increased activity of these analogues comes from a combination of increased binding affinity to receptors on pituitary gonadotrophs and increased hydrophobicity and resistance to enzymatic degradation.

In some species, dopamine acts as a GnRH antagonist to prevent the release of gonadotrophins from the pituitary gonadotrophs. Treatment with dopamine receptor antagonists, such as domperidone, in combination with GnRH-A improves the induction of spawning in some species, such as salmonids, but not in others, such as sciaenids. Syndel Laboratories, Vancouver, Canada, has marketed a formulation of sGnRH-A and domperidone under the name Ovaprim®.

Oestradiol acts by positive feedback to increase the production of gonadotrophins by the pituitary. Injection of testosterone, which is aromatized to oestradiol, has been used to increase the production of gonadotrophins. Oestradiol is also a potent inducer of vitellogenesis in the liver, which provides the components for yolk in the growing oocytes. For more information on the manipulation of spawning in fish, see the reviews by Peter and Yu (1997) and Mylonas and Zohar (2001).

Effects on growth and nutrient utilization

Increasing growth rate and feed efficiency is particularly important in aquaculture, due to the long production cycle of cold-water fish. This cycle normally lasts for years, with a consequently greater risk of loss of fish from disease or accidental release. In contrast, production cycles for poultry, swine and cattle are much shorter. Feed costs account for 50% or more of the cost of production in aquaculture, so improvements in feed efficiency can dramatically reduce operating costs.

Various dietary components can have dramatic effects on hormone levels, but it is difficult to measure these effects since changes in one component of the diet are often linked to changes in other dietary components. Endocrine systems can be regulated at a number of points, including neuroendocrine signalling, synthesis and secretion of the hormone, transport in the blood, metabolism in peripheral tissues, receptor binding, and metabolism and clearance of the hormone (see Chapter 1). Both CCK and NPY (see Section 3.9) have been identified in fish and probably play a major role in how nutritional status affects endocrine function in fish.

Steroid hormones, particularly androgens, can increase growth rate but they are not used commercially. Thyroid hormones are important in larval development in vertebrates. The yolk of newly fertilized eggs from salmonids and striped bass contain high levels of thyroid hormones and supplementing eggs or larvae with additional T_3 improves growth and survival of the larvae and fingerlings. Thyroid hormones are involved in the regulation of growth and energy utilization and they may play a permissive role in growth to potentiate the anabolic effects of insulin and growth hormone. Food deprivation decreases the levels of thyroid hormones and reduces the conversion of T_4 to T_3. Dietary protein and carbohydrate affect thyroid hormone production. Carbohydrate stimulates the secretion of T_4, while protein increases the conversion of T_4 to T_3. Restricted growth rate seen at high stocking densities is accompanied by decreased plasma thyroid hormones.

Fish growth follows a pattern that is related to daylength. Long daylength stimulates growth, in part because the fish has more time to forage for food, but also due to melatonin production by the pineal gland. Melatonin production is regulated by exposure to light and can affect the production of growth hormone and thyroid hormone metabolism. This affects the rate of growth and sexual maturation.

Growth hormone is the most potent growth promoter in teleost fish. Growth hormone acts in synergy with androgens and thyroid hormones to increase growth. Growth hormone release from the pituitary is increased during food deprivation or when low protein diets are fed, and this is thought to be due to direct effects of glucose and amino acids on the pituitary. However, in these situations IGF levels decline, so that growth hormone directly stimulates lipolysis.

Exogenous recombinant growth hormone from chicken or cattle injected into salmon improves growth and feed conversion efficiency. Somatotrophin analogues have also been developed that have increased potency and stability. Somatotrophin and its analogues are also active orally in fish, but the dose required is higher than that given by injection. Immersing trout fry for 30 min in a solution of homologous growth hormone also dramatically improved growth. Transgenic salmon have also been produced that express high levels of growth hormone and have a dramatically increased growth rate. For further information on genetic manipulations in aquaculture, see Hulata (2001).

Insulin and glucagon regulate lipid and carbohydrate metabolism as well as growth in fish. The level of insulin is increased in faster-growing fish, but it is not clear whether insulin stimulates growth or whether the increased level of insulin is a consequence of increased feed intake. Under natural photoperiod and temperature changes, fish show an active period of feed intake and growth rate during times of increased temperature and photoperiod. The level of insulin is highest during the times of rapid growth and lowest in coldest weather, when feeding activity is lowest. During short-term food deprivation, glucagon and glucagon-like peptides stimulate gluconeogenesis to maintain blood glucose levels. Injection of glucose into fed brown trout decreases levels of glucagon but does not affect the level of insulin, although it is increased by the amino acid arginine.

Applications

Periods of feed restriction or 'nutritional stress' can be used to increase feed conversion efficiency and lean growth. Lipolysis and gluconeogenesis are increased during feed restriction as the level of growth hormone increases and insulin and thyroid hormones are decreased. After feed restriction, animals enter a phase of compensatory growth, when levels of anabolic hormones are increased to promote somatic growth rather than storage of energy reserves as fat. This results in improved lean growth and feed efficiency and decreased fat deposition. Compensatory growth after feed restriction is also seen in other species, such as poultry.

Levels of many hormones in fish change with a circadian rhythm. The circadian rhythm of hormone release can also vary with an annual cycle and be dramatically different in winter than in summer. The release of metabolism-related hormones, such as insulin, is also affected by the composition of the diet, including the quality and proportions of protein, carbohydrate, fats and vitamins and minerals. The time of feeding, as well as the amount of feed given, can affect the amplitude of the daily rhythm of thyroid hormone, particularly T_4, as well as insulin and growth hormone. The time of day when fish are fed can thus dramatically affect growth and feed conversion efficiency, due to differences in levels of hormones. It is thus advantageous to determine when the hormone levels in the fish are optimal for growth and how different dietary nutrients affect hormone levels. This will allow the development of feeding strategies to promote anabolic hormone production. The goal is to feed fish at the optimal feeding rate and time of day to provide nutrients when the animal is physiologically best able to deal with them. This approach will achieve the best feed conversion efficiency and growth rate. However, in a practical commercial setting, it may be impossible to feed all fish at the optimal rate and time of day. For more information, see the review by MacKenzie *et al.* (1998).

Stress and effects on the immune system

As with all commercial animal production, poor husbandry practices and suboptimal environmental conditions can cause increased levels of stress. Long-term stress can reduce growth performance and reproductive success and increase the susceptibility to disease. Fish respond to increased stress by activation of the adrenergic system to produce catecholamines, and the hypothalamic–pituitary–interrenal (HPI) axis to produce glucocorticoids. These responses are similar to those described for other commercial animal species (see Section 6.3). Chronic stress reduces the production of sex steroids and impairs gonadal growth, gamete viability and reproductive performance. Levels of growth hormone are increased by stress, but growth rate is decreased due to decreased sensitivity of the target tissues. Chronic stress can also decrease the levels of thyroid hormones and reduce feeding activity, which, together with the catabolic effects of corticosteroids, decreases growth rate.

Physical stress increases the oxygen demand for fish by more than 50%. This becomes more of a problem at high temperatures due to lowered oxygen content of the water and the accumulation of toxic levels of ammonia produced as nitrogenous waste and carbon dioxide from respiration. Respiratory stress can result in high mortality rates in extreme situations. The hormonal responses to stress also reduce the ability of the fish to maintain osmoregulation, which causes a loss of electrolytes in freshwater fish and an increase in electrolytes in marine fish.

Stress also increases the susceptibility of fish to a variety of viral, bacterial, fungal and protozoan pathogens. As occurs in mammals, the endocrine system and immune systems in fish interact with each other (see Section 6.3). Cortisol, growth hormone, prolactin, reproductive hormones, melanotrophins and some pro-opiomelanocortin (POMC)-related peptides influence immune responses in fish.

Cortisol is the major corticosteroid produced by the interrenal gland in teleost fish. Stress or cortisol administration decreases the resistance of fish to infection by bacteria and fungi. The decreased immune response includes decreased phagocytosis and lymphocyte mitogenesis and decreased activity of antibody-producing cells and levels of IgM.

Growth hormone is involved not only in the regulation of somatic growth, but also in osmoregulation and gonadal steroidogenesis in fish. Prolactin is also involved in growth and development, osmoregulation and reproduction. Both prolactin and growth hormone are also involved in the stimulation of immune responses in fish. They stimulate lymphocyte maturation and differentiation, activate phagocytes and natural killer cells and stimulate antibody production.

The major steroids produced by the ovaries of female fish are testosterone and androstenedione, while males produce primarily 11-ketotestosterone and 11-β-hydroxytestosterone. In rainbow trout, oestradiol stimulates, while 11-ketotestosterone decreases lymphocyte proliferation. Testosterone reduces the number of antibody-producing cells, as does cortisol, and these steroids act synergistically to inhibit antibody production. High levels of the gonadal steroids are present in salmonids during their freshwater migration and maturation, and this is linked to immune deficiencies and increased infections during this time.

Skin colour in fish is controlled by the pituitary hormones α-melanocyte stimulating hormone (αMSH) and melanin concentrating hormone (MCH), which act on melanocytes. αMSH darkens the skin by causing dispersion of melanin granules, while MCH concentrates the melanin to decrease skin colour intensity. Both hormones have also been implicated in the regulation of the HPI axis, feeding and osmoregulation in fish. Early studies showed that fish kept in dark tanks are more vulnerable to infections, and this is due to the actions of the melanotrophins. αMSH inhibits fever and inflammation by inhibiting the synthesis of the pro-inflammatory cytokines IL-1, IL-6, IL-8 and TNFα, and stimulating the release of the anti-inflammatory cytokine IL-10. MCH stimulates the proliferation of leucocytes and reduces the inhibitory effects of cortisol on mitogenesis. Other POMC-derived peptides, including ACTH and β-endorphin, have also been implicated in modulating the immune response of fish. For more information on the effects of hormones on the immune system in fish, see Harris and Bird (2000).

Applications

Although it is impossible to completely eliminate stress in commercial aquaculture, the negative effects of stress can be minimized. Appropriate stocking densities should be used and water quality should be maintained. Adequate time (days to weeks) should be allowed for recovery from stress from routine procedures such as netting, grading and transporting, and multiple stresses should be avoided. Manipulations should be done in colder water temperatures. Dilute salt solutions can be used for freshwater fish to limit osmoregulation problems. Withdrawal of food for a few days before the stress minimizes contamination of the water from fecal matter and feed and reduces respiratory stress through increased oxygen availability. Anaesthetics (MS222, phenoxyethanol, etomidate) can be used to minimize the stress response.

The magnitude of the stress response varies dramatically between different breeds and strains of fish. Genetic selection for fish with reduced cortisol response to stress may alleviate some of the negative effects of stress.

Questions for Study and Discussion

Section 5.1 Manipulation of Reproduction in Mammals

1. Outline the role of endocrine factors in the differentiation of gonadal structures and the brain.
2. Describe the differences in the pattern of sexual differentiation in cattle, sheep and pigs.
3. Describe the endocrine factors that regulate meiosis in germ cells.
4. Describe the hormonal changes during the different phases of the oestrous cycle.
5. Describe the hormonal regulation of follicular development.
6. Describe the hormonal changes during pregnancy and parturition.
7. Discuss the effects of seasonality on sexual maturation and breeding.
8. Describe the opportunities for manipulation of the oestrous cycle. What hormone preparations are available for this purpose?
9. Outline methods for detection of oestrus in cattle.
10. Describe methods for synchronizing oestrus in cattle.
11. Discuss methods for maintaining or ending pregnancy.
12. Describe factors that affect the resumption of oestrus postpartum.
13. Outline methods for inducing cycling in seasonal breeders.
14. Discuss the potential for the immunological control of reproduction.

Section 5.2 Endocrine Manipulations in Aquaculture

1. Describe the direct and indirect methods for the use of hormones in sex reversal of fish.
2. Discuss methods for inducing spawning.
3. Describe the potential for the hormonal manipulation of growth and nutrient utilization in fish.
4. Discuss the effects of stress on productivity of farmed fish.

Further Reading

Sexual differentiation and maturation

Albertini, D.F. and Carabatsos, M.J. (1998) Comparative aspects of meiotic cell cycle control in mammals. *Journal of Molecular Medicine* 76, 795–799.

Byskov, A.G., Baltsen, M. and Andersen, C.Y. (1998) Meiosis-activating sterols: background, discovery, and possible use. *Journal of Molecular Medicine* 76, 818–823.

Ford, J.J. and D'Occhio, M.J. (1989) Differentiation of sexual behavior in cattle, sheep and swine. *Journal of Animal Science* 67, 1816–1823.

Gustafsson, J.-A. (1994) Regulation of sexual dimorphism in rat liver. In: Short, R.V. and Balaban, E. (eds) *The Differences between the Sexes*. Cambridge University Press, Cambridge, pp. 231–241.

Haqq, C.M. and Donahoe, P.K. (1998) Regulation of sexual dimorphism in mammals. *Physiological Reviews* 78, 1–33.

Heikkila, M., Peltoketo, H. and Vainio, S. (2001) Wnts and the female reproductive system. *Journal of Experimental Zoology* 290, 616–623.

Koopman, P., Bullejos, M. and Bowles, J. (2001) Regulation of male sexual development by *Sry* and *Sox9*. *Journal of Experimental Zoology* 290, 463–474.

Short, R.V. and Balaban, E. (1994) *The Differences Between the Sexes*. Cambridge University Press, Cambridge.

Regulation of the reproductive cycle

Dhandapani, K.M. and Brann, D.W. (2000) The role of glutamate and nitric oxide in the reproductive neuroendocrine system. *Biochemistry and Cell Biology* 78, 165–179.

Driancourt, M.A. (2001) Regulation of ovarian follicular dynamics in farm animals. Implications for manipulation of reproduction. *Theriogenology* 55, 1211–1239.

Findlay, J.K. (1993) An update on the roles of inhibin, activin, and follistatin as local regulators of folliculogenesis. *Biology of Reproduction* 48, 15–23.

Malpaux, B., Thiéry, J.-C. and Chemineau, P. (1999) Melatonin and the seasonal control of reproduction. *Reproduction Nutrition Development* 39, 355–366.

Peng, C. and Mukai, S.T. (2000) Activins and their receptors in female reproduction *Biochemistry and Cell Biology* 78, 261–279.

Wiltbank, M.C. (1998) Information on regulation of reproductive cyclicity in cattle. In: Williams, E.I. (ed.) *Proceedings of the Thirty-first Annual Conference of the American Association of Bovine Practitioners*. Frontier Printers, Spokane, Washington, pp. 26–33.

Zawilska, J.B. and Nowak, J.Z. (1999) Melatonin: from biochemistry to therapeutic applications. *Polish Journal of Pharmacology* 51, 3–23.

Reproductive manipulations

Amory, J.K. and Bremner, W.J. (2000) Newer agents for hormonal contraception in the male. *Trends in Endocrinology and Metabolism* 11, 61–66.

Day, B.N. (2000) Reproductive biotechnologies: current status in porcine reproduction. *Animal Reproduction Science* 60/61, 161–172.

Day, M.L. (1998) Practical manipulation of the estrous cycle in beef cattle. In: Williams, E.I. (ed.) *Proceedings of the Thirty-first Annual Conference of the American Association of Bovine Practitioners*. Frontier Printers, Spokane, Washington, pp. 51–61.

Diskin, M.G. and Sreenan, J.M. (2000) Expression and detection of oestrus in cattle. *Reproduction Nutrition Development* 40, 481–491.

Fayrer-Hosken, R.A., Dookwah, H.D. and Brandon, C.I. (2000) Immunocontrol in dogs. *Animal Reproduction Science* 60/61, 363–373.

Hoffmann, B. and Schuler, G. (2000) Receptor blockers – general aspects with respect to their use in domestic animal reproduction. *Animal Reproduction Science* 60/61, 25–312.

Marquant-Leguienne, B. and Humblot, P. (1998) Practical measures to improve *in vitro* blastocyst production in the bovine. *Theriogenology* 49, 3–11.

Nagy, P.G.D. and Daels, P. (2000) Seasonality in mares. *Animal Reproduction Science* 60–61, 245–262.

Pratt, P.M. (1998) Manipulation of estrus in the mare. In: *Manual of Equine Reproduction*. Mosby-Year Book Inc., St Louis, Missouri, pp. 15–24.

Prunier, A. and Quesnel, H. (2000) Influence of the nutritional status on ovarian development in female pigs. *Animal Reproduction Science* 60/61, 185–197.

Schillo, K.K., Hall, J.B. and Hileman, S.M. (1992) Effects of nutrition and season on the onset of puberty in the beef heifer. *Journal of Animal Science* 70, 3994–4005.

Thatcher, W.W., Risco, C.A. and Moreira, F. (1998) Practical manipulation of the estrous cycle in dairy animals. In: Williams, E.I. (ed.) *Proceedings of the Thirty-first Annual Conference of the American Association of Bovine Practitioners*. Frontier Printers, Spokane, Washington, pp. 34–50.

Thompson, D.L. (2000) Immunization against GnRH in male species (comparative aspects). *Animal Reproduction Science* 60/61, 459–469.

Wright, P.J. and Malmo, J. (1992) Pharmacologic manipulation of fertility. *Veterinary Clinics of North America: Food Animal Practice* 8, 57–89.

Endocrine manipulations in aquaculture

Harris, J. and Bird, D.J. (2000) Modulation of the fish immune system by hormones. *Veterinary*

Immunology and Immunopathology 77, 163–176.

Hulata, G. (2001) Genetic manipulations in aquaculture: a review of stock improvement by classical and modern technologies. *Genetica* 111, 155–173.

MacKenzie, D.S., VanPutte, C.M. and Leiner, K.A. (1998) Nutrient regulation of endocrine function in fish. *Aquaculture* 161, 3–25.

Mylonas, C.C. and Zohar, Y. (2001) Use of GnRHα-delivery systems for the control of reproduction in fish. *Reviews in Fish Biology and Fisheries* 10, 463–491.

Peter, R.E. and Yu, K.L. (1997) Neuroendocrine regulation of ovulation in fishes: basic and applied aspects. *Reviews in Fish Biology and Fisheries* 7, 173–197.

Piferrer, F. (2001) Endocrine sex control strategies for the feminization of teleost fish. *Aquaculture* 197, 229–281.

6
Effects on Animal Behaviour, Health and Welfare

6.1 Control of Broodiness in Poultry

Broodiness, also known as incubation behaviour, occurs when birds spend an increasing amount of time in the nest to incubate eggs. Food and water consumption decrease dramatically and egg production ceases due to regression of the ovary and oviduct. Broodiness is decreased in battery-type cages and increased in 'welfare friendly' production systems that allow full nesting behaviour to be expressed, such as modified cages, aviary and free-range systems. Broodiness is a natural behaviour that allows the hen to incubate and hatch out a clutch of eggs. However, broodiness is a problem in modern production systems, since when a hen becomes broody it ceases to lay eggs due to the regression of the reproductive system.

Broodiness can be controlled by manipulation of the environment to remove the stimuli that encourage nesting behaviour. This includes regular removal of eggs and use of uniform lighting to discourage birds from nesting in dim corners. Adequate numbers of nest boxes should be provided and these should be closed at night. The arrangement or appearance of the nest boxes should also be changed regularly to prevent the birds from becoming attached to a familiar nest site. Hens that are in the early stage of persistent nesting can be transferred to a pen designed to make nesting uncomfortable, such as one with wire floors and strong updrafts of cold air.

Altering hormone function can be used to control broodiness (Fig. 6.1). Nesting behaviour begins just before an egg is laid and is stimulated by the interaction of oestrogen and progesterone. After birds have been in lay for a few weeks, nesting behaviour may become prolonged and advance to become broodiness. An extensive brood patch can develop on the breast, designed to improve heat transfer from the hen to the developing embryos. Prolactin acts in concert with the ovarian steroids to increase vascularization and feather loss in the brood patch. Stimulation of the brood patch by the eggs or a familiar nest site maintains the high levels of prolactin release, and once broodiness is established oestrogen and progesterone are no longer needed. Prolactin also decreases the secretion of LH, which results in regression of the ovary and the cessation of egg laying.

The secretion of prolactin by the anterior pituitary gland is controlled by the avian prolactin releasing hormone, a 28-amino-acid peptide originally isolated from the gut and also known as vasoactive intestinal peptide (VIP). VIP is released from the hypothalamus and binds to specific receptors on lactotroph cells in the anterior pituitary. VIP release is increased by serotonin or dopamine, and turnover of these neurotransmitters is increased in the hypothalamus of broody birds. Dopamine may also act directly on the anterior pituitary to inhibit prolactin secretion.

The prolactin receptor gene has been suggested as a candidate gene for the control

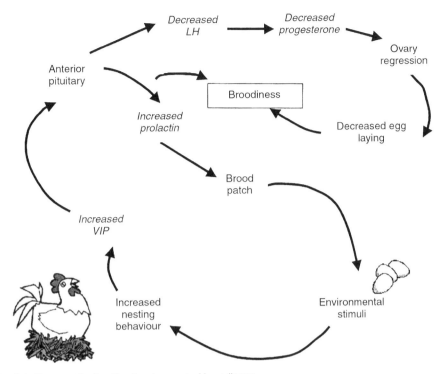

Fig. 6.1. Factors affecting the development of broodiness.

of broodiness and it has been mapped to the Z chromosome. However, there are no differences in the amounts of the prolactin receptor between White Leghorns, which do not become broody, and bantam hens, which commonly develop broodiness behaviour. It may be that differences in the expression of the various isoforms of the receptor are important.

Applications

Broodiness occurs where the birds are kept on the floor, and affects broiler breeders, especially the dwarf strains, egg-layer multiplier flocks and especially turkeys. Broodiness is affected by genotype, and is present in most species of laying fowl except White Leghorn.

Administration of pharmacological doses of progesterone or oestrogen disrupts broodiness, but this delays the return to lay. Treatment with anti-oestrogens (clomiphene citrate and tamoxifen) produced inconsistent results. Further treatments to disrupt broodiness have been directed at either increasing gonadotrophin secretion or decreasing prolactin secretion. Treatment with gonadotrophins returns broody turkeys to lay but is too expensive to be of practical use. Treatment with synthetic GnRH is also impractical since the anterior pituitary becomes non-responsive to GnRH after prolonged treatment. Treatment with a dopamine receptor blocker, pimozide, does not prevent broodiness. Decreasing the synthesis of serotonin with p-chlorophenyl alanine (PCPA) decreases the production of prolactin, but does not return turkeys quickly back into lay. Use of the dopamine agonist, bromocryptine, delays the resumption of egg laying.

Passive immunization against prolactin in bantam broody hens results in increased LH. Subsequent cloning of chicken prolactin allowed the active immunization of hens against recombinantly derived prolactin coupled to β-galactosidase. This procedure decreases broodiness without affecting egg production, and is more effective if it is initiated before egg laying begins. The peptide sequences of chicken prolactin and turkey prolactin are practically identical, so this should also be effective in turkeys. Alternatively, hens can be immunized against VIP. Injection of antibodies against VIP decreases the prolactin level to baseline and decreases broodiness. Active immunization of turkeys with synthetic turkey VIP conjugated to keyhole limpet haemocyanin (KLH) decreases the prolactin level and broodiness. Active immunization protocols to decrease broodiness require regular booster injections at 3–5-week intervals to maintain sufficient antibody titres. For more information, see the review by Sharp (1997).

6.2 Applications of Pheromones

The use of chemical signals is an important method for many diverse species to communicate between and within a species, and between a species and its environment. An appreciation of the use of chemical signals in animals began in the 19th century, but it was not until the 1930s that experimental evidence showed that volatile chemicals were involved. For further information on pheromones, see the *Scientific American* text by Agosta (1992) and the symposium proceedings by Johnston *et al.* (1999). For an interesting discussion of the evolution of pheromones and their receptors, see Stoka (1999).

Naturally occurring compounds that are used by organisms to perceive and communicate with other organisms and the environment are collectively known as semiochemicals. They include pheromones, allomones and kairomones. The word pheromone, used by Karlson and Butenandt in 1959, is derived from the Greek *pherein*, 'to transfer', and *hormon*, 'to excite', and refers to a chemical messenger released to the exterior of one individual to stimulate a response in another individual of the same species. Pheromones are thus hormones that are used external to the animal. Allomones are emitted by a species to control other species for its advantage. For example, when different species compete for the same resources, repellant allomones emitted by one species can reduce the number of individuals present from another species. Kairomones are emitted by a species to its disadvantage and typically as a result of efforts to overcome stresses. For example, pears release ethyl-2,4-decadieoate, which is a highly potent attractant to the codling moth, *Cydia pomonella*, which is a serious insect pest of walnuts, apples and pears.

The field of chemical communication and semiochemicals has concentrated mainly on insects, although several vertebrate pheromones have now been identified. Current research on pheromones is involved in determining receptor function and signal transduction mechanisms, and developing natural product analogues.

Types of Pheromones

Pheromones are divided into two types, depending on their effect on the recipient. **Signalling pheromones** transfer specific information to induce changes leading to a prompt behavioural response in the recipient. They include sex attractants, which either lead to aggression or are involved in courtship or copulatory behaviour. Alarm pheromones prompt organisms to evacuate an area. Aggregation pheromones bring others to a food source or a suitable habitat, or bring others to a sexual partner. Dispersion pheromones maintain optimal separation between animals and maintain separation between territorial social groups. **Priming pheromones** induce physiological changes that may have a long-term influence on the recipient, such as the induction of puberty or the termination of anoestrus.

Chemistry of pheromones

The chemical structures of several pheromones are known, but in most cases the existence of a pheromone is implied from behavioural studies when the chemical structure of the pheromone has not yet been established. Most of the work comes from studies on insects, with much less work on pheromones from vertebrates. Pheromones are organic compounds or mixtures of organic compounds. They may contain functional groups, such as carbon–carbon double bonds, or carbonyl, hydroxyl, carboxyl and ester groups. Most volatile pheromones are molecules with 5–20 carbon atoms and include hormones or their breakdown products.

There are several examples of very similar, or even identical, compounds being active as pheromones in different species. In insects, it is common for closely related species to use similar mixtures of compounds as pheromones, but the ratios of the components are different or other components are included to give species specificity. In other cases, the same pheromone is used for similar species that have different timing of their reproductive cycle, so there is no confusion. In other cases, widely different species utilize the same pheromones. For example, (Z)-7-dodecenyl acetate is a sex attractant for Asian elephants as well as the turnip looper and cabbage looper and many species of butterflies and moths.

A family of low molecular weight soluble proteins known as lipocalins (Fig. 6.2) acts as carriers for pheromones and are important for their delivery and detection. Members of the lipocalin family include odorant-binding proteins (ODPs) that are produced by glands in the nasal cavity to concentrate volatile odorant molecules and improve their detection. Major urinary proteins (MUPs) present in the urine and saliva of rodents act to stabilize and control the release of volatile pheromone deposits. Aphrodisin is a pheromone-related protein in the vaginal discharge of hamsters. These lipocalin proteins have a common three-dimensional structure called a β-barrel, which is made of eight antiparallel β-sheets enclosing the hydrophobic binding pocket that binds the pheromone. In addition to binding pheromones, some of the pheromone-binding proteins have been shown to have pheromonal properties of their own, whether or not a volatile pheromone is bound. Other lipocalins bind lipophilic compounds such as retinol and retinoic acid. For more information on lipocalins, see the review by Flower (1996).

Pheromone production and release

The communication system for pheromones requires a mechanism for emitting the pheromone, a medium through which pheromones can be transmitted and a mechanism for receiving the pheromone. Pheromones are usually emitted by glandular organs equipped with specialized structures for their release. In insects, most pheromone glands are composed of groups of modified epidermal cells. The glands may be complex and associated with internal reservoirs.

Chemical signals can be quite volatile and carried for a long distance in air or water by diffusion and passive transport. These signals would provide an immediate message that would last for a short time. For example, a sex attractant released into the environment would indicate that a member of the opposite sex is available for mating. In insects, wing fanning disperses the highly volatile pheromones. In male sheep and goats, a primer sex pheromone produced in the skin evaporates from the hair or wool. In fish, free steroids are released from the gills. Many species secrete pheromones in the urine. Some chemical signals can be persistent, stable and non-volatile. This can occur by binding a volatile compound to a protein or using more stable chemical signals of larger molec-

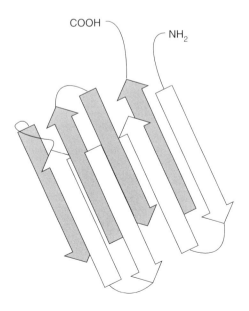

Fig. 6.2. Structure of lipocalins.

ular weight. These signals would be deposited and remain after the animal has left to provide a signal for the longer term. This would be useful for marking territory or indicating the location of a food source.

Detection of pheromones

The perception of pheromones is through taste and smell receptors. Insects detect pheromones through different hair-like projections (sensilla) on the antennae. Chemosensory systems can be extremely sensitive. For example, the antennae chemoreceptors of the male silkworm moth, *Bombyx mori*, can detect concentrations of the female sex pheromone bombykol as low as 1 molecule in 10^{17} in air. This allows a male moth to locate a female that is several miles downwind, at night. Similarly, a steroid that induces moulting in crabs, and also acts as a sex attractant, is active at 10^{-13} M. Pheromones are active over a limited concentration range and are ineffective when the levels are too high.

In mammals, odorants are detected in the olfactory sensory neurones in the epithelium that lines the nasal cavity. Most mammals also have a second olfactory organ known as the vomeronasal organ (VNO) that is involved in pheromone detection. The VNO is a tubular cavity in the nasal septum, lined with sensory epithelium and vomeronasal glands, which secrete a fluid into the lumen. The lumen of the VNO is connected to the nasal cavity by narrow ducts, which allows the movement of secretions between the two compartments. During the 'lip curl' Flehmen response in male courtship behaviour in certain mammals, including cats, ungulates, bats and marsupials, fluid containing female pheromones is sucked into the VNO and the pheromones are detected. A 70 kDa odorant-binding glycoprotein (vomeromodulin) is expressed in the VNO and presumed to be involved in the binding of pheromones.

There are three distinct families of olfactory receptor molecules that are all G-protein-coupled receptors. The odour receptor (OR) gene family, which is expressed in the olfactory epithelium, consists of about 1000 genes that code for receptor proteins that vary dramatically in amino-acid sequence and can bind a wide variety of ligands. The other two families of receptors are expressed in the VNO; these are the V1R family of about 35 members and the V2R family of 150 members, which are presumably involved in detection of pheromones. All of the olfactory receptors have seven transmembrane domains and the V2Rs have a very large N-terminal extracellular domain. Each olfactory neurone expresses only one OR and each OR can recognize multiple odorants. The detection of odours involves the combined response of a large number of signals from individual neurones with different receptor molecules. This allows the discrimination of a very large number of odorants. On the other hand, it is hypothesized that the V1Rs and V2Rs only respond to individual pheromones. It has been suggested that V1Rs react with lipophilic and volatile odorants, while the V2Rs react with proteins. The activation of specific receptor molecules initiates the endocrine and behavioural response to pheromones. This mechanism provides for the specific recognition of a defined number of pheromones.

In mammals, ligand binding to odour receptors (ORs) stimulates adenylyl cyclase to increase cAMP and opens cyclic nucleotide-gated cation channels, causing membrane depolarization and generating an action potential in the neurone. Alternatively, a second pathway exists for V1R and V2R receptors and ion channels that are specific to the vomeronasal organ. In this system, phospholipase C is activated to increase IP_3 and trigger the opening of an ion channel, causing an action potential (see Section 1.3). Insect pheromones are bound by specific pheromone-binding proteins to interact with a G-protein-coupled odorant receptor and increase the IP_3 second messenger. Phosphorylation of the receptors and high levels of cGMP reduce the signal transduction mechanism. For more information on odour and pheromone receptors, see Buck (2000).

Vertebrate pheromones

In vertebrates, various bodily secretions, such as vaginal secretions and urine, have pheromonal effects. Pheromones have been implicated as indicators of the reproductive status of females and in the induction of behaviours resulting in mating, in aggressive behaviour towards intruder males, in the inhibition of oestrus cycling in group-housed females and in the induction of oestrus by exposure to males. An individual or group can use pheromones or chemical signals to mark its territory. These signals are deposited from urine, or specialized anal or suprapubic glands.

There is also evidence that individuals can be identified based on odour that is directly related to the composition of the major histocompatibility complex (MHC) genes. Odorants may be bound directly by proteins encoded by MHC genes, released into serum and concentrated in urine. It is possible that recognition of MHC types may be used to prevent inbreeding in communal populations. It may also be the basis of the Bruce Effect, in which pregnancy is terminated in rodents by exposure of the female to a male that is genetically different from the inseminating male.

Due to the flexibility of mammalian behaviour and its dependence on a number of factors, behavioural responses to olfactory signals may be less obvious than in insects. For example, a mammal could respond quite differently to a pheromonal signal depending on its emotional state. Various tactile, visual or auditory cues from the male or female may also affect the response. Repeated exposure to the pheromone can cause the animal to become habituated and then not respond to the pheromone. These factors can make the identification of the pheromone, and the role that pheromones play in vertebrate behaviour, very difficult to determine.

Rodents

Most detailed studies of reproductive pheromones have been done with rodents. The female Syrian golden hamster in oestrus leaves a pheromone trail to attract a male. The attractant pheromone is dimethyl disulphide, but the later stages of courtship and copulation are stimulated by a protein pheromone known as aphrodisin, which is a member of the lipocalin family. Aphrodisin, present in the vaginal secretions of the female, is licked by the male and detected in the vomeronasal organ.

The major urinary protein (MUP) in the mouse is another member of the lipocalin family and is analogous to α-2u globulin in the rat. MUP is thought to function in pheromone transport in male urine but also has pheromonal properties of its own. In fact, the hexapeptide N–Glu–Glu–Ala–Arg–Ser–Met, which is a truncated form of MUP, accelerates puberty in female mice. Male mouse urine contains *sec*-butyldihydrothiazole and dehydro-*exo*-brevicomin (Fig. 6.3) bound to MUP, and these compounds stimulate aggression in other males, female attraction, and synchronization and acceleration of oestrus. The pheromonal property of these compounds is also illustrated by the fact that a similar compound, *exo*-brevicomin, is a male attractant produced by the female western pine beetle.

Six small organic compounds have also been identified in female mouse urine that delay the onset of female puberty. Rats

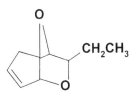

Fig. 6.3. Pheromonal compounds in male mouse urine.

produce a maternal pheromone, deoxycholic acid, in the faeces that attracts the young back to the maternal nest.

Pigs

Oestrous sows detect pheromones released by boars. These pheromones are the steroids, 5α-androstenone and 3α-androstenol (Fig. 6.4), which are members of the 16-androstene steroid family, responsible for boar taint (see Section 3.3). These steroids are synthesized in the testis and subsequently removed from the bloodstream by pheromone-binding proteins (lipocalins) in the salivary gland. Arousal of the boar results in profuse salivation and release of the pheromones and binding proteins with the saliva. 5α-Androstenone is one of the first vertebrate pheromones available as an aerosol preparation (Boar Mate®).

The boar pheromones cause the sows to 'stand' for mating, so the sow resists forward pressure (lordosis) when in oestrus. These pheromones may also establish dominance hierarchies and decrease fighting of males in regrouping situations. In addition to these signalling actions, boar pheromones have a priming action and increase the onset of puberty in prepubertal gilts. Boar exposure should occur after 140 days of age, since earlier boar exposure actually increased the age at puberty. This may be due to a habituation of the boar presence when the gilt is physiologically unable to respond. Androstenone or boar exposure also decreases the time from weaning to first oestrus in sows.

There is some evidence to suggest that the 16-androstene steroids are active in humans and are found in trace amounts in human sweat. *In vitro* synthesis of the steroids has been demonstrated in human testis microsomes. The 16-androstene steroids have also been found in various tissues in camels. Androstenone is also found in truffles, and pigs have traditionally been used to hunt for these underground fungi. Androstenone is also present in caviar, celery and young parsnips.

There is some evidence that pheromones that signal alarm are released in the urine of pigs undergoing stress, but the chemical nature of these pheromones has not been investigated (Amory and Pearce, 2000). Sows also produce a maternal pheromone that stimulates feeding behaviour and post-weaning weight gain in piglets (McGlone and Anderson, 2002).

Cattle

The cervicovaginal mucous and urine from oestrous cows stimulates sexual activity in bulls. Higher levels of *n*-propyl-phthalate and 1-iodoundecane have been found in oestrous urine, but these compounds have not been proven to act as pheromones. Detection of olfactory cues as a method for detecting oestrus in cows would benefit the dairy and beef industries (Section 5.1). The olfactory cues could potentially be detected using trained animals (e.g. sniffer dogs), and recently electronic sensors have been used experimentally for this purpose. There is also evidence that cows produce pheromones that regulate the oestrous cycle of other cows.

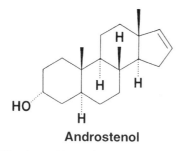
Androstenol

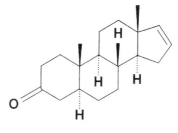
Androstenone

Fig. 6.4. Structures of boar pheromones.

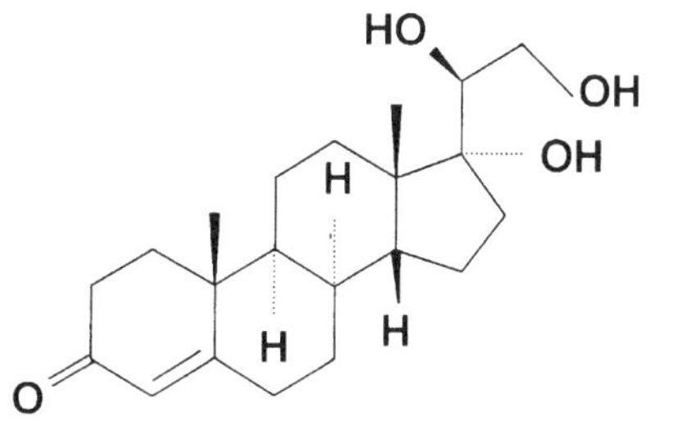

17α, 20β, 21-Trihydroxy-4-pregnen-3-one 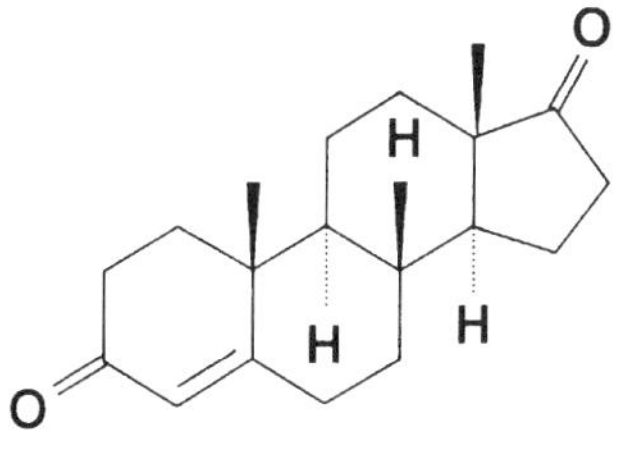 4-Androsten-3,17-dione

Fig. 6.5. Primer pheromones produced by female goldfish.

Exposure to bull urine has been shown to accelerate puberty and shorten the calving season in heifers. Bull pheromones could therefore be used to increase the pregnancy rate of first-season heifers and concentrate the calving season in beef heifers. Exposure of cows to a bull also decreases the time to conception after calving.

Sheep and goats

Rams sniff odours from vaginal secretions to detect oestrus in ewes. Oestrous ewes also seek out rams, suggesting that rams emit a signalling pheromone. The introduction of a male also aids in the termination of seasonal anoestrus and synchronization of oestrus in sheep and goats. A primer pheromone is produced in the sebaceous glands of male sheep and goats; the production of this pheromone is induced by testosterone. The pheromone is released from the male hair or wool and stimulates the neuroendocrine system in females to increase the frequency of pulsatile LH release and induce ovarian function and cycling.

Fish

The existence of chemical signals for fish has been demonstrated since 1932, but the compounds responsible have only been identified in the past 10 years. Most of the work has been done with goldfish as a model system. About 10 h before spawning, female goldfish produce large quantities of di- and tri-hydroxy-4-pregnen-3-one, testosterone and 4-androsten-3,17-dione (Fig. 6.5). These steroids act first as hormones to promote oocyte maturation in the female then are released into the water to act as primer pheromones, increasing the production of gonadotrophins in males. This stimulates the formation of milt (sperm and seminal fluid) and increases aggression in males. The steroids are released as conjugates with glucuronide and sulphate in the urine (see Section 1.2) and free steroids are released from the gills.

Prostaglandin $F_{2\alpha}$ and its metabolite 15-keto-$PGF_{2\alpha}$ are associated with follicular rupture and ovulation in females. When they are released into the water, they stimulate aggression among males and courtship behaviour towards females. This is an unusual application for prostaglandins, as they normally have only local hormonal effects. Thus, for fish the distinction between hormones and pheromones is very blurred.

Injured fish release an alarm pheromone to trigger anti-predator behaviour in other fish. In ostariophysian (order Cypriniformes) fish, the alarm pheromone has been proposed to be hypoxanthine-3-oxide (Fig. 6.6), which is produced by alarm system cells in the skin. However, this result has been questioned due to the high levels of the pure compound that are needed to produce the alarm response. Nothing is known about the chemical structure of alarm pheromones in non-ostariophysian fishes.

Other

Female Asian elephants release (Z)-7-dodecenyl acetate in their urine to signal

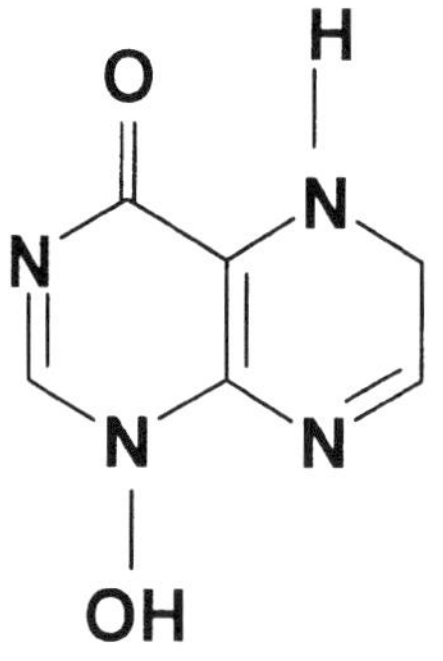

Hypoxanthine-3-oxide

Fig. 6.6. Alarm pheromone in ostariophysian fish.

that they are ready to mate. Female garter snakes use a mixture of long-chain fatty acids, ranging from 29 to 37 carbon atoms, as a mating pheromone. Squalene is used as a male recognition pheromone in garter snakes. The foul-smelling musk of skunks is produced by scent glands near the anus. The active ingredients are *trans*-2-butene-1-thiol, 3-methyl-1-butanethiol and *trans*-2-butenyl methyl disulphide.

Nepetalactone is the active component of catnip that induces behaviour similar to the mating ritual in domestic cats, lions and tigers. A synthetic feline facial pheromone product (Feliway®) has been marketed. It is claimed to reduce stress in cats and thus reduce urine spraying and territorial marking in this species.

Insect pheromones

Much more is known about the chemical nature of insect pheromones than pheromones in vertebrates. Bombykol (Fig. 6.7), the sex attractant produced by the female silkworm, was the first pheromone to be chemically identified by A. Butenandt and co-workers in 1959. The limited equipment and analytical techniques available at that time made this a very difficult task. Milligram quantities of pure material were required for structural determination, and sensitive bioassays needed to be developed to guide the purification work, so the structural identification took over 20 years to accomplish. Since this work was completed, the structures of many other insect pheromones have been elucidated. The majority of the lepidopteran female sex pheromones are long-chain alcohols, acetates and aldehydes.

Insects use sex pheromones to attract a mate, aggregation pheromones to indicate to others the site of a good food source, and alarm pheromones to indicate attack by a predator. In some cases, insects do not synthesize pheromones *de novo*, but use plant-derived compounds as starting materials for the synthesis. For example, the sex attractant pheromone of danaid butterflies, danaidone, is made by male butterflies from pyrrolizidine alkaloids obtained from plants. Similarly, aggregation pheromones of dark beetles are composed of mixtures of compounds such as exo-brevicomin, which are derived from the plants they infest.

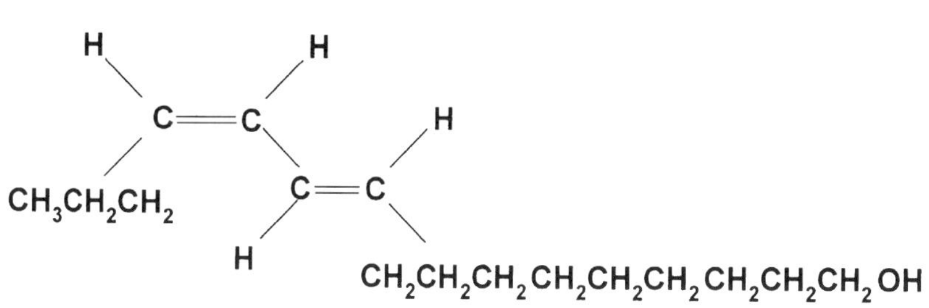

Bombykol

Fig. 6.7. Sex pheromone of the female silkworm.

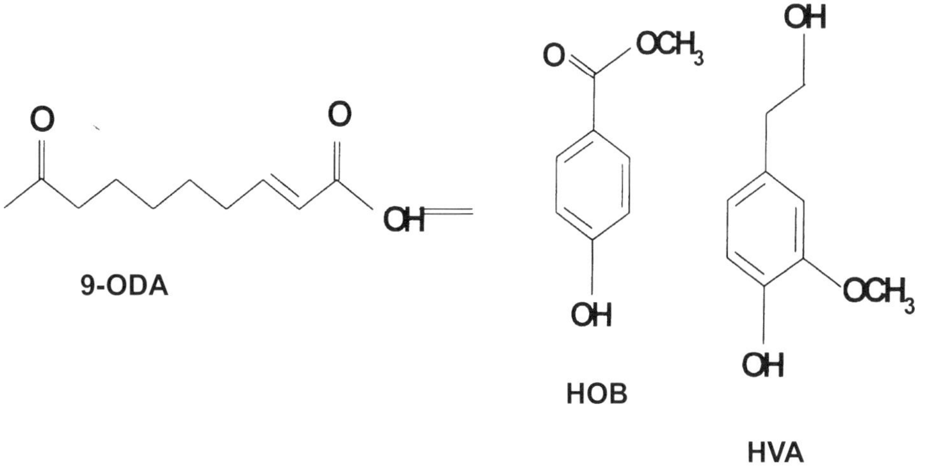

Fig. 6.8. Components of the queen bee pheromone.

The queen bee produces queen mandibular pheromone (QMP), which maintains order in the hive, suppresses ovary development in worker bees and prevents the rearing of new queen bees. It also stimulates worker bees to build new comb, rear more bees, forage and store food. Queen bee pheromone is a mixture of several compounds (Fig. 6.8), mainly 9-oxodec-2-enoic acid (9-ODA) and 9-hydroxydec-2-enoic acid (9-HDA). QMP also contains methyl p-hydroxybenzoate (HOB) and 4-hydroxy-3-methoxyphenylethanol (HVA); levels of HOB and HVA differ among different bee species.

Worker bees release an attractant pheromone (Fig. 6.9) from the Nasonov gland to keep the colony of bees together and help disoriented workers find the nest. It is released when swarming or forming a cluster, when marking the location to the nest or marking the location of water. The attractant pheromone is comprised of seven closely related compounds derived from the essential oils of fragrant plants. The most biologically active of these is citral, which is a component of the fragrance of oranges. Oleic acid from dead and decomposing bees acts as a chemical message to eliminate the dead bee from the hive.

Alarmed bees produce alarm and

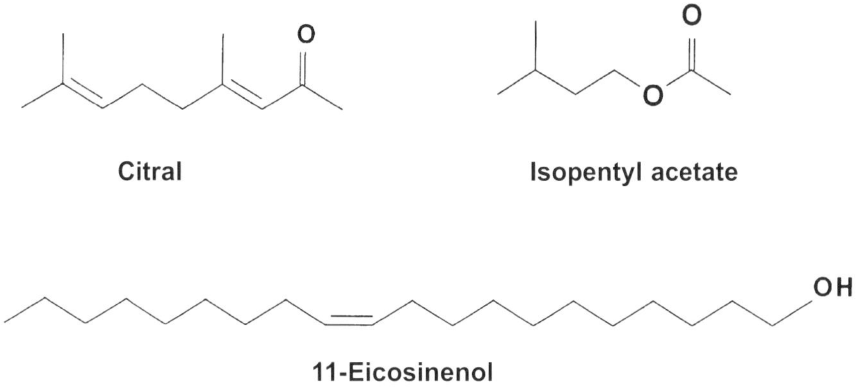

Fig. 6.9. Attractant, alarm and aggression pheromones from bees.

aggression pheromones to elicit a colony defence response from other workers. A major component of the alarm pheromone is isopentyl acetate, which is a component of the sting material as well as acting as a targeting pheromone to guide other bees to the sting site. 11-Eicosinenol also induces stinging and, mixed together with isopentyl acetate, is as effective as natural sting extract in causing a bee to engage in stinging behaviour.

Ants also have alarm and attractant pheromones as well as trail pheromones that they use to mark the trail to a food source. Different species of ants use different chemicals as trail pheromones, which reduces competition for the food.

Applications

Pheromone research has applications in pest control, insect management (e.g. honey bees) and the manipulation of reproduction in domestic animals and humans. Once the chemical composition of a pheromone is known, it can be synthesized chemically in amounts that can be used commercially. For reviews on the commercial use of pheromones and other semiochemicals, see Jones (1998) and Copping and Menn (2000).

Pest control

Agricultural damage from insects amounts overall to a loss of about 13% of the potential crop, with much greater losses in some instances. The use of chemical insecticides to control insect pests results in water pollution, pesticide residues in food and damage to non-target species, some of which are beneficial for agriculture. Insects also carry viruses, bacteria and parasites that cause diseases such as malaria, bubonic plague, yellow fever, typhus and others. The use of specific agents to effectively control targeted insect pests would be of great benefit and is an important part of integrated pest management (IPM) programmes.

In the USA, pesticides have to be approved by the Environmental Protection Agency (EPA), and pheromone-based pesticides have to pass a much simpler set of guidelines than conventional chemical pesticides. This saves a lot of money and time in the approval of pheromones. This is not the case in the EU, where the same criteria apply to chemical pesticides, plant extracts and pheromones. Pheromones are natural products and therefore cannot be patented. They are effective in small amounts, and individual pheromones are specific for one particular species. This makes pheromones less attractive commercially compared to broad-spectrum chemical insecticides, which are typically patentable and needed in large quantities.

Pheromones are generally used as a blend of several chemical components. Some 45 pheromones are listed as biopesticides and these are used for mass trapping, as monitoring agents, mating disrupters and aggregation pheromones, and as the mite alarm pheromone. Some plant-derived compounds attract insects to a good food source and these can be used together with attractant pheromones from insects.

POPULATION MONITORING. The most important use of pheromone-baited traps is for population monitoring. Pheromone-baited traps can be used qualitatively and quantitatively to detect the presence and density of insects. Traps work at low population levels, and thus can provide information on whether pests may be infecting an area or a particular crop. This information is then used to regulate the amount of chemical pesticides that are used and to ensure that they are being used at the most effective times. For example, the pea moth pheromone (dodecadienol acetate, Fig. 6.10) is usually used in traps designed to monitor the size of the moth population, so that conventional insecticides can be used during the vulnerable larval stages. Traps can also be used after spraying to determine whether the treatments were effective. Population monitoring using pheromone-baited traps is also useful for ecological studies and surveys, and for quarantine detection of non-indigenous insects.

MATING DISRUPTION. Pheromones are used at high concentrations for mating disruption. This high concentration of pheromone masks

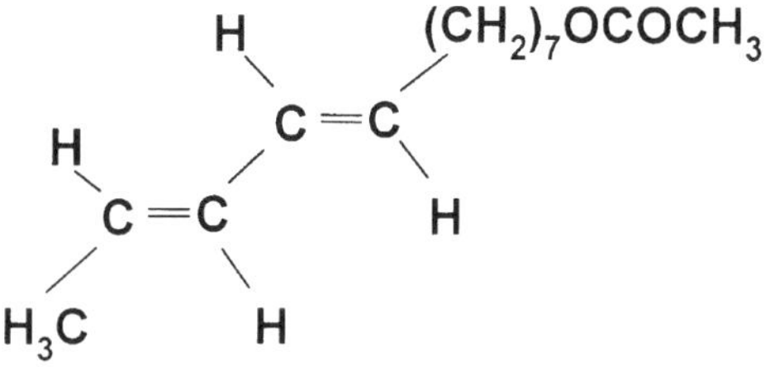

Dodecadienol acetate

Fig. 6.10. Structure of pea moth pheromone.

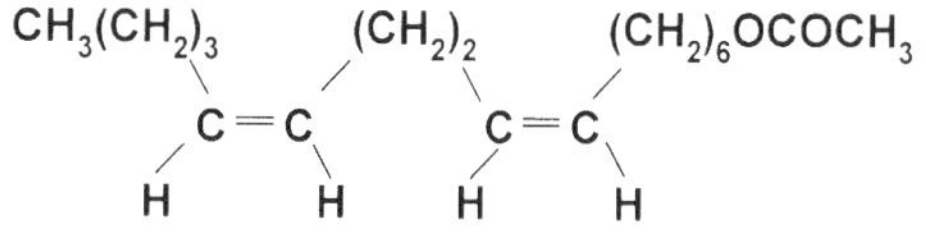

Gossyplure

Fig. 6.11. Structure of pink bollworm sex pheromone.

natural trails left by the species and may divert insects from natural sources, so they are unable to find a mate as effectively. High levels of pheromones may also cause adaptation of antennae receptors and habituation of the CNS, so insects no longer respond to the pheromone. About 30 pheromone-based products that act as mating disruptors are registered by the US EPA to control 11 lepidopterous pest species. For example, high levels of the pink bollworm sex pheromone (gossyplure, Fig. 6.11), placed indiscriminately, make it impossible for the males to follow the pheromone trails left by the females and thus mating is disrupted.

MASS TRAPPING. Mass trapping is a powerful and highly specific method for pest control, particularly for those insects that respond to aggregation pheromones. The appropriate number of traps per unit area must be used to achieve control. Generally, it is not effective enough to completely eliminate insect infestations. It is also possible to add an insecticide to the trap to kill the insects that are attracted to the pheromone. For example, the Japanese beetle pheromone is used in traps designed for lure-and-kill. In addition to the pheromone, which attracts the male beetles, a floral scent is included to attract the females. These traps are not 100% effective, but they do reduce beetle populations to more acceptable levels. Another example is the use of the aggregation pheromone of the spruce dark beetle (Fig. 6.12) to attract the beetles to about four baited trees per hectare. These trees are then attacked and the heavily infested trees are then removed and destroyed. Another example is the use of the mite alarm pheromone (farnesol) along with an insecticide. This reduces feeding of the mites and increases their activity and susceptibility to the insecticide.

To summarize pheromone use for pest control: pheromones account for much less than 1% of the total pesticide market. They have key roles in pest monitoring systems for speciality crops where the timing of insecticide use is important, as with boring insects. Pheromones have been used successfully for mating disruption, but they are specific for

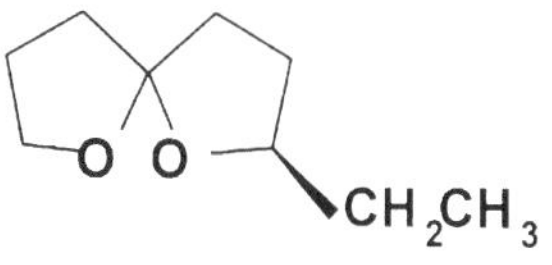

Chalcogran (spruce dark beetle)

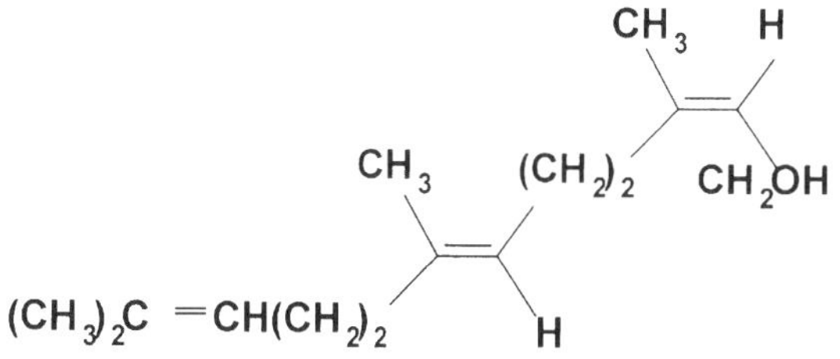

Farnesol (two-spotted mite)

Fig. 6.12. Two pheromones used for mass trapping.

one species and this is a major problem if more than one insect species is involved. Mass trapping is not effective to remove all of the insects and it is usually only the males that are trapped. Aggregation pheromones are useful in specialized areas such as forestry, where part of the crop can be sacrificed for the remainder, but there are not many suitable crops and insect pests for other applications.

Insect management

Pheromones can also be used as part of integrated resource management (IRM) programmes for beneficial organisms. For example, queen mandibular pheromone (QMP) can be used in the management of honeybee colonies. It can be used to create a pseudoqueen when shipping queenless packaged bees and to improve queen rearing success inside the hive. QMP can also be sprayed on flowering crops (fruit trees) to attract honeybees and improve pollination. The honeybee Nasonov orientation pheromone can be used to reduce the loss of bees due to swarming and to trap bees in food processing plants and other areas where they are undesirable.

Reproduction control in mammals

Signalling pheromones produced by females can be present at oestrus and pre-oestrus and can serve as attractants or inducers of sexual activity. Male pheromones can be used to detect oestrus for correct timing of AI to increase conception rates (e.g. Boar Mate®). Primer pheromones can be used to induce puberty in young females, end seasonal anoestrus, and improve conception rates (see Section 5.1).

Groups of females tend to be synchronized with respect to ovulation. This has been described in rodents, humans and also potentially in cattle. Pheromones may have applications in contraception and in the treatment of infertility. Compounds such as the 16-androstene steroids are also being marketed and are advertised to act as aphrodisiacs in humans to increase the sex-drive of potential partners.

6.3 Effects of Stress

What is stress?

A stressor is any environmental change that disrupts homeostasis, and stress refers to a state in which the homeostasis of the animal is threatened or perceived to be threatened. Neutral stress results in responses that are neither harmful nor beneficial to the animal. Distress causes the animal to respond in a way that can negatively affect its well-being or reproduction and which may cause pathological damage. The response to stress may have evolved as a mechanism for coping with threats to the survival and well-being of the animal by adjusting several systems within the body to maintain homeostasis. When these adjustments fail to compensate for the stress, or when the response is excessive or inappropriate, pathological changes and damage to the animal can occur.

Stress cannot be totally avoided in a complex real world. However, not all stress is harmful and some stimuli are necessary and beneficial to the animal. The well-being of animals in a captive environment can actually be better than in a natural state, since in the natural state predators, starvation, disease and natural disasters can threaten animals. The conditions in the captive environment should be adjusted to maximize the well-being of the animals and not simply to mimic the natural environment.

The effects of stress are ultimately controlled at the level of the hypothalamus, and are mediated by changes in behaviour and neuroendocrine effects via the sympathetic nervous system and the hypothalamic–pituitary–adrenal axis. These effects are summarized in Fig. 6.13.

Stress can cause changes in behaviour that may simply involve the animal moving away from a threat. When this is not possible, the animal may fight to remove the threat or try to cope by hiding or developing tonic immobility or other coping behaviours. If behaviour is to be used to assess animal well-being, a thorough knowledge of species-specific behaviours and the normal behaviours of the animals is necessary. Coping or abnormal behaviours are recognized as being not goal oriented and of no obvious benefit to the

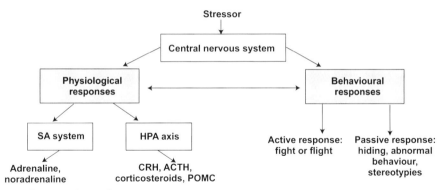

Fig. 6.13. Scheme of overall responses to stress.

animal. They include behaviours that are seen in captivity but not in natural settings, such as self-mutilation or repetitive stereotypic behaviours, as well as behaviours that are seen in the natural setting but are performed with abnormal frequency or in an abnormal setting.

Hormonal responses to stress

Hormone signalling plays a vital role in regulating homeostasis, and virtually every endocrine system responds in some manner to specific stressors. This includes hormonal regulation of metabolism, reproduction, growth and immunity. Rapid endocrine responses are mediated by the sympathetic nervous system activation of the adrenal medulla (SA system). Longer-term effects are due to changes in hypothalamic signalling and pituitary function, with the best known of these being the activation of glucocorticoid production by the adrenal cortex (HPA axis). For more information on the neuroendocrine responses to stress, see Matteri *et al.* (2000).

The adrenal gland in mammals consists of two separate organs, the adrenal medulla on the interior of the gland, and the cortex, which surrounds the medulla on the outside of the gland. The medulla produces the catecholamines, adrenaline, noradrenaline and dopamine. The cortex is comprised of the zona glomerulosa, which produces mineralocorticoids (predominantly aldosterone), and the zona fasciculata and zona reticularis, which produce glucocorticoids and androgens.

SA system

Efferent motor neurones carrying information from the central nervous system can be divided into two main systems, the somatic or voluntary system that controls voluntary movements in skeletal muscle, and the autonomic system that controls smooth and cardiac muscle and various glands. Neurones of the autonomic nervous system belong to either the sympathetic or parasympathetic pathway, which generally act in opposition to each other. The neurones of the parasympathetic pathway contact the target organs via cholinergic receptors that use acetylcholine as the neurotransmitter. The sympathetic neurones use noradrenaline as the neurotransmitter at the target organs and thus have adrenergic receptors. There are two types of adrenergic receptors, the α-receptors, which are selectively blocked by the drug phenoxybenzamine, and β-receptors, which are blocked by propanolol (see Section 3.5).

The sympathetic and parasympathetic pathways act in opposition to one another, with the balance between these two systems regulating body systems. The parasympathetic pathway predominates in the relaxed state to lower the heart rate and maintain housekeeping tasks such as digestion. When the animal is threatened, the sympathetic pathway predominates (Fig. 6.14) and stimulates the release of adrenaline from the adrenal medulla (see Sections 3.5 and 3.12). The locus ceruleus region in the brainstem contains nerve fibres that secrete noradrenaline (LUC-NE) and are also stimulated by stress. Adrenaline and noradrenaline inhibit the

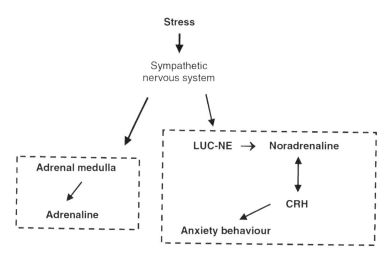

Fig. 6.14. Sympathetic nervous system response to stress.

storage of glucose and fatty acids, inhibit protein synthesis and stimulate the release of glucose, amino acids and free fatty acids from muscle, fat tissue and liver. Heart rate is increased, with blood flow redistributed to the skeletal and heart muscles, and anabolic processes, such as digestion, growth, reproduction and immune function, are suppressed. This is a rapid hormonal response that occurs in seconds to support the animal in the 'fight or flight' response (see Section 3.5.)

HPA axis

Neuroendocrine hormones regulate reproduction, shift metabolism, influence growth, affect immunity and alter behaviour. The neuroendocrine hypothalamic–pituitary–adrenal (HPA) axis response involves the release of corticotrophin releasing hormone (CRH, a 41-amino-acid peptide also known as corticotrophin releasing factor, CRF) by the hypothalamus (Fig. 6.15). CRH stimulates the

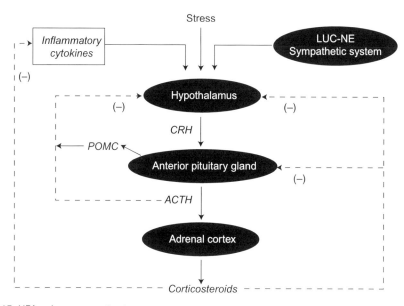

Fig. 6.15. HPA axis response to stress.

anterior pituitary to release adrenocorticotrophic hormone (ACTH) and subsequently cause the release of glucocorticoids by the adrenal cortex. Vasopressin also increases the release of glucocorticoids by potentiating the effects of CRH on the pituitary to increase the release of ACTH and also release peptides derived from pro-opiomelanocortin (POMC), such as β-endorphin. These opioid peptides have an analgesic effect and reduce the stress response by inhibiting the release of CRH. The major glucocorticoid produced in humans, pigs, cattle, sheep and fish is cortisol, while in rats and birds the major glucocorticoid is corticosterone.

Cortisol acts by negative feedback on the hypothalamus and pituitary to decrease the production of CRH and ACTH. The HPA response occurs more slowly (minutes to hours) and has a more general effect on the animal than the SA response. ACTH is secreted in a pulsatile manner throughout the day and is thought to enhance attention, motivation, learning and memory retention. Both ACTH and CRH suppress reproduction.

The LUC-NE system is also involved in the stress response, as noradrenaline increases general brain activity and stimulates the release of CRH (Fig. 6.14). CRH, in turn, stimulates the release of noradrenaline, which acts both locally and systemically. CRH receptors are present in several areas of the brain involved with cognitive function and emotion. Injection of CRH into the brain of rats produces anxiety behaviour and wakefulness and decreases appetite and sexual receptivity. Elevated levels of CRH in the hypothalamus have been linked to depression in humans. CRH thus acts as a neurotransmitter and plays a role in the behavioural response to stress. The response to CRH depends on previous experience and type of stressor as well as the genetic disposition of the animal. For more information, see the review by O'Connor et al. (2000).

CRH AND CRH RECEPTORS. The effects of CRH on behaviour can be assessed using CRH receptor antagonists. The first CRH receptor antagonists used were the peptide fragments of CRH, namely α-helical CRH_{9-41} and D-Phe-CRH_{12-41}. These peptide antagonists do not cross the blood–brain barrier and had to be injected into the brain to block stress-induced behaviours. CRH is a member of a family of neuropeptides that includes the mammalian CRH analogues, urocortin and the amphibian peptides, sauvagine and urotensin I (Table 6.1; see also the review by Lovejoy and Balment, 1999). The peptide antagonists antisauvagine-30 and astressin have been developed based on the structure of this family of peptides (Fig. 6.16). A number of non-peptide receptor antagonists that readily cross the blood–brain barrier have also been developed (see reviews by Christos and Arvanitis, 1998; McCarthy et al., 1999).

CRH acts via two classes of receptors, CRHR1 and CRHR2 (α,β, and γ forms), all of which are G-protein-coupled receptors. There is also a CRH-binding protein that is thought to regulate the concentration of free extracellular CRH. CRHR1 is linked to increased anxiety-like behaviour and is localized in the brain cortex and cerebellum. CRHR2α is localized in the lateral septum and paraventricular and ventromedial nuclei of the hypothalamus; CRHR2β is found in heart, skeletal

Table 6.1. Naturally occurring CRH-like peptides.

Hormone	Species/source	Function
Urotensins	Fish	Ion balance
Sauvagine	Amphibians	Osmoregulation in skin, binds to CRHR1 and CRHR2
Urocortin	Lateral hypothalamus, peripheral tissues (lymphocytes, GI tract)	40-fold higher affinity for CRHR2, decreased gastric emptying and food intake
CRH	Neuropeptide	Release of ACTH by anterior pituitary, behavioural changes

Amino-acid sequence of CRH from human/rat

NH$_2$-I-I-E-M-L-K-N-S-H-A-Q-Q-A
L-Q-E-A-R-A-M-E-L-V-E-R-L-L-H
F-T-L-D-L-S-I-P-P-E-E-E-S-COOH

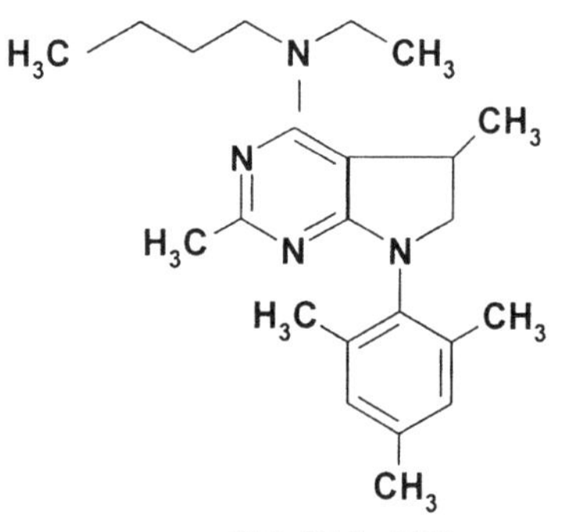

CP 154, 526
CRHR1 receptor antagonist

Fig. 6.16. Structure of CRH and CRHR1 receptor antagonist

muscle, and in the choroid plexus and cerebral arterioles in the brain; and CRHR2γ is localized in the amygdala. These receptors have been cloned and expressed in mammalian cells and these cells have been used in a bioassay system to screen for non-peptide antagonists for specific receptors.

Sauvagine (a CRH-like peptide isolated from frog skin) binds to both CRHR1 and CRHR2 receptors. Specific CRHR1 receptor antagonists include NBI 27914, CP 154,526 from Pfizer (Fig. 6.16), and antalarmin, which differs from 154,526 only by a methyl group on the pyrollopyrimidine moiety. These orally active compounds reduce anxiety behaviour, act as antidepressants and decrease the production of stress hormones, and can delay parturition. Other antagonists include DMP-696 by Dupont and NBI-30775 from Neurocrine Biosciences, which reduce stress behaviour but do not affect the HPA axis. NBI-30775 has a high affinity for CRHR1 but not CRHR2 and reduces the severity of depression in humans without affecting the production of cortisol, and would therefore be unlikely to cause serious endocrine side-effects, such as adrenal insufficiency.

Urocortin is a CRH-related peptide produced by the lateral hypothalamus as well as by peripheral tissues, including lymphocytes and the gastrointestinal tract. It binds CRHR2 with 40-fold greater affinity than CRHR1 and is thus considered the natural ligand for CRHR2. Antisauvagine-30 is a specific peptide antagonist for CRHR2 but not CRHR1.

Studies in transgenic mice have linked activation of CRHR2 to reduced signs of anxiety, suggesting that CRHR1 and CRHR2 have opposite effects on stress behaviour. This is important for **allostasis**, which is the adaptive response to stress by changes in hormones and autonomic responses. CRHR1 present in the pituitary gland is critical for the rapid initiation of corticosteroid release in response to stress, while CRHR2 is thought to be involved in the proper recovery of the response that may be independent of the negative feedback by glucocorticoids (Coste *et al.*, 2001). For example, injection of CRH into the brain results in increased blood pressure and heart rate, as is seen during stress. In contrast, CRH or urocortin given systemically causes vasodilation and decreases blood pressure. Urocortin also decreases gastric emptying and food intake (Wang *et al.*, 2001). For more information on the action of CRH analogues, see the review by Eckart *et al.* (1999).

Role of various hormones in stress responses

Glucocorticoids stimulate gluconeogenesis in the liver by increasing the synthesis of the enzymes involved in converting amino acids, glycerol and lactate into glucose and increasing the mobilization of amino acids from muscle. Glucocorticoids also decrease the rate of glucose transport and utilization by cells. This increases blood glucose levels by as much as 50% above normal. Cellular protein levels are also decreased by glucocorticoids in all cells except the liver, due to increased protein catabolism and decreased protein synthesis. Although labile proteins are first depleted, prolonged exposure to high levels of glucocorticoids can cause muscle weakness, decreased size of lymphoid tissue and impaired immune function. However, the rate of synthesis of liver protein and plasma proteins made in the liver increases as part of the acute-phase response, which occurs in response to infection. Glucocorticoids increase the mobilization of fatty acids from adipose tissue and increase their utilization for energy. Long-term exposure to high levels of glucocorticoids may suppress growth and accelerate the ageing process by increasing protein catabolism and causing hyperglycaemia.

Glucocorticoids affect the expression of many genes involving various aspects of metabolism and development. They are permissive for the actions of many hormones, including potentiating the synthesis and actions of catecholamines in increasing hepatic gluconeogenesis. Glucocorticoids have immunosuppressive and anti-inflammatory effects. They also increase gastric secretion, which can lead to the development of gastric ulcers. Glucocorticoids also decrease reproductive efficiency by blocking the effects of LH on the gonads. For further information, see Sapolsky *et al.* (2000).

DHEA (dehydroepiandrosterone) and DHEA sulphate are also produced by the adrenal cortex in response to ACTH and are normally secreted in synchrony with cortisol. However, in critical illness and acute trauma, such as burns, levels of DHEA and DHEAS decrease while cortisol levels increase. The stress of increased physical exercise tends to increase levels of DHEA and DHEAS. The role of these steroids in the stress response is not clear, but DHEA is thought to enhance immune function and to be correlated with a sense of well-being in humans. Further information can be found in a review by Kroboth *et al.* (1999).

Short-term stress increases the secretion of GH in non-rodents, but reduces the secretion of IGF-I, thus diverting energy from growth to survival. GH acts as an antagonist to insulin in peripheral target tissues to spare blood glucose, while decreased production of IGF-I reduces growth. Stress also increases levels of IGF-binding proteins. The increased GH may be caused by increased glucocorticoids. The mechanism of decreased IGF-I during stress is not established, but lower numbers of GH receptors and decreased signal transduction in the liver have been demonstrated during undernutrition or exposure to cold temperatures. Under long-term stress, growth hormone release is suppressed.

Levels of prolactin increase within a few minutes in response to acute stimuli and then decrease. Passive, rather than active, coping responses to stress increase prolactin. The increased prolactin may be due to production of β-endorphin, which decreases levels of dopamine, which normally acts as prolactin inhibitory factor (PIF) to inhibit prolactin release. Vasoactive intestinal peptide (VIP) and thyrotrophin releasing hormone (TRH) also directly stimulate prolactin release. The role of prolactin in the stress response is not clear, but it has been suggested to modulate immune function.

Acute psychological stress increases TSH secretion from the pituitary and increases release of T_3 and T_4 from the thyroid in non-rodents, to increase the metabolic rate. Cold temperature increases the activity of thyroid hormones, while chronic undernutrition decreases thyroid activity. This is due to changes in hypothalamic TRH release, pituitary thyrotroph function, thyroid hormone production and peripheral thyroid hormone receptors. The reduction in metabolic rate when food supply is limited would decrease energy usage and improve survivability. Chronic stress reduces thyroid hormone function and elevated levels of glucocorticoids inhibit the conversion of T_4 to T_3.

Renin is released from the kidney in response to various forms of distress. It acts on angiotensinogen to produce active angiotensin, which is a powerful vasoconstrictor. The activation of α-adrenergic receptors on pancreatic β cells by the sympathetic nervous system also suppresses insulin release. Glucagon release is also increased by aversive states and these effects on insulin and glucagon elevate the plasma glucose level during stress.

In summary, exposure to distress increases levels of adrenaline, noradrenaline, corticosteroids and, in some species, growth hormone and thyroxine. These hormones have a catabolic function and mobilize energy reserves to allow the animal to deal with the adverse situation. Conversely, levels of anabolic hormones (insulin, androgens and oestrogens) decrease with distress. Prolonged exposure to aversive situations thus results in decreased thyroid activity, reduced body growth and decreased reproductive activity.

Assessment of stress

Behavioural and physiological measures

Animals can respond to stress by changes in behaviour, changes in the autonomic nervous system and catecholamine production, neuroendocrine changes in the hypothalamic–pituitary system and changes in immune function (Table 6.2). However, not all of these systems are altered by any one stress and there is no non-specific stress response that applies to all stressors. It is thus important to monitor a number of systems to assess the degree of stress in animals. For further information on the concept of stress and methods for assessing stress, see the series of review papers by Clark *et al.* (1997a,b,c).

Animals respond to stress with species-specific behaviours as well as learned behaviours that are specific to the individual animal. Species-specific behaviours have allowed animals to survive in the wild, but may result in a harmful response in captive animals. Individual animals may react differently to the same stimulus, depending on their genetic constitution and previous experiences. Circadian rhythm and environmental effects, as well as age, sex, physiological state and population density, can also affect the reactions of individuals to stress.

Useful behaviours for monitoring well-being include level of activity, posture, vocalization, aggressiveness, movement patterns, feed and water intake and sleep patterns. Animal preference tests, in which an animal can choose between different situations or stimuli, can also be used to assess whether a particular stimulus is undesirable. The amount of effort that an animal is willing to use to avoid the stimulus is used as a measure of the degree of averseness of the stimulus. In this way, preference testing may be used to rank the suitability of housing systems, food and other environmental factors.

Physiological responses to stress include changes in heart rate, blood pressure, gonadal function and immune function that could be monitored to assess the level of stress (Table 6.2). Chronic stress results in decreased growth, increased disease, reduced fertility, gastric ulcers and hypertension.

Hormonal measures

The activity of the hypothalamic–pituitary–adrenal axis, as measured by the levels of ACTH and glucocorticoids, has been widely used as an index of stress. However, changes in levels of these hormones do not necessarily indicate that the animal is in distress. For example, surgical trauma can result in release of ACTH even though the animal is anaesthetized and presumably unaware of what is happening. The magnitude of the ACTH response does not always match the degree of

Table 6.2. Summary of potential methods for assessing stress.

Behavioural/physiological	Endocrine	Metabolic systems
Activity/sleep patterns	Catecholamines	Immune function
Posture/stereotypes	ACTH/CRH, glucocorticoids	Disease state
Feed and water intake	Gonadotrophin/sex steroids	Growth performance
Heart rate and blood pressure	Cytokines	Reproductive performance
	β-Endorphin, renin and prolactin	

painfulness or unpleasantness of the treatments. Glucocorticoid levels also depend on the behavioural response that occurs due to the stress. Other factors, such as the availability of food and water, can also affect the magnitude of this response. If an animal perceives an event as threatening, it will respond whether the event is actually threatening or not. Glucocorticoid levels are also increased in situations that are not unpleasant, such as during coitus, voluntary exercise, in anticipation of regular deliveries of food, and simply exposure to novel environments. There is a large diurnal variation in the pulsatile release of glucocorticoids. Sampling at the same time every day can minimize this variation.

Repeated regular exposure to a stress, in which the animal can anticipate when the stress will occur, can result in habituation and dampening of the corticosteroid response. For example, electric shocks delivered at irregular intervals produce a higher corticosteroid response than electric shocks given at regular intervals. In the same manner, if the animal has a sense of control and can predict the stress, the glucocorticoid response will be lessened. In contrast, for highly painful and aversive stressors, such as dehorning, the response may increase when the stress is repeated. The endocrine response to stress also depends on the individual coping responses of the animals. If the animal reacts in an active manner in an attempt to control the threat, it will exhibit aggression and status control behaviours. If it is driven by anger, the endocrine responses will be increased levels of noradrenaline and sex steroids, while if it is driven by fear the levels of glucocorticoids and catecholamines will increase and sex steroids will decrease. If the animal reacts in a passive manner due to a loss of control, it will exhibit defeat behaviours, resulting in increased glucocorticoids and decreased sex steroids.

Glucocorticoid levels return to near normal in animals under long-term stress, because of negative feedback of glucocorticoids on ACTH release. However, there may be changes in the response of the adrenal glands to ACTH in chronic stress. Repeated challenge with ACTH increases the production of glucocorticoids, so administering ACTH or providing an acute stress to chronically stressed pigs (a challenge test) may result in an increased glucocorticoid production compared with that of non-stressed pigs.

In addition to glucocorticoids, levels of β-endorphin, renin and prolactin are affected by stress, and plasma levels of these hormones have been suggested as possible indicators of animal well-being.

In summary, a number of different variables should be measured to accurately assess animal well-being, since animals respond in a variety of different ways to aversive situations. Behaviour can be assessed non-invasively and provide information on the nature of the distress and the coping strategy that the animal has adopted. Differences in behaviour among animals kept in different environments may be used to indicate that an environment is causing distress. However, some changes in behaviour may be difficult to interpret. Additional criteria need to be measured to assess the effects of long-term or subtle aversive stimuli.

Issues related to sampling

Measures of physiological, metabolic or endocrine changes in blood samples tend to be invasive and thus induce confounding reactions in the animal. This can be more severe in semi-domesticated or wild animals. Levels of catecholamines can increase within seconds after restraint, while the cortisol level increases within a few minutes after restraint and then remains elevated for several hours. Sedating the animals or habituating them to the blood sampling procedures may reduce the effects of handling, but may result in atypical responses. Habituation is not suitable, for example, for procedures involving the stress of handling. Alternatively, less invasive methods that reduce the degree of handling can be used. Indwelling catheters can be placed in blood vessels and animals can be trained to cooperate with blood sampling procedures. Remote blood sampling equipment attached to the free-moving animal can be used for non-invasive blood sampling of large animals. However, implanting catheters requires surgery and it can be difficult to keep the catheters patent.

Alternative non-invasive methods include measurements of cortisol in readily accessible body fluids, such as saliva, milk, urine and faeces. Salivary cortisol level is generally correlated with blood cortisol but is present at only about 10% of the level present in blood. Saliva can be collected non-invasively by having the animal chew on a gauze sponge. There can be problems with contamination of the sponges with feed and with rumen fluid in cattle, but this method has been successfully used to assess cortisol status in swine. Cortisol is also transported from plasma to milk, so that milk cortisol concentration can be used as an indicator of stress in dairy cattle. Urine contains both free cortisol and cortisol metabolites, but levels are quite low. There is also a lag time between cortisol release and excretion. For a quantitative estimate of cortisol status, a urine sample should be collected over the entire day. Faecal cortisol may be useful for studies of stress in wild animals, but factors such as lag time, volume of faeces and rate of passage need to be considered.

In plasma, there is an equilibrium between glucocorticoids bound to corticoid-binding protein (CBP, transcortin) or albumin and free, unbound steroids. Only the free steroids are thought to have biological activity, so increasing the concentration of CBP would lower the activity of the glucocorticoids. The ratio of free to bound steroids can be altered by changes in the amount of binding protein, particularly during chronic stress when protein catabolism occurs. There is, therefore, some controversy over whether the total amount of glucocorticoids, or simply the free steroids, should be measured to assess the stress response. All cortisol in the saliva is present in the free form and not bound to protein.

Effects of stress on the immune system and disease resistance

Overview

The immune system contains a number of different leukocytes (white blood cells) that play integral roles in the overall response to an immune system challenge. This includes

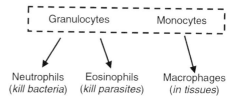

Fig. 6.17. Innate immunity (bone marrow).

granulocytes and monocytes, which are formed in the bone marrow and protect the body by ingesting invading organisms, by the process of phagocytosis, to destroy them. This is called innate immunity (Fig. 6.17). Monocytes mature in tissues to become large macrophages, which can combat foreign pathogens without prior exposure to them through disease or vaccination. The granulocytes include neutrophils, which attack bacteria by phagocytosis, and eosinophils, which attach to parasites to kill them.

Other types of white blood cells include lymphocytes and plasma cells, which are produced in the lymphoid tissues and are involved in the immune reactions. They are involved in acquired immunity (Fig. 6.18), since the response is directed specifically towards foreign agents that have been present previously. Humoral immunity is a type of acquired immunity that refers to the production of antibodies by B-cell lymphocytes (see Section 2.1), while cell-mediated immunity is acquired immunity carried out by activated T cells. There are three groups of T-cell lymphocytes, helper T cells, cytotoxic T cells and suppressor T cells. The helper T cells (Th) produce lymphokines (also known as cytokines), which are peptide hormones that include interleukins (IL) and interferons (IFN) that act in a paracrine or autocrine manner to stimulate other immune cells in the process of inflammation and immune reactions. There are two major types of Th cells that are mutually inhibitory. Th-1 cells produce IL-2, IFN-γ and TNF-β to promote cellular immunity and inflammation. Th-1 cells are inhibited by IL-4 and IL-10, which are the major anti-inflammatory cytokines. Th-2 cells secrete IL-4, IL-10 and IL-13 to promote humoral immunity. Th-2 cells are inhibited by IL-12 and IFN-γ, which are the major

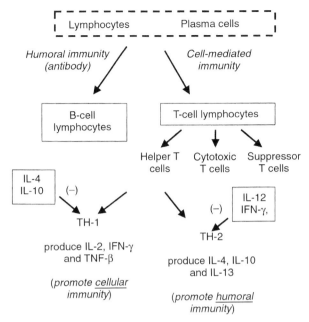

Fig. 6.18. Acquired immunity (lymphoid tissue).

pro-inflammatory cytokines. Thus, the products of Th-1 cells inhibit the Th-2 cells and vice versa. Cytotoxic killer T cells attack and kill foreign cells or cells that have been invaded by viruses. The suppressor T cells regulate the other cells to prevent severe immune reactions that could damage the animal's own tissues.

Stress from traumatic injury resulting in bacterial or viral infection causes an inflammation response and production of cytokines by macrophages, vessel endothelial cells and fibroblasts. Bacterial infections induce the production of IL-1, IL-6, IL-8 and TNF-α, while viruses induce the production of type I interferons, such as IFN-α and -β. These distinctions are not absolute, but cytokines can be used as markers of infection and for discriminating between bacterial and viral infections (Fossum, 1998).

Cytokines induce a sickness response, causing fever, fatigue, loss of appetite and decreased libido. They also stimulate an acute-phase response, which includes the production of hepatic acute-phase proteins, including protease inhibitors, coagulation factors, complement factors, transport proteins and scavengers that are released into the circulation. In addition, synthesis of acute-phase-regulated intracellular proteins, including transcription factors, intracellular enzymes and cell surface receptors is also increased. The role of the acute-phase proteins is to restore homeostasis. For example, C-reactive protein (CRP, named for its ability to bind the C protein of pneumococci), along with other complement proteins, binds to bacteria in the process of opsonization and tags the bacteria for destruction by macrophages. Tests are available for the rapid determination of CRP in blood, and these tests are used to diagnose inflammation and determine the effectiveness of treatments.

Inflammation is the complex of changes that occurs in injured tissue. It is characterized by enhanced local blood flow, increased permeability of the capillaries, clotting of the fluid in the interstitial spaces and swelling of the tissue. Inflammation serves to wall off the damaged or infected area from the remaining tissues to contain the damage. These changes cause the redness, heat, swelling and pain that are associated with tissue injury. Large number of granulocytes and monocytes migrate to the area and are activated by lymphokines and tissue products such as histamine, bradykinin, serotonin and prostaglandins to begin their phagocytic actions.

Stress effects on the immune system

There is a lot of evidence demonstrating that prolonged exposure to aversive conditions reduces the health of animals. Many intensive livestock production systems reduce the immune response of animals and the resistance to disease. For example, the stress of shipping calves to feedlots increases their susceptibility to bovine respiratory disease (shipping fever). This is due in part to the decrease in neutrophil release from bone marrow caused by glucocorticoids. In contrast, shipping pigs decreases body weight but does not reduce immune function, although pigs with lower social status have generally lower immune function compared to dominant pigs.

The nervous system can directly modulate the activity of the immune system to decrease disease resistance. This occurs by the direct innervation of lymphoid tissue by autonomic nerves, by paracrine action of noradrenaline released from nearby nerves on immune cells, and through the HPA axis.

Glucocorticoids affect the immune response in animals to lower the resistance to infection and decrease inflammation (Fig. 6.19). Glucocorticoids affect cell-mediated immunity by decreasing the number of lymphocytes and eosinophils, suppressing natural killer (NK) cell activity and decreasing the amount of lymphatic tissue. Increased levels of glucocorticoids can destroy thymus cells and negatively affect T cells. There are also negative effects on B cells, causing an inhibition of antibody formation. Adrenalectomy prevents the negative effects of stress on the immune response in laboratory animals.

Glucocorticoids modulate immune responses by inhibiting the production of type 1 pro-inflammatory cytokines, which stimulate cellular immunity. These cytokines, particularly interleukin 1 (IL-1), also act on the hypothalamus to increase the secretion of CRH, which acts on the pituitary to release ACTH and subsequently increase the production of glucocorticoids from the adrenal cortex. CRH is also produced by immune cells in the spleen or in nerve endings and acts on immune cells at the site of inflammation. The glucocorticoids and CRH then act by negative feedback on the immune system to increase the formation of neutrophils and decrease the formation of macrophages and lymphocytes. The production of pro-inflammatory leukotrienes and prostaglandins by immune cells is also decreased and the activity of pro-inflammatory transcription factors, such as nuclear factor-κB (NF-κB) and activating protein-1 (AP-1), is inhibited by glucocorticoids. A polymorphism has been identified in the human glucocorticoid receptor gene that increases the susceptibility to glucocorticoids.

Both glucocorticoids and catecholamines also increase the production of the type 2 anti-inflammatory cytokines IL-4, IL-10 and IL-13 by Th-2 cells. These cytokines promote humoral immunity and protect against extracellular bacteria and parasites and soluble toxins and allergens. Further information on the effects of stress on the immune response can be found in the review by Elenkov and Chrousos (1999).

Thus cortisol reduces inflammation and speeds up the healing process. In addition to inhibiting the production of cytokines, cortisol stabilizes the lysosomal membranes and reduces the release of proteolytic enzymes

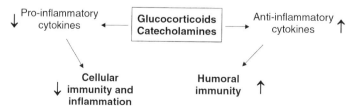

Fig. 6.19. Effects of glucocorticoids and catecholamines on immune function.

from lysosomes. It also decreases the permeability of capillaries and reduces the loss of plasma and the migration and activity of white blood cells. This blocks the further release of inflammatory materials from cells. The reduction in cytokine production also reduces fever. The activation of the HPA axis and the production of glucocorticoids thus keep the immune response in check and prevent it from becoming hypersensitive and therefore harmful.

The effects of stress on immune function depend on the nature of the stressor as well as the immunological function being measured. Immune system function can be evaluated by *in vitro* measures, such as the ability of natural killer cells to destroy virus-infected cells and the proliferation of T and B cells after exposure to mitogens in culture. *In vivo* measures of immune function are antibody formation after exposure to antigens and T-cell-mediated delayed hypersensitivity and graft rejection. The effects of stress on resistance to disease can vary. Social stress in chickens decreases the resistance to viral diseases such as Newcastle disease or Marek's disease, but increases the resistance to bacterial infections. This may be due to the action of glucocorticoids in suppressing inflammation from bacterial infection.

The reduction in cellular immunity, and the accompanying inflammation, pain and sickness behaviour, allow the animal to respond more effectively to stressful situations in the short term. However, the chronic depression of cellular immunity by long-term stress decreases the ability of the animal to fight infection by viruses, bacteria, fungi and protozoa, and increases the possibility of illness. Thus, long-term stress is undesirable and makes the animal more susceptible to disease. Long-term aversive stress can also lead to organ damage, such as cardiovascular disease and gastric ulcers. Increased release of free fatty acids can result in the formation of fatty deposits on arterial walls and can lead to atherosclerosis and hypertension. Other changes due to long-term stress include adrenal hypertrophy, haemorrhage, skeletal muscle degeneration and reduced weight of other organs, such as the spleen and thymus.

The levels of anabolic hormones, including GH and IGF-I, are decreased during disease, so supplementing with anabolic hormones might reduce the catabolic effects of disease. GH treatment of critically ill patients increases protein synthesis, stimulates the immune system and improves wound healing. Use of insulin and anabolic steroids also increases nitrogen retention. Treatment of growing cattle with GH improves the immune response to experimental challenge with bacterial lipopolysaccharide (LPS), but not to infection with parasitic coccidia. However, treatment with the anabolic steroids oestradiol and progesterone improves the response to coccidia infection. It is thus apparent that treatment with anabolic hormones can improve animal responses to disease. Further research is needed in this area.

Effects on reproduction

Animals that are suffering from chronic stress do not have the same reproductive success as non-stressed animals. Acute stressors may also impair reproductive function during critical periods of the reproductive cycle, such as ovulation, early pregnancy and lactation. Stress decreases the secretion of GnRH by the hypothalamus and subsequent production of LH and FSH by the pituitary and sex steroids by the gonads (Fig. 6.20). This causes decreased libido and fertility rates, lack of implantation of the fertilized ovum and retarded growth of the embryo. Perinatal stress during sensitive periods can demasculinize males and delay puberty in females.

Glucocorticoids, ACTH and CRH, as well as vasopressin and opioids such as β-endorphin, reduce GnRH secretion. IL-1 also inhibits the hypothalamic–pituitary–gonadal axis. Glucocorticoids may also directly interfere with gonadotrophin secretion by the pituitary, inhibit the production of gonadal steroids and reduce the sensitivity of target tissues to sex steroids. The initial rapid release of LH is mediated via arachidonic acid and its metabolites, while the prolonged release of LH is regulated by protein kinase C-dependent mechanisms. Glucocorticoids reduce the release of LH by inhibiting the

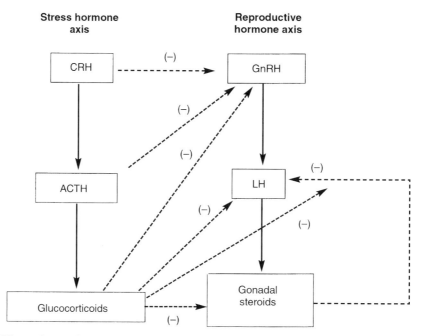

Fig. 6.20. Effects of stress hormones on gonadal function.

hydrolysis of phospholipids to release arachidonic acid. Glucocorticoids also affect the feedback of gonadal steroids on the pituitary gonadotropes. In rats, glucocorticoids prevent oestrogen from sensitizing the pituitary to GnRH and thus prevent the preovulatory release of LH.

Glucocorticoids interact with reproductive hormones during early development of the neuroendocrine system. Elevated corticosteroid levels negatively affect pregnancy rates in gilts and the rate of sexual development of boars. Sexual behaviour depends on the production of gonadal hormones and may be suppressed when a stressor interferes. Sexual activity of low-ranking males is depressed when dominant animals are present, due to low production of testosterone. The stress of low social rank in females can interfere with LH production and prevent ovulation and oestrous behaviour. Paradoxically, acute stress, such as mixing with unfamiliar gilts, can increase LH production and is used for oestrus synchronization and induction of early puberty. The first few days of early pregnancy until implantation is thought to be a particularly stress-sensitive period.

Stress also negatively affects lactation. Catecholamines and opioids produced during stress reduce oxytocin production, which negatively affects milk ejection and yield.

Nutritional stress delays the onset of puberty, interferes with normal cycling in females and results in hypogonadism and infertility in males. Leptin plays a role in this effect, and adequate levels of leptin are required for gonadotrophin secretion (see Section 3.8). Heat stress also inhibits gonadotrophin secretion, due to decreased production of GnRH and decreased sensitivity of the pituitary to GnRH stimulation. Long-term exposure to cold temperature also reduces LH and FSH secretion. This may be due in part to inadequate nutrition and lack of leptin signalling due to the increased metabolic demand at cold temperatures.

The effects of stress and glucocorticoids are not always negative. Glucocorticoids prepare the fetus for birth by stimulating the production of lung surfactants and the maturation of the gut. Parturition is initiated by a surge of cortisol from the fetus, which is stimulated by the production of CRH (see Section 5.1). Infusion of the CRHR antagonist antalarmin into the fetus has been shown to delay parturition in sheep (Chan et al., 1998).

Effects on growth performance

Chronic treatment of grower pigs with glucocorticoids decreases growth rate. Glucocorticoids are catabolic and adversely affect growth performance by increasing gluconeogenesis and decreasing protein incorporation into tissues. The overall body response to stress results in decreased growth rate, decreased efficiency of nutrient utilization for growth and increased energy requirements for maintenance (Fig. 6.21). In young animals, a loss of growth efficiency from stress can be offset somewhat by increased efficiency of nutrient use for thermogenesis. In addition to endocrine responses, stress reduces nutrient availability by decreasing appetite, gut motility and nutrient absorption, and by affecting the activity levels of the animal. Nutrient use is also directed towards fever thermogenesis and away from growth. While poor growth and reproduction likely indicate that the animal is in a poor state of well-being, the reverse is not always true.

One cannot assume that if an animal appears to perform well, it is not under some form of stress.

Stress effects on the metabolism of different tissues depend on the responsiveness of the cells to the different hormones produced during stress. GH and IGF-I produce anabolic effects, and these hormones are generally inhibited during disease. ACTH and the glucocorticoids produced during stress cause catabolic effects. Thyroid hormone regulates the basal metabolic rate and affects nutrient uptake by cells. In addition, changes in blood flow to different tissues affect nutrient availability and the exposure of the tissues to hormones. Blood flow changes from vasoconstriction and vasodilation are caused by arachidonic acid metabolites (prostaglandins, prostacyclins and thromboxanes) and nitric oxide (NO).

Catabolism of glycogen in liver and muscle provides glucose, resulting in transient hyperglycaemia and then hypoglycaemia as the glycogen is depleted. Catabolism of adipose tissue provides fatty acids for energy production. Muscle protein catabolism due to TNF-α and IL-1 provides

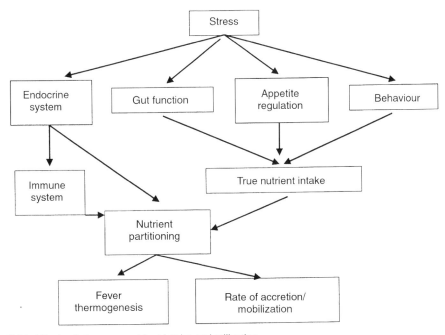

Fig. 6.21. Effects of stress on nutrient intake and utilization.

amino acids such as glutamine that are required by immune cells. There is a priority of nutrient use among different tissues, so that different body structures are differentially affected by stress. For example, the locomotor and fast-twitch rectus femoris muscle is relatively unaffected, but psoas major, which is a postural muscle with slow-twitch fibres, is catabolized in cattle during stress. Severe stress results in the overproduction of stress hormones and cytokines, causing dramatic changes in metabolism, leading to catabolism, tissue wasting and pathological conditions. IL-1 induces fever and reduces feed intake. Blocking the response to the cytokines TNF-α and IL-1 using receptor antagonists decreases the severity of weight loss and muscle wasting due to disease stress.

Exposure to disease organisms stimulates the immune system response and diverts nutrients away from growth. The stress of infections also reduces appetite. This may be due in part to the action of urocortin. Intravenous injection of urocortin has been shown to delay gastric emptying and reduce food intake in rodents and marsupials. These effects are blocked by CRHR2 receptor antagonists, but not by selective antagonists for CRHR1. Low levels of infection cause increased production of acute-phase proteins, reduced synthesis of muscle proteins, increased maintenance energy requirements and decreased feed intake. Subtherapeutic levels of antibiotics are sometimes included in animal feed to reduce the activation of the stress response by subclinical infections with bacteria. Alternatively, the health status in a herd can be improved by increasing sanitation and limiting public access. For pigs, the effects of disease can be minimized with specific-pathogen-free (SPF) herds or selecting for animals with a high immune response. For further information on the effects of stress on performance, see von Borell (1995) and Elsasser *et al.* (2000).

Summary

Stress occurs when the homeostasis of an animal is threatened, and this can have dramatic negative effects on the immune system, reproduction and the growth and performance of animals. These responses are due to the production of stress hormones, including the HPA and SA axis (CRH, ACTH, glucocorticoids and catecholamines) and cytokines, as well as decreases in anabolic hormones (IGF-I, GH, thyroid hormones and sex steroids). These hormonal changes, as well as behavioural and physiological changes, can be used to assess the degree of stress that an animal is experiencing. Altering these hormonal responses can potentially reduce the effects of stress. In particular, the use of specific CRH receptor antagonists (for example, see Deak *et al.*, 1999) shows promise to alleviate some of the negative aspects of stress without adversely affecting other endocrine functions. Further research in this area that produces a more complete understanding of the endocrine and metabolic responses to stress may lead to additional treatment strategies.

6.4 Endocrine Applications in Toxicology

Endocrine disruptors or modulators

The terms 'endocrine disruptor' or 'endocrine modulator' refer to chemicals that have direct endocrine effects or indirectly affect the normal endocrine systems of animals. The primary concern has been xenobiotic chemicals in the environment, including persistent organohalogen compounds, such as pesticides, industrial chemicals including surfactants, and organometals. However, other naturally occurring compounds, such as phyto-oestrogens, are also potent endocrine modulators.

Environmental pollutants have been studied for their lethal and carcinogenic effects, but many also have endocrine disruptor activities. High levels of these compounds in the Great Lakes have been linked to thyroid dysfunction, decreased fertility and hatching success, metabolic and behavioural abnormalities, altered sex ratios and compromised immune systems in various animals in this ecosystem. High levels of dioxins were linked to decreased fish populations, and reproductive losses and mortality in mink that were fed fish from the Great Lakes. Mothers with high levels of polychlorinated biphenyls (PCBs) and dioxin had babies with statistically decreased mental development and increased incidence of learning disorders that could be detected by skilled psychologists. The developing fetus may be more sensitive to the effects of endocrine disruptors than is an adult animal. Associations between decreased sperm counts in men and the increased human exposure to synthetic chemicals over the period 1938–1998 have also been claimed, but a cause and effect relationship has not been demonstrated conclusively. For more information on this issue, see Colborn and Thayer (2000).

Natural, plant-derived compounds such as phyto-oestrogens can also have endocrine-modulating effects that are at least as potent as those of man-made xenobiotic chemicals. Many plant and herbal extracts have been used as health aids for some time and their use is generally unregulated. Most natural compounds that affect endocrine function are biodegradable, but exposure to these compounds through the diet can be high.

Since it is impossible to perform chemical analysis for all potential endocrine disruptors, certain 'indicator species', such as rats, have been used as bioindicators to assess the impact of endocrine-modulating chemicals in the environment. While this approach measures actual effects of complex mixtures of chemicals on biological systems, the results obtained with one species have often been extrapolated to other, unrelated species. However, there are dramatic differences between species in the activities of different hormones and receptors, the ability to detoxify contaminants and in other factors that control attributes, such as sex ratios. Very little is known about the comparative endocrinology of many species. Thus, the effects of a potential endocrine disruptor seen in one species may not be applicable to other species.

Assessment of endocrine disruptor activity

Researchers have used a number of different assays to demonstrate the effects of potential endocrine disruptor chemicals (EDCs). Public concern about the effects of endocrine disruptors on wildlife and human populations has led to the development of government testing programmes. The US Environmental Protection Agency (EPA) and the European Organization for Economic Cooperation and Development (OECD) are developing standardized testing procedures for endocrine disruptors. This will provide standardized protocols that produce reproducible results that can be compared across many different laboratories. A number of different testing protocols with different end-points are being validated. Testing starts with the *in vitro* assessment of endocrine disruption activity, followed by *in vivo* tests to determine the effects on different development stages of the animal model. In addition to tests with mammals (rodents), tests using various wildlife species (fish, birds, reptiles, amphibians and invertebrates) are being considered. Testing procedures and the endocrine end-point being measured need to be defined carefully and appropriate reference chemicals that produce defined effects are required to standardize the procedures. For information on the validation of assay methods for endocrine disruptors, see Ashby (2000).

The first set of tests (Fig. 6.22) includes *in vitro* screens to determine whether the test compound has the potential to interact with the oestrogen receptor, androgen receptor or the thyroid receptor. This does not confirm toxicity, but allows for priority setting for further testing and determination of the potential mechanism of action of the chemical. The next level of testing is short-term *in vivo* assays, such as the effect on the development of male and female sex organs and tests in prepubertal animals, to determine effects on thyroid activity, as well as effects on neonatal animals from *in utero* exposure. Finally, in order to fully assess the potential risk of endocrine disruptors, multigenerational *in vivo* studies should be conducted to assess the effects on reproduction and development. Differences among species, dose–response effects that cover the potential range on exposure, persistence in the environment, route of exposure and potential for exposure during critical periods of development need to be considered. Further details on the testing programmes for EDCs being developed by the US EPA are summarized by Fenner-Crisp *et*

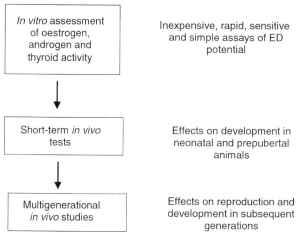

Fig. 6.22. Components of testing for EDCs.

al. (2000) and the OECD programme is described by Huet (2000).

Endocrine disruptors can potentially act by:

1. Binding to hormone receptors to act as hormone agonists or antagonists;
2. Altering the synthesis, transport or metabolism of hormones, thus affecting hormone levels; and
3. Interfering with signalling between different glands, such as the hypothalamus–pituitary–endocrine gland system.

The testing procedures are designed to investigate one or more of these potential mechanisms.

In vitro *assays*

In vitro assays for estimating oestrogenic activity have been developed. In one example, a cell line is transfected with a recombinant oestrogen receptor and a reporter gene (*lac-z* for β-galactosidase, *luc* for luciferase or *gfp* for green fluorescent protein) linked to an oestrogen-responsive element. Treatment with compounds that activate the oestrogen receptor increases the expression of the reporter gene. Competitive binding assays for the oestrogen receptor, androgen receptor, progesterone receptor and thyroid hormone receptor have also been developed, but these assays do not determine whether inhibition or activation of the hormone is occurring. Oestrogenic compounds can also stimulate cell division and expression of oestrogen-responsive genes, such as c-*myc*, *pS2* and the progestin receptor in oestrogen-responsive cells. The human breast tumour cell line MCF-7, T-47D or Bg-1 cells have been used for this purpose. Oestrogenic compounds can also stimulate synthesis of sex hormone binding globulin in HepG2 liver cancer cells. Oestrogen receptor antagonists, such as ICI 182,780 and ICI 164,384, can be used to block the oestrogenic effects of the test compounds and demonstrate that specific binding to the oestrogen receptor is occurring.

In vivo *assays*

In vivo bioassays for oestrogenic activity include determining the effect of treatment with the test compound on immature (22–23-day-old) female rats. Oestrogenic compounds cause premature vaginal opening and increases in ovarian and uterine weight. The induction of oestrogen-responsive progesterone receptor and peroxidase genes, overall DNA synthesis in the uterus and histological examination to measure thickening and mitotic activity of the vaginal epithelium can be compared to an oestradiol-positive control.

Oestrogen induces the production of vitellogenin in the liver of juveniles or males of oviparous species, although only females normally produce vitellogenin. Test compounds can be administered *in vivo* by intraperitoneal injections or added to *in vitro* cultures of liver cells from fish. Other effects of endocrine disruptors in fish are reduced gonad size, fecundity, egg size, hatching success, and expression of secondary sex characteristics and increased time to sexual maturity due to decreased levels of plasma sex hormones. The effects on plasma hormone levels and sperm quality and quantity can be measured in males. The Hershberger assay, using castrated male rats, is used to detect chemicals that interact with the androgen receptor and cause proliferation of androgen-dependent tissues.

Thyroid-active substances can be detected using a frog metamorphosis assay with *Xenopus laevis*. Measurement of tail resorption and mortality, as well as thyroid receptor mRNA and tissue levels of T_3 and T_4, can be made.

The induction of specific behavioural responses (e.g. lordosis in female rats and mounting behaviour in males) can also be measured, keeping in mind that behaviours vary among different species. For more information on assays for endocrine disruptors, see the reviews by Oberdörster and Cheek (2000) and Lister and Van der Kraak (2001).

Sources of endocrine disruptors

Particular interest has been expressed in chemicals that affect the actions of gonadal steroids and thus affect reproduction. This concern may be due, in part, to the experience with the use of diethylstilbestrol (DES). This was a synthetic oestrogen that was used to prevent miscarriages in pregnant women, but it was later found to have teratogenic effects in the fetus. A large number of xenobiotic chemicals and plant-derived compounds have been shown to act as agonists of steroid hormones, particularly oestrogens. These phyto-oestrogens and xeno-oestrogens are typically several orders of magnitude lower in activity than natural oestrogens. Phyto-oestrogens are the most potent source of environmental oestrogens, based on *in vitro* activity measurements. There was concern that exposure to these oestrogenic compounds in the environment was causing reproductive problems in wildlife. Fears that they were also causing decreased sperm counts in men and breast cancer in women have not been substantiated.

Phyto-oestrogens can also have anti-oestrogen effects, and foods and dietary supplements rich in phyto-oestrogens are commonly used as a preventive measure for breast and prostate cancer. For further information on endocrine disruptors in the food chain and environment, see Nilsson (2000).

Plant-derived endocrine modulators

A wide variety of oestrogenic and anti-oestrogenic compounds has been identified in plants (Fig. 6.23). The flavonoids are the most widely distributed group of phyto-oestrogens and include isoflavones in soy and clover, as well as flavone, flavanone and chalcone. Coumestrol is derived from flavonoids and is one of the most potent oestrogenic compounds in plants. It is thought to be responsible for 'clover disease' in sheep, which results in lesions in the reproductive organs and infertility. The isoflavonoids genistein and daidzein are present in high amounts in soy products, with 60 g of soy protein providing 45 mg of isoflavones. This dosage can significantly affect the menstrual cycle in women. Genistein is also a specific inhibitor of tyrosine protein kinases. Lignans such as enterolactone are present in vegetables and in high levels in flaxseed. They have weak oestrogenic effects and may act as anti-oestrogens and inhibit the conversion of androstenedione to oestrone by the aromatase enzyme. Sterols such as β-sitosterol are also important phyto-oestrogens. They are present in wood and can be released from bleached kraft pulp mills at levels high enough to affect the endocrine status of fish. Bacterial degradation of plant sterols present in kraft mill effluents can also produce androgenic steroids. The oestrogenic effects of the mycotoxin zearalenone have been known for some time as a cause of infertility and constant oestrus in pigs fed *Fusarium*-contaminated maize.

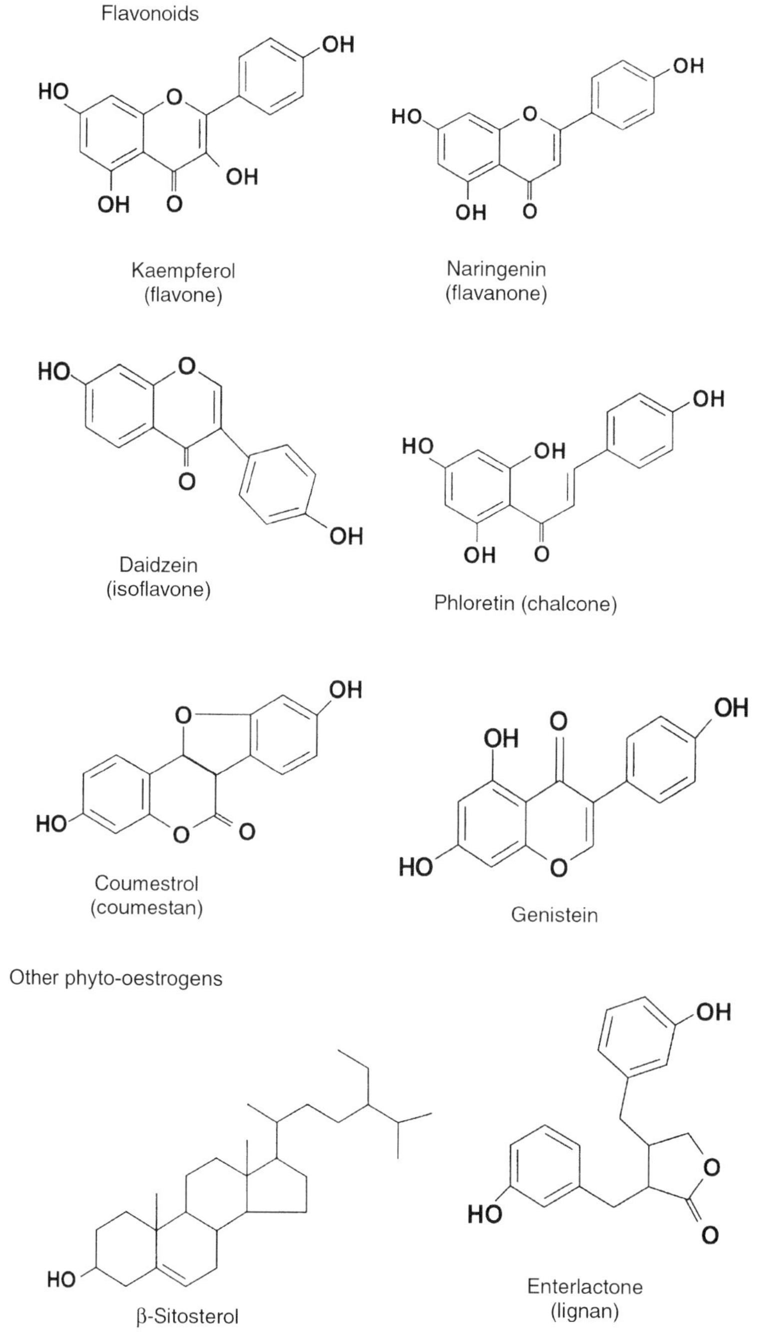

Fig. 6.23. Oestrogenic and anti-oestrogenic compounds from plants.

Thus, there is considerable exposure to endocrine-modulating substances in feeds and water. This can affect animal production systems as well as interfere with rodent bioassays designed to measure endocrine-modulating compounds. For more information on phyto-oestrogens, see the reviews by Safe and Gaido (1998) and Santti et al. (1998).

Xenobiotic endocrine modulators

Xeno-oestrogens include organochlorine pesticides such as PCBs, dichlorodiphenyltrichloroethane (DDT) and the related chemical methoxychlor, and bisphenol A (Fig. 6.24). Industrial chemicals such as nonylphenol, which is a breakdown product of alkoxyphenol detergents, also have oestrogenic effects. Most oestrogenic compounds are phenols or, like methoxychlor, are activated to phenols. Methoxychlor is 200,000 times less potent, nonylphenol is 300,000 less potent and PCBs are 1 million times less potent oestrogens than oestradiol. The oestrogenic effects from organochlorines in the diet have been estimated to be 2.5×10^8 less than the oestrogen equivalents from isoflavonoids in foods. The most potent xenobiotic endocrine disruptors in mammals are the 2-phenyl-substituted cyclic tetrasiloxanes, such as 2,6-*cis*-diphenylhexamethyl cyclotetrasiloxane, which has a potency of one-tenth that of DES.

Tributyltin (TBT) is used as an antifoulant in paints for the hulls of ships and has dramatic effects on marine invertebrates. It results in masculinization (imposex) of females in several species of marine snails and interferes with reproduction in bivalves, including oysters and other molluscs. TBT becomes associated with nerves and ganglia and may affect the production of neurohormones. TBT also interferes with shell growth through effects on calcium channels. The resultant decline in gastropod populations has been reversed following restrictions on TBT use.

Insecticides such as juvenile hormone analogues (methoprene), which act by interfering with moulting and metamorphosis, can also interfere with the development of important marine crustaceans, including lobster, crab and shrimp.

Municipal wastewater can contain significant levels of oestrone, oestradiol and ethinyl oestradiol. These oestrogens can come from pregnant women, use of oral

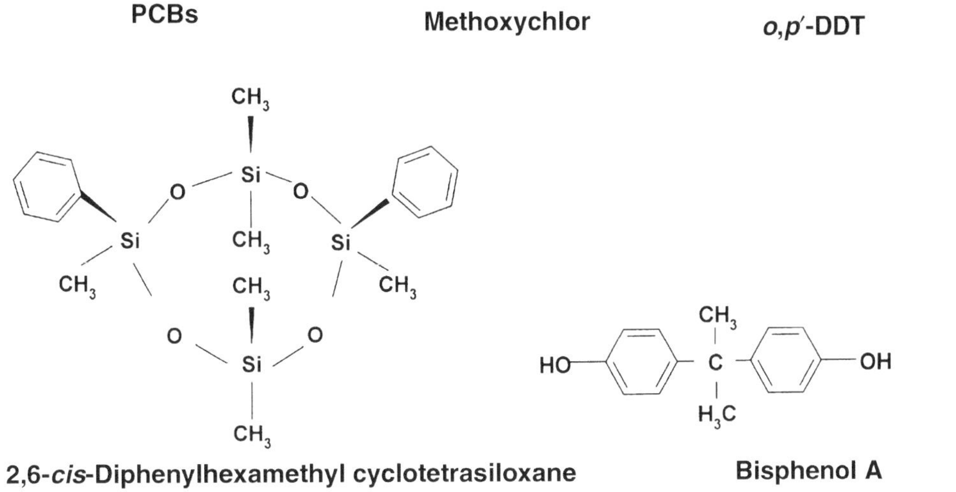

Fig. 6.24. Xenobiotic endocrine modulators.

contraceptives and hormone replacement therapy. There may also be significant release of steroid hormones in the manure from swine production and other intensive animal-holding facilities.

Indirect mechanisms of action

Effects on hormone metabolism

Xenobiotic chemicals can affect the activities of the cytochrome P450 (CYP) enzymes that are involved in the metabolism of steroid hormones. CYP enzymes in families 1–4 are responsible for the metabolism of both natural endogenous steroids and xenobiotic chemicals to more polar compounds for excretion. These families of enzymes can be induced by exposure to xenobiotics. CYP enzymes in families 11, 17, 19, 21 and 27 are involved in the synthesis of steroid hormones and these enzymes can be inhibited by xenobiotics. Thus, exposure to xenobiotic chemicals can decrease the synthesis and increase the degradation of endogenous steroid hormones, resulting in decreased levels of these hormones. Birds fed diets containing DDT, dieldrin or PCBs showed induction of hepatic CYP enzymes and increased metabolism of steroids to more polar products. Exposure of blue heron embryos to tetrachlorodibenzo-p-dioxin (TCDD) increased testosterone hydroxylase activity and 7-ethoxyresorufin-O-deethylase (EROD) activity (a measure of CYP1A1) measured *in vitro*. However, TCDD treatment did not affect plasma levels of testosterone or oestradiol *in vivo*, suggesting that the animal's endocrine system adapts to changes in steroid hormone degradation through feedback mechanisms. Ergosterol biosynthesis inhibiting fungicides (EBI) are widely used in agriculture and can inhibit CYP enzymes. However, *in vivo* treatment of partridges with a potent EBI did not affect levels of plasma steroids.

Organochlorine pesticides such as DDT are persistent in the environment and accumulate in animals at the top of the food chain. They can cause direct toxic effects at high levels and act as endocrine disruptors at lower levels. Fish from sites with high levels of PCBs have abnormal gonad development and decreased fecundity. Toxic effects have been shown in birds and alligators and in mammals such as seals and mink that eat these fish. However, it should be noted that these animals might also be exposed to high levels of other toxic compounds, such as methyl mercury. A well-known example of the effects of organochlorine pesticides is the major spill that occurred in Lake Apopka in Florida, which resulted in distorted sexual development and reproductive function in alligators (Guillette *et al.*, 1994). In birds, organochlorine pesticides are thought to cause eggshell thinning, resulting in increased embryo mortality from breakage or increased permeability of the shell. This is thought to be the cause of the decline in a number of raptor species in North America and Europe. The major cause of this problem is thought to be p,p'-DDE, which is a major metabolite of p,p'-DDT. Commercial DDT is made up of several isomers, with 75–80% being p,p'-DDT. The o,p'-DDT isomers are relatively unstable and are rarely found in the environment. These compounds can act as oestrogens themselves, as well as inducing CYP enzymes that metabolize oestrogen and thus decrease oestrogen activity. The p,p'-DDT isomer has also been shown to be an androgen receptor antagonist. The levels of organochlorines in the environment have decreased somewhat since the use of these chemicals was phased out in the 1970s and early 1980s and eggshell thickness in birds has increased. However, levels of PCBs and TCDD are still significantly elevated in the Great Lakes and feeding mink with carp from Lake Michigan has been shown to adversely affect reproduction and survival of the young.

Xenobiotic chemicals can inhibit the activity of the oestrogen-specific 17β-hydroxysteroid dehydrogenase (17β-HSD) type I that converts oestrone to the more biologically active oestradiol. A variety of flavonoid compounds inhibit 17β-HSD type I, at concentrations from 0.12 to 1.2 mM. Some flavones also inhibit the 17β-HSD type II that catalyses the 17β-oxidation of testosterone to androstenedione and oestradiol to oestrone, producing less active steroids. Flavonoids also inhibit aromatase activity. This anti-oestrogen activity of flavonoids may explain the reduced incidence of breast and prostate cancer attributed to diets high in phyto-oestrogens.

Effects on thyroid function

Thyroid hormones are necessary for the development of neurones and they affect metabolism, growth and reproduction in all vertebrates. They are also involved in the smoltification (conversion from a freshwater to a saltwater habitat) of salmonid fish and metamorphosis in flounder to form the asymmetrical adult. They are also required in the development of amphibians and the metamorphosis from the tadpole stage to the adult frog, which is the basis for a popular bioassay for thyroid hormone activity.

There are three major mechanisms for the disruption of thyroid function:

1. Inhibition of the iodoperoxidase enzyme of the follicle cells in the thyroid gland;
2. Effects on the transport of thyroid hormones by thyroid-binding globulin, thyroxine-binding prealbumin (transthyretin) and albumin; and
3. Effects on the catabolism of thyroid hormones and conjugation by the UDP-glucuronyl transferase.

Cabbage and a number of other *Brassica* species contain potent anti-thyroid compounds. The most important of these is goitrin (vinyl-2-thioxazolidine), which is as active as the drug propylthiouracil that is used for treating hyperthyroidism. Glucosinolates and cyanogenic glucosides present in foods can also be metabolized by the myrosinase enzyme present in brassica plants, as well as by intestinal bacteria to form thiocyanate, which is goitrogenic. Synthetic thionamides, aromatic amines and polyhydric phenols inhibit the synthesis of thyroid hormones. There are dramatic species differences in the susceptibility of thyroid function to sulphonamides, with rats, mice, hamsters, dogs and pigs susceptible, while primates, guinea-pigs and chickens are resistant. This can be explained in part by the inhibition of thyroid peroxidase by sulphamonomethoxine in the rat but not in monkeys. Thyroid activity is also higher in rats, due to high levels of TSH and the lack of thyroid-binding globulin, which increases the turnover of thyroid hormone. Sulpha drugs also induce thyroid neoplasia in the rat but are not human carcinogens. The relevance of thyroid disruption in a rodent test model to endocrine disruption in other species is thus questionable.

There is evidence that levels of thyroid hormone can be reduced in rats by exposure to high levels of PCBs, polybrominated biphenyls and dioxins. These effects are seen mainly with T_4, with no effect on active T_3, suggesting that the deiodinase activity increases to maintain T_3 levels. The decrease in T_4 is not due to decreased synthesis of T_4, but may be due to increased activity of the T_4-UDP-glucuronyl transferase or reduced binding of T_4 to the transthyretin binding protein in the blood. However, these effects on thyroid hormones are not seen in birds.

Effects on adrenal function

An example of EDCs that affect adrenal function is the consumption of liquorice. Liquorice can affect the synthesis of adrenal steroids and result in retention of sodium, excretion of potassium and hypertension due to an imbalance in the renin–angiotensin–aldosterone system. Liquorice contains glycyrrhizic acid, which is deconjugated to the free steroid glycyrrhetic acid (Fig. 6.25) in the intestine. Glycyrrehetic acid is absorbed from the intestine and acts as an inhibitor of the 11β-hydroxysteroid dehydrogenase in the kidney, which catalyses the conversion of cortisol to cortisone in the synthesis of the adrenal corticoids. A regular intake of 50 g of licorice provides about 100 mg of glycyrrhetic acid, which is sufficient to induce clinical signs of excess mineralocorticoid activity in sensitive individuals.

Effects on CNS function and behaviour

Exposure to endocrine disruptors can affect the function of the central nervous system

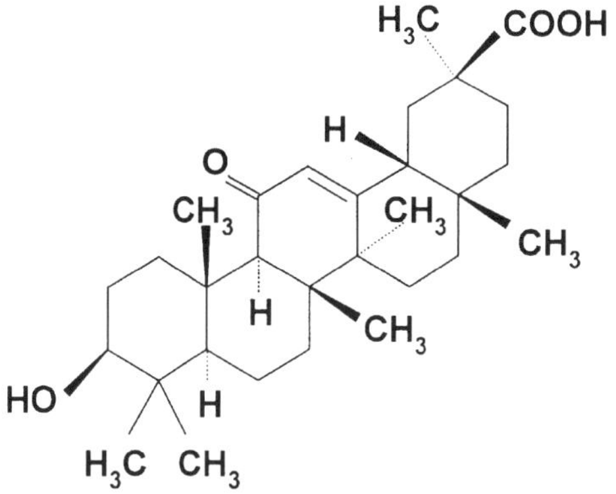

Fig. 6.25. Glycyrrehetic acid.

(CNS) and alter behaviour. This can occur during early development and neonatal life, during puberty and breeding or after the behaviour is established. Hormones direct the organization of the CNS and reproductive tract during early development and are then involved in the activation of sexual dimorphic behaviours via the hypothalamus–pituitary–gonadal axis at puberty and the maintenance of behaviour during adulthood (see Section 5.1).

Testosterone (T), dihydrotestosterone (DHT) and oestradiol (E_2) are involved in male programming of the CNS and development of the reproductive tract, with the relative importance of each steroid varying among species. Treatment of neonatal female hamsters or rats with oestrogenic compounds induces male mounting behaviour when these animals were later treated with testosterone as adults. Masculinization of the CNS in this way increases the size of the sexually dimorphic nucleus in the preoptic area of the hypothalamus. Perinatal exposure to oestrogenic or anti-androgenic compounds produces developmental abnormalities in the male reproductive tract. Exposure of immature female rats to the oestrogenic pesticide methoxychlor at doses from 2 to 200 mg kg^{-1} day^{-1} induces precocious puberty within a few days of treatment. In adult rats, xeno-oestrogens reduce food consumption, weight gain and sexual behaviours.

Some pesticides act by disrupting nerve function, such as organophosphorous insecticides that act as acetylcholinesterase inhibitors and pyrethroid insecticides that affect sodium flux. These might be expected to act as endocrine disruptors, but this has not been well documented. There is evidence that endocrine disruptors impair the stress response in fish, but not in birds. For more information on the effects of endocrine disruptors on the CNS, see Gray and Ostby (1998).

Summary

In summary, particular indicator species have been used for monitoring endocrine disruptors in the environment. However, even though endocrine-modulating chemicals can have significant effects with *in vitro* systems, many *in vivo* studies have shown that the effects may not overwhelm the normal homeostatic control mechanisms to cause adverse effects in animals. The potential adverse effects of endocrine-modulating compounds depend on the potency, levels of exposure and time during development when exposure to the compounds occurred. Possible effects on wildlife populations are sometimes difficult to determine due to a lack of baseline data and the presence of other environmental effects. For example, declines in fish populations can be due to habitat destruction or competition from other species. The exposure of humans in terms of oestrogenic equivalents from synthetic endocrine disruptors is estimated to be much less than that from oestrogens used for oral contraceptives or natural oestrogenic compounds in foods.

Questions for Study and Discussion

Section 6.1 Control of Broodiness in Poultry

1. Describe factors that lead to broodiness in poultry. How can broodiness be controlled?

Section 6.2 Applications of Pheromones

1. Describe the physiological roles of pheromones.
2. Describe the nature of pheromone-binding proteins and pheromone receptors.
3. Describe sex pheromones in rodents, pigs, cattle, and sheep and goats.
4. Describe pheromones used by fish.
5. Give examples of sex pheromones, aggregation pheromones and alarm pheromones in insects.
6. Describe the use of insect pheromones in pest control.

Section 6.3 Effects of Stress

1. Discuss the nature of stress. What are the behavioural and physiological responses to stress?
2. Describe the role of the SA and HPA systems in the response to stress.
3. Describe the different roles of CRH and CRH receptors in the stress response.
4. Discuss the role of other hormones in the stress response.
5. Describe behavioural, physiological and endocrine methods to assess the level of stress.
6. Describe the systems involved in innate and acquired immunity. What is the effect of stress on these systems?
7. Describe the effects of stress on reproduction.
8. Describe the effects of stress on nutrient utilization and growth performance.

Section 6.4 Endocrine Applications in Toxicology

1. Discuss the nature of endocrine disruptor chemicals.
2. Outline *in vitro* and *in vivo* tests for endocrine disruptors.
3. Give examples of plant-derived and xenobiotic endocrine disruptors.
4. Describe the potential effects of endocrine disruptors on hormone metabolism, thyroid and adrenal function, and CNS function and behaviour.

Further Reading

Control of broodiness

Sharp, P.J. (1997) Immunological control of broodiness. *World's Poultry Science Journal* 53, 23–31.

Applications of pheromones

Agosta, W.C. (1992) *Chemical Communication: the Language of Pheromones*. Scientific American Library, New York.
Amory, J.R. and Pearce, G.P. (2000) Alarm pheromones in urine modify the behaviour of weaner pigs. *Animal Welfare* 9, 167–175.
Buck, L.B. (2000) The molecular architecture of odor and pheromone sensing in mammals. *Cell* 100, 611–618.
Copping, L.G. and Menn, J.J. (2000) Biopesticides: a review of their actions, applications and efficacy. *Pest Management Science* 56, 651–676.
Flower, D.R. (1996) The lipocalin protein family: structure and function. *Biochemical Journal* 318, 1–14.
Johnston, R.E., Müller-Schwarze, D. and Sorensen, P.W. (1999) *Advances in Chemical Signals in Vertebrates*. Kluwer Academic/Plenum Publishers, New York.
Jones, O.T. (1998) The commercial exploitation of pheromones and other semiochemicals. *Pesticide Science* 54, 293–296.
McGlone, J.J. and Anderson, D.L. (2002) Synthetic maternal pheromone stimulates feeding behaviour and weight gain in weaned pigs. *Journal of Animal Science* 80, 3179–3183.
Stoka, A.M. (1999) Phylogeny and evolution of chemical communications: an endocrine approach. *Journal of Molecular Endocrinology* 22, 207–225.

Effects of stress

Chan, E.-C., Falconer, J., Madsen, G., Rice, K.C., Webster, E.L., Chrousos, G.P. and Smith, R. (1998) A corticotropin-releasing hormone type I receptor antagonist delays parturition in sheep. *Endocrinology* 139, 3357–3360.

Christos, T.E. and Arvanitis, A. (1998) Corticotrophin-releasing factor receptor antagonists. *Expert Opinion on Therapeutic Patents* 8, 143–152.

Clark, J.D., Rager, D.R. and Calpin, J.P. (1997a) Animal well-being II. Stress and distress. *Laboratory Animal Science* 47, 571–579.

Clark, J.D., Rager, D.R. and Calpin, J.P. (1997b) Animal well-being III. An overview of assessment. *Laboratory Animal Science* 4, 580–585.

Clark, J.D., Rager, D.R. and Calpin, J.P. (1997c) Animal well-being IV. Specific assessment criteria. *Laboratory Animal Science* 4, 586–597.

Coste, S.C., Murray, S.E. and Stenzel-Poore, M.P. (2001) Animal models of CRH excess and CRH receptor deficiency display altered adaptations to stress. *Peptides* 22, 733–741.

Deak, T., Nguyen, K.T., Ehrlich, A.L., Watkins, L.R., Spencer, R.L., Maier, S.F., Licinio, J., Wong, M.-L., Chrousos, G.P., Webster, E. and Gold, P.W. (1999) The impact of the non-peptide corticotropin-releasing hormone antagonist antalarmin on behavioral and endocrine responses to stress. *Endocrinology* 140, 79–86.

Eckart, K., Radulovic, J., Radulovic, M., Jahn, O., Blank, T., Stiedl, O. and Spiess, J. (1999) Actions of CRF and its analogs. *Current Medicinal Chemistry* 5, 1035–1053.

Elenkov, I.J. and Chrousos, G.P. (1999) Stress hormones, Th1/Th2 patterns, pro/anti-inflammatory cytokines and susceptibility to disease. *Trends in Endocrinology and Metabolism* 10, 359–368.

Elsasser, T.H., Kahl, S., Rumsey, T.S. and Blum, J.W. (2000) Modulation of growth performance in disease: reactive nitrogen compounds and their impact on cell proteins. *Domestic Animal Endocrinology* 19, 75–84.

Fossum, C. (1998) Cytokines as markers for infections and their effect on growth performance and well-being in the pig. *Domestic Animal Endocrinology* 15, 439–444.

Kroboth, P.D., Salek, F.S., Pittenger, A.L., Fabian, T.J. and Frye, R.F. (1999) DHEA and DHEA-S: a review. *Journal of Clinical Pharmacology* 39, 327–348.

Lovejoy, D.A. and Balment, R.J. (1999) Evolution and physiology of the corticotropin-releasing factor (CRF) family of neuropeptides in vertebrates. *General and Comparative Endocrinology* 115, 1–22.

Matteri, R.A., Carroll, J.A. and Dyer, C.J. (2000) Neuroendocrine responses to stress. In: Moberg, G.P. and Mench, J.A. (eds) *The Biology of Animal Stress: Basic Principles and Implications for Animal Welfare*. CAB International, Wallingford, UK, pp. 43–76.

McCarthy, J.R., Heinrichs, S.C. and Grigoriadis, D.E. (1999) Recent advances with the CRF1 receptor: design of small molecule inhibitors, receptor subtypes and clinical indications. *Current Pharmaceutical Design* 5, 289–315.

O'Connor, T.M., O'Halloran, D.J. and Shanahan, F. (2000) The stress response and the hypothalamic–pituitary–adrenal axis: from molecule to melancholia. *Quarterly Journal of Medicine* 93, 323–333.

Sapolsky, R.M., Romero, L.M. and Munck, A.U. (2000) How do glucocorticoids influence stress responses: integrating permissive, suppressive, stimulatory, and preparative actions. *Endocrine Reviews* 21, 55–89.

von Borell, E. (1995) Neuroendocrine integration of stress and significance of stress for the performance of farm animals. *Applied Animal Behaviour Science* 44, 219–227.

Wang, L., Martínez, V., Rivier, J.E. and Taché, Y. (2001) Peripheral urocortin inhibits gastric emptying and food intake in mice: differential role of CRF receptor 2. *American Journal of Physiology Regulatory Integrative Comparative Physiology* 281, R1401–R1410.

Endocrine applications in toxicology

Ashby, J. (2000) Validation of *in vitro* and *in vivo* methods for assessing endocrine disrupting chemicals. *Toxicologic Pathology* 28, 432–437.

Colborn, T. and Thayer, K. (2000) Aquatic ecosystems: harbingers of endocrine disruption. *Ecological Applications* 10, 949–957.

Fenner-Crisp, P.A., Maciorowski, A.F. and Timm, G.E. (2000) The endocrine disruptor screening program developed by the U.S. environmental protection agency. *Ecotoxicology* 9, 85–91.

Gray, L.E. Jr and Ostby, J. (1998) Effects of pesticides and toxic substances on behavioral and morphological reproductive development: endocrine versus nonendocrine mechanisms. *Toxicology and Industrial Health* 14, 159–184.

Guillette, L.J. Jr, Gross, T.S., Masson, G.R., Matter, J.M., Percival, H.F. and Woodward, A.R. (1994) Developmental abnormalities of the gonad and abnormal sex hormone concentrations in juvenile alligators from contaminated and control lakes in Florida. *Environmental Health Perspectives* 102, 680–688.

Huet, M.-C. (2000) OECD activity on endocrine disrupters test guidelines development. *Ecotoxicology* 9, 77–84.

Lister, A.L. and Van Der Kraak, G.J. (2001) Endocrine distruption: why is it so complicated? *Water Quality Research Journal of Canada* 36, 175–190.

Nilsson, R. (2000) Endocrine modulators in the food chain and environment. *Toxicologic Pathology* 28, 420–431.

Oberdorster, E. and Cheek, A.O. (2000) Gender benders at the beach: endocrine disruption in marine and estuarine organisms. *Environmental Toxicology and Chemistry* 20, 23–36.

Safe, S.H. and Gaido, K. (1998) Phyto-oestrogens and anthropogenic estrogenic compounds. *Environmental Toxicology and Chemistry* 17, 119–126.

Santti, R., Makela, S., Strauss, L., Korkman, J. and Kostian, M.-L. (1998) Phytoestrogens: potential endocrine disruptors in males. *Toxicology and Industrial Health* 14, 223–237.

Index

acceptable daily intake (ADI) 60, 74
acquired immunity 212
actinomycin D 39
active/passive immunization 61
activin 159
acute phase proteins 213
additive effects of growth promoters 95
adenohypophysis 27
adenylate cyclase 14–15
adjuvant 41
adrenal gland 205
adrenaline 117, 205
adsorption chromatography 45
affinity chromatography 41, 46
aggregation pheromone 200–201
alarm pheromone 199, 201
allometric growth 124
allomones 194
allostasis 208
allosteric regulator 15
anabolic effects 69
anagen 145
androgenic effects 70
androst-16-ene steroids 77, 198
androstenone 77
antagonists 39
antibodies (Ab) 40
antigens (Ag) 40
antimicrobials 108–109
aquaculture 182–189
aromatase 9, 72, 155
arteriovenous differences 68
ascites 102
assay
    bioassay 43
    calibration 42–43
    chemical 44
    competitive binding 47
    enzyme-linked immunosorbent (ELISA) 48
    radioimmunoassay (RIA) 48
    solid-phase 48
association constant (K_a) 49
autocrine 1, 127, 130
autoradiography 42

B lymphocytes 40, 212
beta adrenergic agonist receptors 91–93, 205
beta adrenergic agonists 90–94
blood–brain barrier 31
boar taint 68, 77–83
bombykol 196
brood patch 192
broodiness 192–193
bST and milk production 130–132

calbindin 144
calcitonin 134
calcium metabolism 134–136, 143–145
calcium release channel (CRC) 114
calmodulin 16
cAMP-responsive-element binding protein (CREB) 16, 27, 92
capillary electrophoresis 47
carrier proteins 4
catagen 145
catecholamines 90, 117
cDNA library 54
cell-mediated immunity 40, 212
chemical inactivation 36
cholecystokinin CCK 107, 186
chromatin 23–24
chromium 109–112
circadian rhythm 187
clover disease 221
coefficient of variation 43
colchicine 39
concerted/additive effects 3
conjugated linoleic acid (CLA) 102–105

corpus luteum 161, 164
corticotrophin releasing hormone (CRH) 206–207, 216
corticotrophin releasing hormone receptors 207–208, 218
cortisol 72, 117
    in saliva 212
C-reactive protein (CRP) 213
cross-talk 18
cyanosis 114
cycloheximide 39
cyclooxygenase (COX) 10, 102, 164
cystic ovarian disease 178
cytochrome P450 8, 13, 80–81, 156, 224
cytokine receptors 20, 88
cytokines 145, 147–151, 213
cytotoxic T lymphocytes 40

dark, firm, dry (DFD) meat 113
deafferentation 36
defleecing 145
dehydroepiandrosterone (DHEA) 209
desensitization 87
dichlorodiphenyltrichloroethane (DDT) 223
diethylstilbestrol (DES) 52, 221
displacement experiment 50
dissociation constant (K_d) 49
disulphide bonds 13
DNA-binding domain 22
dopamine 163, 181, 192
dressing percentage 66

ectopic site 36
Edman degradation 53
eggshell formation 141–145
eggshell matrix 142
eicosanoids 9
electrophoresis 47
electrophoretic mobility shift assays 26
endocrine disruptor chemical (EDC) 218
endocrine disruptor chemical testing 219–221
endocrine gland 1
endopeptidases 12
energy balance 179
enzyme-linked immunosorbent assay (ELISA) 48
epidermal growth factor (EGF) 18, 128, 147, 149–151
epitope 40
erythropoiesis 69
exocrine 4
exocytosis 7
exopeptidases 12
extracellular receptors 13

feed efficiency 66
feedback, positive/negative 2
fertility control 170

fetoneonatal oestrogen binding protein (FEBP) 156
fibroblast growth factor 148
FLAG sequence 57
fluorescence immunoassay (FIA) 48
follicular development 138–141, 159–161
follistatin 159
footprinting 26
forskolin 39
freemartin 156
fusion products 57

G protein 14, 16
galactopoiesis 124,128
gamma linolenic acid (GLA) 101–102
gas chromatography 46
gastrin 107
ghrelin 87
glucagon 110
glucocorticoid 117, 208–209, 214–216
glucose tolerance factor 109
glucose transport proteins (GLUT) 110
glycoproteins 7
glycyrrhetic acid 225
growth efficiency 67
growth hormone binding protein (GHBP) 88
growth hormone releasing hormone (GHRH) 85
growth hormone releasing peptides (GHRP) 86
growth hormone secretagogue receptor (GHS-R) 186
growth performance 66

habituation 211
half-life 4
hapten 40, 48
heat shock protein 21
Hershberger assay 221
high performance liquid chromatography (HPLC) 45
histones 23
homeostasis 1
homologous recombination 62
hormone-responsive elements (HREs) 22
human chorionic gonadotrophin (hCG) 165
humoral immunity 40, 212
hybridoma 41
hydrophobic interaction 45
hyperplasia 36, 66
hypertrophy 36, 66
hypophysectomy 36
hypothalamic–pituitary portal system 27
hypothalamic–pituitary–adrenal (HPA) axis 117, 204
hypothalamic–pituitary–gonadal axis 160, 226
hypoxia 102

idiotypes 61
immortalized cell lines 38

immune system 40, 212
immunoassay 48
immunocytochemistry 42
immunoenzyme histochemistry 42
immunoneutralization 61, 83, 181–182
in situ hybridization 42
indicator species 219
inflammation 213
infundibulum 141
inherent capacity 66–67
inhibin 140, 159
inhibitors 39
innate immunity 212
insulin 109, 187
insulin-like growth factor (IGF) 89, 128, 147–148, 160
interferon 212–213
interleukin 212
intracellular receptor 20–25
ion exchange 45
ionophores 39
isolated cells 38
isthmus 141

JAK (Janus kinase) 20, 127

kairomones 194
ketosis 134
knockouts 62

lactoferrin 126
lactogenesis 124, 128
leptin 87, 105–107, 164
leukotrienes 101
ligand-binding domain 22
lipid peroxidation 102
lipocalin 195
lipogenesis 93
lipolysis 93
liposomes 40
liquid chromatography 45
locus ceruleus 205
luteolysis 161

magnum 141
major histocompatability complex (MHC) 197
major urinary protein (MUP) 195, 197
MALDI-TOF 53
malignant hypothermia 116
mammary-derived growth inhibitor 127
mammary gland involution 126
mammogenesis 124, 127
mass spectrometry 46
mass trapping 202
maternal pheromone 198
mating disruption 202
meiosis activating substance (MAS) 158

melanin 188
melatonin 163
metabolic inhibitors 39
milk composition 132–133
milk fever 134–136
milk production 124
milk removal 139
mimosine 146
mineralocorticoids 8
model systems 35–39, 126–127, 147, 220–221
monoclonal antibodies 41
Mullerian and Wolffian ducts 138, 155
multiple ovulation and embryo transfer (MOET) 175–177

n-3/n-6 fatty acids 100
neofemale 184
neomale 184
neurocrine 1
neuroendocrine 1
neurohypophysis 27
neuropeptide Y (NPY) 106, 163, 186
nitric oxide 163–164, 217
no observed effects limits (NOEL) 74
non-additive effects 3
non-specific binding 49
non-steroidal anti-inflammatory drug (NSAID) 10
noradrenaline 205
nuclear matrix 24–25
nuclear receptor 71

odour receptors 196
oestrous cycle 159–164
    hormones for manipulation 164–171
    mammary gland development 126
oestrus (heat) 161
    detection 171–172, 198–199
    induction and synchronization 172–175, 204
osteoblast 144
osteoclast 143
osteopontin 142
ouchterlony immunodiffusion 41
ovocleidin-17 142
ovulation 161
oxytocin 28

pale, soft, exudative (PSE) meat 112
palindromes 22
parabiosis 36
paracrine 1, 127
parasympathetic nervous system 205
parathyroid hormone 134, 143
pars distalis 97
parturition 162
    induction 177
peptide sequencing 53–54
peptide synthesis 54–57

peptide YY (PYY) 108
peptidomimetics or peptoides 55
perfused organ 36
permissive effects 3
peroxisome proliferator-activated receptor (PPAR) 100
phorbol ester 39
phosphodiesterase inhibitors 39
phosphorylation/dephosphorylation 3
phyto-oestrogens 219, 221–222
placental lactogen 127
poikilotherms 2
polychlorinated biphenyls (PCBs) 218, 223
polyclonal antibody 40
polymerase chain reaction (PCR) 54
polyunsaturated fatty acid (PUFA) 99–105
population monitoring 202
porcine stress syndrome (PSS) 114–118
postpartum interval 178
potential daily intake (PDI) 60
pregnancy 162
pregnancy maintenance 177
pregnant mare serum gonadotrophin (PMSG) 165
prehormone 5
preprohormone 7, 28
primary cell culture system 38
primary immune response 40
priming pheromones 194
proanagen 145
progesterone-releasing intrauterine device (PRID) 58
prohormone 5
prolactin 30, 127, 192, 209
promoter 26
prostaglandins 11, 101
protein kinase C 16
puberty 162
puberty induction 180
pulsatile hormone release 31, 58, 87
puromycin 39

quantitative trait locus (QTL) 82
queen mandibular pheromone 201, 204

receptor 4
    agonists 39
    cooperativity 50
    dimerization 18
    serine kinase 20
    tyrosine kinase 18
recombinant proteins 56
releasing hormones 30
restriction fragment length polymorphism (RFLP) 116
reverse transcription 54
rifampicin 39

saturation-binding curve 49
sauvagine 207
Scatchard analysis 48
seasonality 162–163, 180–181
second messenger 4, 14
secondary immune response 40
selective androgen receptor modulator (SARM) 71
semiochemicals 194
sex determination 137, 154, 182
sex hormone binding globulin (SHBG) 26
sex pheromone 198–200
sex reversal 182–185
sexual development 136–138, 154–157, 226
    labile period 183
SH_2 domain 18
sham-operated 36
shipping fever 214
signal peptides 5
signal transducers and activators of transcription (STATS) 20
signalling pheromones 194
size exclusion or gel filtration 46
skatole 77
somatostatin (SS) 85
somatotrophin (ST) 83–90, 130–132, 187, 209
spawning 185–186
standard curve 42
steroid acute regulatory protein (StAR) 164
steroid hormones 7, 13, 51, 76
stress 204
    assessment 210–212
    behaviour 210
    growth 217–218
    hormonal responses 205–212
    immune function 188, 214–215
    reproduction 179, 215–217
subcellular fractions 39
supermale 184
suprachiasmatic nucleus 31
sustained release of hormones 57
    osmotically controlled systems 59
    surface eroding 59
sympathetic nervous system 90
    activation of adrenal medulla (SA) 117, 205
synergistic effects 3

target tissue 1
telogen 145
testis determining factor 157
thyroglobulin 96
thyrotrophin (TSH) 11, 97, 209
thyroxine 96, 186, 225
tissue slices 38
tonic inhibition 31
transcortin 212

transcription 5
transcription factors 20, 71
transforming growth factor 128, 149
transgenic animals 62–64
translation 5
translocation 20
tributyltin 223
triiodothyronine 96
trophic hormone 11

uncoupling proteins 106
urocortin 207

vasoactive intestinal peptide (VIP) 192
vasopressin 28
vitamin D 134, 143
vomeronasal organ (VNO) 196

Western blotting 47
whole animal model 35
    see also model system

xeno-oestrogens 223

zinc fingers 22